Cure of Thermosetting Resins
Modelling and Experiments

J.-M. Vergnaud and J. Bouzon

Cure of Thermosetting Resins

Modelling and Experiments

With 240 Figures

Springer-Verlag
London Berlin Heidelberg New York
Paris Tokyo Hong Kong
Barcelona Budapest

J.-M. Vergnaud, PhD
J. Bouzon, PhD
Université de Saint-Etienne, Laboratoire de Chimie des Matériaux et Chimie Industrielle, Faculté des Sciences et Techniques, 23 rue du Docteur Paul Michelon, 42023 Saint-Etienne Cédex 2, France

ISBN-13:978-1-4471-1917-3 e-ISBN-13:978-1-4471-1915-9
DOI: 10.1007/978-1-4471-1915-9

British Library Cataloguing in Publication Data
Vergnaud, Jean-Maurice
Cure of thermosetting resins : modelling and experiments.
I. Title II. Bouzon, Jean
668.412
ISBN-13:978-1-4471-1917-3

Library of Congress Cataloging-in-Publication Data
Vergnaud, J.-M.
Cure of thermosetting resins : modelling and experiments / Jean-Maurice Vergnaud and Jean Bouzon.
p. cm.
Includes index.
ISBN-13:978-1-4471-1917-3
1. Thermosetting plastics–Curing. I. Bouzon, Jean, 1929–
II. Title.
TP1180.T55V47 1992 91-40808
668.4′22–dc20 CIP

Softcover reprint of the hardcover 1st edition 1992

Typeset by Keytec Typesetting Ltd, Bridport Dorset
69/3830-543210 Printed on acid-free paper

Preface

The process of cure of thermosets is rather complex, and good knowledge of the various steps and different problems is necessary for the user. For instance, the following basic facts characterise the cure of thermosets:

1. In the same way as rubbers, thermosets are generally polymerised and processed in a simple operation which involves the irreversible transformation of a low molecular weight resin in viscous liquid state into a solid network polymer. The process of cure is thus much more important for thermosets or rubbers than for thermoplastics, because if something goes wrong during the cure process of thermosets, the final products may have undersirable properties and will be of no use or value, while the thermoplastic material can be melted again to make a new material.
2. In contrast with rubbers, a high exothermic cure reaction is the aspect of fundamental importance in the cure process for thermosets. This high enthalpy of cure associated with a rather low thermal conductivity can give rise to an excessively high temperature which may cause discoloration and degradation of the material, and also to substantial temperature gradients. The material is thus heterogeneous during the process of cure, and these temperature–time histories in the resin may have some effects on the properties of the final material.
3. Moreover, the increase in production following the reduction in time of the cure cycle necessitates the use of a higher mould temperature. The rate of heat evolved from the cure reaction varies with temperature by following an Arrhenius equation. Thus a higher temperature in the mould is responsible for higher temperature gradients and a higher increase in temperature, which in turn increase the rate of cure reaction.

On the whole, the process of cure of a thermoset consists of two main stages:

1. The heating period of the liquid resin (either pure or in the form of a composite with fillers) in the mould. The resin is thus heated up to a temperature at which the cure reaction starts. Because of the low thermal conductivity of the resin or composite, some gradients

of temperature are thus developed within the sample, with a higher temperature on the surface of the resin.

2. The cure reaction. Generally, the reaction starts on the surface of the resin where the highest temperature exists. But, because of the internal heat generated by the cure reaction associated with the low thermal conductivity, other gradients of temperature are developed within the resin with the highest temperature at the centre of the resin. The temperature–time history at the centre passes through a maximum, and decreases regularly down to the mould temperature.

The process of cure of thermosets thus covers two different fields: one where heat is transferred by conduction through the resin and the mould, by convection (free or forced) at the surface of the mould when it is immersed in a heated fluid; the other where the cure reaction and the rate of the heat evolved from the overall reaction can be described by an Arrhenius equation with one or two pre-exponential factors and one or two activation energies. When each of these two phenomena occurs separately, i.e.

1. Heat transfer by conduction within the solid and simple boundary and initial conditions
2. Heat evolved from the overall cure reaction at constant temperature

an analytical solution can generally be found by mathematical treatment. But when these two phenomena take place simultaneously, no analytical solution exists, and the problem must be treated by numerical methods with finite differences.

Modelling the process of cure is of great interest either from a theoretical point of view or for practical and industrial applications. The numerical models based on methods with finite differences are very flexible and can be applied to all cases, as they are capable of taking into account all the known facts: especially the initial and boundary conditions and also the heat transfer by conduction (or even convection) through the solid, either with constant, temperature- or state-of-cure-dependent thermal parameters, as well as the heat generated by the overall cure reaction. Of course, building a model necessitates a precise view and knowledge of the process by the researcher or the engineer, because the various stages of the process must be set forth. But when the model is established, the process can be simulated with the help of computers, and comparisons made between the experimental and calculated results are able to test not only the validity of the model but also the accuracy of the various data which are used, such as the thermal and kinetic parameters. Finally, the model is able to afford further insight into the nature of the process of cure. Simulation provides the user with more information than experiments do: for instance, the profiles of temperature as well as the profiles of state of cure developed within the sample at any time.

These simulations of the process are very powerful and informative not only for researchers but also for those who are in charge of the industrial applications. They can enable the user to evaluate the effect of each parameter on the process in a quantitative manner, however great their number, without having to carry out further new experiments, some of them being tedious and highly time-consuming. In conclusion, simulating the process is necessary to determine precisely, without using rule of thumb, the best operational conditions when the process of cure must be developed on a larger scale than the laboratory scale for which the model was tested. This information is of great concern, as the main objective in industry is to reduce the price of the final material and to improve its quality, at the same time. Moreover, the efficacy of techniques able to reduce the time of the cure cycle can be evaluated easily with the help of the model. For instance, in the case of the injection moulding system, the typical conditions of the system such as the temperature at which the resin is stored and injected in the mould can be predetermined, in the same way as the best operational conditions for the mould, e.g., time and temperature.

An important problem must be considered, namely the selection of a kinetic equation of the overall cure reaction and the determination of the kinetic parameters. A simple Arrhenius expression with only one activation energy was chosen in these studies for describing the rate of heat evolved from the cure reaction, as it can match the experimental results with high accuracy. Calorimetry is of course the best technique for measuring the rate of heat evolved from the cure. In calorimetry techniques, enthalpy changes accompanying chemical events in the sample are monitored as a function of temperature or time. The result is a profile of the rate of enthalpy change either as a function of temperature when the sample is heated at a known linear rate, in differential scanning calorimetry (DSC), or as a function of time, the calorimeter being held at constant temperature, in differential calorimetry (DC). Two problems of great importance thus appear, namely the dimension of the sample, and the way in which the calorimeter works.

The question of the dimension of the sample used in calorimetry is of great concern, especially in the case of thermosets. Thermosets are often obtained by mixing two viscous components with other solid additives, fillers, loose fibres, woven fibres, etc. The small sample size used in calorimetry makes it necessary that all ingredients in the sample be very well dispersed. The question which always arises is how representative of the homogeneity of a large batch a 1–5 mg sample is. A good answer to this question, in our opinion, is to choose a calorimeter working with a fairly large sample, about 150–200 mg of material, located for instance in a cylindrical holder heated radially.

Making a choice between the DC or the DSC mode for determining the kinetics of heat evolved from the cure reaction is certainly also a problem.

For the supposed constant temperature operation in the DC mode, the calorimeter is stabilised at the desired temperature, and the sample in the holder at room temperature is introduced. A period of time is thus required for the sample and calorimeter to establish thermal equilibrium after the introduction of the sample, over which very steep temperature gradients are developed within the sample, and a very high endothermic heat flux results from this strong heating. At the end of this heating period, the applied temperature which is registered is the set calorimeter temperature and not the true sample temperature during the cure because of local heating arising from the exothermic reaction. In a few words, it was once said that the sample and calorimeter are at the same temperature at two times only: when the heat flux reaches the zero value at around the end of the heating stage, and when the reaction is over. Moreover, for thermosets, a temperature window of about 20 °C width is found over which meaningful data can be obtained.

In DSC, the sample is scanned at a selected heating rate up to a temperature at which the reaction exotherm is complete. After the sample has been given this precure, it is quench-cooled and then scanned under conditions identical to the initial scan, giving the reference baseline. Of course, the important parameter is the heating rate. Too high a value is responsible for a large increase in temperature within the sample which in turn produces a higher rate of enthalpy change. When the heating rate is too low, the rate of enthalpy change (or the heat flux) is too small, and the limitation is the sensitivity of the calorimeter.

The general plan chosen for the book is as follows:

In the first part, the analytical solutions obtained by the mathematical treatment of the equation of heat transfer by conduction are given when they exist, i.e., when no reaction takes place and for simple initial and boundary conditions, by considering various shapes of the sample.

In the second part, the numerical treatment by methods based on finite differences is described for various shapes of the sample. This part is written in a didactic way in order to make the numerical analysis very comprehensive and attractive to the reader.

In the third part, many examples of modelling are described, when the resin is cured either in a calorimeter or in various moulds.

The first part with six chapters presents a general overview of the state of the mathematical treatment of heat conduction within a solid with simple initial and boundary conditions. In spite of the lack of analytical solutions in the general problem of heating and cure, it is necessary for researchers and engineers to have good background knowledge of the mathematical treatment of heat conduction for at least two reasons: to know when the analytical solutions are available and the conditions at which they are obtained, and to know how to determine the values of the thermal parameters such as the thermal conductivity or the coefficient of surface heat transfer. Special attention is focused on the assumptions made before calculating each analytical solution, because it is important to know exactly the conditions upon which the solution is found. Two kinds of results are given: the profiles of temperature developed through the sample, as well as the kinetics of heat transferred.

The six chapters are described as follows:

Chap. 1 summarises the principles of heat transfer by considering heat conduction within a solid, heat convection in the fluid next to the surface of the solid, and heat radiation.

Chap. 2 presents the principal methods of solution of the equation of heat conduction through a solid when mathematical treatment can be used, i.e., in the case of heat conduction with constant thermal conductivity and without reaction: the method of separation of variables, the method of reflection and superposition, and the method of the Laplace transform.

Chap. 3 considers a thin plane sheet of solid with conduction heat through its thickness in either a transient or a stationary state, with various initial and boundary conditions. Some of the cases are worth noting: constant temperature on the surface, finite coefficient of surface heat transfer, sandwich form, and where the sheet is located in a surrounding medium of infinite or of finite volume.

Chap. 4 studies a sphere, as this shape can be of interest in a few applications. The same initial and boundary conditions as chosen for the sheet are considered.

Chap. 5 presents interesting cases of the cylinder, either of infinite or finite length, solid or hollow, as all these cases display various applications. In the case of the cylinder of finite length, the solutions are expressed as the product of the solutions obtained with radial heat transfer and the Bessel functions, and with the one-direction transfer through a sheet.

Chap. 6 looks at the rectangular parallelepiped, which is also of great interest, and the solutions are expressed as the product of the solutions obtained with the simple problem of the sheet.

The second part of the book is concerned with the concept of numerical analysis. Explicit numerical methods are mainly developed, because they are easy to understand for non-specialists such as engineers, physicists or chemists, and they are well adapted for microcomputers. In order to accustom the reader to this new way of working, special attention is given to presenting these new techniques in a didactic way, starting with very simple examples and leading progressively to difficult problems. For instance, simple problems of heat transfer through a sheet which have an analytical solution are still treated by using numerical analysis, so that the reader may test the validity of his simple model. On the whole, some emphasis is placed upon presenting these new methods in such a way that they appear friendly and attractive to the reader.

Chap. 7 presents examples of numerical analysis in the plane sheet, with heat conduction and with or without cure reaction, while the thermal parameters are either constant, or temperature- or state-of-cure-dependent. Various cases are considered for the surface which can be either the resin–mould interface or the external surface of the mould surrounded by a fluid.

Chap. 8 describes the numerical treatment of solid or hollow cylinders of infinite of finite length, with heat conduction and with or without reaction. Various cases are studied for the surface of the cylinder, as well as for the thermal parameters which may be either constant or temperature-dependent.

Chap. 9 deals with the numerical analysis for solid spheres, with radial heat conduction and with or without reaction, the thermal parameters being either constant, or temperature- or state-of-cure-dependent.

Chap. 10 traces the development of numerical treatment for a rectangular parallelepiped, applicable to resin surrounded by a mould or the mould surrounded by a fluid.

In the third part of the book various approaches to studies of different problems dealing with the cure of thermosets are examined. Chapters 11–19 deal with each of these problems, from the difficulties encountered in experiments and calculation, finally leading to conclusions of interest either from a theoretical or a practical point of view.

Chap. 11 examines the techniques which are used to determine the properties of the resin after or during cure. A comparison made between the techniques used earlier for rubbers and more recently for thermosets may be informative. Special attention is given to the size of the sample in the calorimetry, as this small sample taken from a larger batch must be representative of the homogeneity of the batch.

Chap. 12 shows the complexity of the process of cure in calorimetry working under isothermal conditions. An application of the numerical model to this technique by taking into account the heat transfer and heat generated by the cure reaction gives informative results, enabling one to conclude that the process of cure in this case is not quite as simple as may be thought.

Chap. 13 studies the process of heating and cure in calorimetry in scanning mode. Modelling of the process lays emphasis on the importance of the heating rate, whose value must be selected by considering not only the sample size, but also the cure enthalpy and the sensitivity of the rate of cure to temperature.

Chap. 14 deals with the problems of the homogeneity of composites, and in particular the effect of the distribution of the components in a binary mixture of epoxy resin on the kinetic parameters of the cure.

Chap. 15 describes the process of cure by modelling and experiments when the resin, either in a pure state or as a composite, is placed in a cylindrical mould heated by air. The effect of parameters, such as temperature and mould size, on the process is determined precisely, as well as a slight change in the kinetic and thermal properties, as a lack of accuracy for these parameters can be observed.

Chap. 16 studies the process of cure of epoxy resin and composites in a cylindrical mould heated by silicone oil. The effect of various parameters on the process is evaluated with the help of the model.

Moreover, these moulds can easily be used to simulate the injection moulding technique, and the effect of the temperature of the resin at injection on the cure cycle can be calculated precisely.

Chap. 17 looks at the association of the value of the state of cure with a mechanical property such as the resistance to torsion during the process of cure of an impregnated resin.

Chap. 18 focuses on the problem of cure of polyesters with metallic moulds heated by oil, and especially the effect of parameters such as temperature and the coefficient of surface heat transfer on the process.

Finally, in Chap. 19, an example of cure of a thin epoxy resin coating is given where the temperature is fairly low, and the hardness of the coating is correlated with the state of cure.

In this third part, where several models are used to describe the process of cure in various cases, studies are carried out by simultaneously making experiments and using models, in order to test the validity of the models.

Saint-Etienne J.M. Vergnaud
May 1991 J. Bouzon

Acknowledgements

A large part of this book (Chaps 11–19) deals with various applications and industrial problems. Several colleagues and students have by their work brought contributions worth noting.

We are grateful for the collaboration of our colleagues, A. Accetta, J.B. Rochette, M. Hivert and B. Touchard in the field of the cure of rubber. Deep gratitude is extended to the colleagues working in the field of thermosets: M. Dursapt, R. Jones, H. Liu, Y. Robert and especially to R. Granger.

Our best appreciation is given to our students: M. Abdul, C. Ferradou, Y. Khatir, A. Khouider and A. Sadr for their work in the cure of rubber; J.Y. Armand, G. Chataing for their efficient help in calorimetry; K. Azaar, M. Chater, A. El Brouzi, S. Laoubi for their very active contribution in the field of thermoset cure. Many thanks to D. Berthet for his kind help in calculation and for his drawings.

We appreciate the contributions of people from industry, and especially of P. Le Parlouer from SETARAM, and D. Lalart and F. Michel-Dansac from CDF-Chimie (Orkem).

Special thanks to our colleague Norton who read the initial drafts. Many thanks to C. Çervantes and C. Marcon for their competent typing of the manuscript.

Contents

Chapter 1

Principles and General Equations

Thermosets are polymerised and processed in a single operation which involves the irreversible transformation of a resin with a low molecular weight into a network polymer. The crosslinking reaction is dependent upon the type of resin, the relative concentrations of the curing agent, and also upon the temperature. One aspect of fundamental importance in this process is the exothermic nature of the curing reaction which can give rise to substantial temperature gradients and even to excessively high temperatures, which may cause degradation of the material.

The laws of heat transmission either within the resin and the mould or out of the mould surface are thus of very great importance in the cure operation.

1.1 Modes of Heat Transmission

Heat is transmitted by three possible means: conduction, convection and radiation. Each mode is considered separately, even if in most cases heat is transferred by more than one of these three modes.

1.1.1 Conduction

Heat is transferred by conduction in a homogeneous solid from one part to another, under the influence of a temperature gradient. As conduction involves only the transfer of kinetic energy from one molecule to an adjacent molecule, no appreciable displacement of the material is considered. When the material is transparent, some energy may be transmitted by radiation as well as by conduction. With a fluid, conduction may be supplemented by convection.

1.1.2 Convection (Natural or Forced)

Heat convection involves transfer by mixing one part of a fluid with another. Two kinds of motion of the fluid are considered:

Natural convection. The motion of the fluid is due to the differences of density throughout the fluid resulting from temperature differences at various places. The motion may be either laminar or turbulent, depending on various factors.

Forced convection. The motion of the fluid is produced by mechanical means, such as stirring.

Energy is also transferred by molecular conduction and, in transparent materials, by radiation.

1.1.3 Radiation

Radiant energy is emitted in all directions from a hot body. When radiant energy strikes another body, a part is transmitted through this body, while another part may be reflected.

Radiant energy is emitted at a rate, and of a quality, depending primarily on the temperature of the body. When two bodies, one hotter than the other, are placed in an enclosure, a continuous interchange of energy between them takes place. As the hotter body radiates more energy than it absorbs, the colder body absorbs more than it radiates. When the temperature is not very high, as in the case of thermoset curing, heat radiation is often neglected.

1.2 General Laws of Heat Transmission

The general laws of heat transmission are given in the three cases of conduction, convection and radiation.

1.2.1 Conduction (Non-steady and Steady State)

The flux of heat at a point M, at time t, along the direction of the $0x$ axis (Fig. 1.1), is defined by the relation

$$\phi_{0x}(M, t) = \frac{dQ}{dAdt} \tag{1.1}$$

where dA is the area of a small plane surface located at M and normal to the $0x$ direction, and dQ is the amount of heat passing through this surface during a short interval of time of duration dt. dQ is positive if the heat moves in the positive direction of the $0x$ axis, and negative in the opposite direction.

If the material is isotropic and if three perpendicular axes $0x$, $0y$, $0z$ are used, the flux of heat convection along the $0x$ axis is given by Fourier's relation [1]:

$$\phi_{0x}(x, y, z, t) = -\lambda \frac{\partial T}{\partial x} \tag{1.2}$$

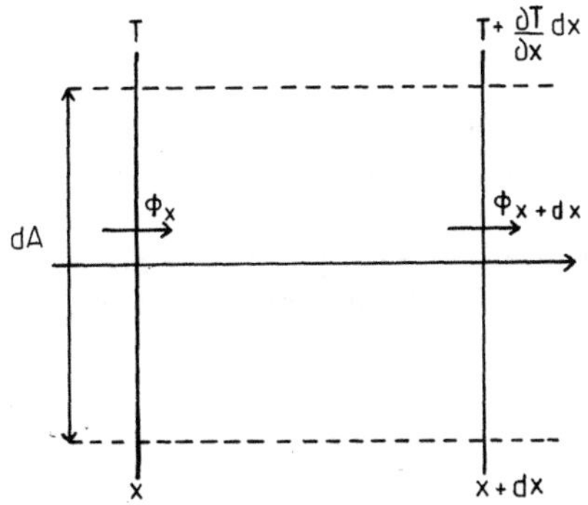

Fig. 1.1. Flux of heat ϕ across the surface dA taken at right angle to the direction of the $0x$ axis.

where $\partial T/\partial x$ is the temperature gradient along the $0x$ direction and λ is the thermal conductivity of the material, which is a physical property of this material. λ is a positive number (independent of the direction as the material is isotropic). The minus sign in Eq. (1.2) means that heat is transferred from a place at a high temperature to a place at a lower temperature.

Three heat fluxes ϕ_{0x}, ϕ_{0y}, ϕ_{0z} are obtained along the three perpendicular axes $0x$, $0y$, $0z$, respectively. These three fluxes are the coordinates of a vector $\vec{\phi}$, called the flux vector, at the point considered:

$$\vec{\phi} = -\lambda\, \overrightarrow{\text{grad}}\,(T) \tag{1.3}$$

λ has the dimensions $MLt^{-3}T^{-1}$. λ is often expressed in terms of $\text{kcal}\,\text{h}^{-1}\,\text{m}^{-1}\,°\text{C}^{-1}$ or $\text{Btu}\,\text{h}^{-1}\,\text{ft}^{-1}\,°\text{F}^{-1}$.

The differential form given in Eq. (1.2) is general for unidirectional transfer and can be applied in all cases, whatever the temperature gradient variation with time and with the location of the point considered.

When the solid is anisotropic, the thermal properties are different in different directions. In this case, the direction of heat flow at any point is not normal to the surface of constant temperature through the point. Eq. (1.2) must be replaced by the assumption:

$$\left.\begin{aligned} \phi_{0x} &= -\lambda_{11}\frac{\partial T}{\partial x} - \lambda_{12}\frac{\partial T}{\partial y} - \lambda_{13}\frac{\partial T}{\partial z} \\ \phi_{0y} &= -\lambda_{21}\frac{\partial T}{\partial x} - \lambda_{22}\frac{\partial T}{\partial y} - \lambda_{23}\frac{\partial T}{\partial z} \\ \phi_{0z} &= -\lambda_{31}\frac{\partial T}{\partial x} - \lambda_{32}\frac{\partial T}{\partial y} - \lambda_{33}\frac{\partial T}{\partial z} \end{aligned}\right\} \tag{1.2'}$$

The previous axes of coordinates can be changed, so that with the new axes the equations (1.2′) become

$$\left.\begin{aligned} \phi_{0x} &= -\lambda_{1}\frac{\partial T}{\partial x} \\ \phi_{0y} &= -\lambda_{2}\frac{\partial T}{\partial y} \\ \phi_{0z} &= -\lambda_{3}\frac{\partial T}{\partial z} \end{aligned}\right\} \tag{1.2''}$$

These new axes of coordinates are called the principal axes of heat conduction, and λ_1, λ_2, λ_3 the principal thermal conductivities.

Heat conduction in a non-steady state. When temperature at any position varies with time, the process of heat flow is called "heat conduction in a non-steady state", or "transient heat conduction".

Heat conduction in a steady state. When temperature at each point does not vary with time, heat is conducted in a steady state. The temperature gradient, as well as heat flow, is thus independent of time.

1.2.2 Convection

The rate of heat convection from the surface of a solid to a fluid is given by the equation:

$$\frac{dQ}{dt} = h_m A(T_s - T_f) \tag{1.4}$$

where

h_m is the mean coefficient of heat transfer by convection from the solid surface to fluid.

A is the area of the surface of contact.

T_s is the surface temperature of the solid.

T_f is the temperature of the fluid far from the solid surface (this distance depends on the dimensions of the solid).

The coefficient of heat transfer h_m depends on various factors, such as certain physical properties of the fluid, the dimensions of the surface as well as its position (vertical, horizontal), the difference in temperatures $(T_s - T_f)$ called sometimes the temperature potential ΔT, and especially the velocity of the fluid over the surface.

Depending on whether the heat convection is natural or forced, the values of the coefficient of heat transfer h are calculated in different ways. The coefficient of heat transfer h between the surface at T_s and the surrounding atmosphere at T_f is expressed in terms of kcal h^{-1} m^{-2} °C^{-1} or Btu h^{-1} ft^{-2} °F^{-1}.

1.2.3 Radiation

Kirchhoff's Law

When the surface of a body at a given temperature receives a heat radiation of power W and given wavelength from a given direction, this surface absorbs the fraction α and reflects the complement fraction $1 - \alpha$. This fraction α is the absorptivity of the body under these conditions. When $\alpha = 1$, whatever the conditions (direction, wavelength and temperature), the body is called a black body.

The ratio of the emissive power W of a heated surface to the emissive power from a black body at the same temperature, for the same wavelength and direction, is the emissivity of this heated surface under these conditions. Kirchhoff's law expresses that the emissivity is equal to the absorptivity for each body, under the same conditions.

Black-Body Radiation Law

When a body is heated, radiant energy is emitted at a rate and a quality which depends on the temperature of the body. The emissive power per unit area of a black body depends on its temperature only, the emissive power being proportional to the fourth power of the absolute temperature, as shown in the Stefan–Boltzmann law [2, 3]:

$$W_{\mathrm{B}} = A\sigma T^4 \tag{1.5}$$

where σ is the Stefan–Boltzmann constant ($4.96 \times 10^{-8}\ \mathrm{kcal\,m^{-2}\,h^{-1}\,K^{-4}}$) and A is the area of the black body.

The black body radiation is distributed in a spectrum, following Planck's law:

$$W_\lambda = 3.22 \times 10^{-8} \frac{\lambda^{-5}}{\exp\left(\dfrac{1.438}{\lambda T}\right) - 1} \tag{1.6}$$

W_λ being expressed in $\mathrm{kcal\,m^{-2}\,h^{-1}}$, per centimetre of wavelength (λ being the wavelength in centimetres).

When $\lambda T < 0.21$, Planck's law reduces to

$$W_\lambda = 3.22 \times 10^{-8}\,\lambda^{-5} \exp\left(-\frac{1.438}{\lambda T}\right) \tag{1.6'}$$

Radiation between the Surfaces of Solids Separated by a Non-absorbing Medium

For two infinite parallel plane surfaces facing each other, with emissivity or absorptivity ε_1 and ε_2, the net radiation interchange between the two surfaces is given per unit area by

$$\sigma(T_1^4 - T_2^4)\left(\frac{1}{\varepsilon_1} + \frac{1}{\varepsilon_2} - 1\right)^{-1} \tag{1.5'}$$

When the solid of surface A_1 at temperature T_1 is located in an enclosure of very large dimensions whose surface is at temperature T_2, the radiation emitted from the body is

$$A_1\sigma(T_1^4 - T_2^4)\varepsilon_1 \tag{1.5''}$$

1.3 Equations of Heat Conduction in Isotropic Solids

Basic equations for heat conduction are developed for homogeneous bodies of various shapes.

1.3.1 Sheet

Consider a sheet with a distribution of temperature at time t given by

$$T = f(x, t) \tag{1.7}$$

where x is a coordinate along the axis perpendicular to the sheet. Heat conduction takes place along the $0x$ axis, and the planes perpendicular to this axis at positions x and $x + dx$ are isothermal surfaces (Fig. 1.1). The flux of heat conduction is respectively ϕ_x at the x axis, and ϕ_{x+dx} at $x + dx$.

The heat balance is calculated during the increment of time dt within the solid located between the parallel planes at positions x and $x + dx$, with the area A through which heat is transferred. This heat balance is obtained by considering, on the one hand, the difference in the amounts of heat entering and leaving the volume $A dx$ during the increment of time dt, and on the other hand, the change in temperature in this volume resulting from the difference in these amounts of heat:

$$(\phi_x - \phi_{x+dx}) A\, dt = A\, dx \rho C dT \tag{1.8}$$

where

ρ is the density of the solid at temperature T.
C is the specific heat of the solid at temperature T.
dT is the increase in temperature during the increment of time dt resulting from unsteady heat conduction.
ϕ_x is the flux of heat entering the solid, at time t.
ϕ_{x+dx} is the flux of heat leaving this solid at the same time.

The difference in the flux is

$$\phi_x - \phi_{x+dx} = -\frac{\partial \phi}{\partial x} dx \tag{1.9}$$

and the heat balance becomes

$$-A \frac{\partial \phi}{\partial x} dx\, dt = A\, dx \rho C\, dT \tag{1.10}$$

or more simply

$$\rho C \frac{\partial T}{\partial t} = -\frac{\partial \phi}{\partial x} \tag{1.11}$$

As the flux of heat conduction is expressed in terms of heat conductivity and of the gradient of temperature,

$$\phi = -\lambda \frac{\partial T}{\partial x} \tag{1.2}$$

the following general equation is obtained, by combining Eqs (1.2) and (1.11):

$$\rho C \frac{\partial T}{\partial t} = \frac{\partial}{\partial x}\left(\lambda \frac{\partial T}{\partial x}\right) \tag{1.12}$$

In the simple case where thermal conductivity λ is constant, Eq. (1.12) is reduced to

$$\frac{\partial T}{\partial t} = \alpha \frac{\partial^2 T}{\partial x^2} \tag{1.13}$$

by writing the thermal diffusivity of the solid as

$$\alpha = \frac{\lambda}{\rho C} \tag{1.14}$$

The diffusivity α has the dimensions $L^2 t^{-1}$ as shown in Eq. (1.13). It is expressed in square centimetres per second, or in square metres per second, in square metres per hour, or square feet per hour.

1.3.2 Parallelepiped

The element of volume in the form of a rectangular parallelepiped whose sides are parallel to the axes of coordinates is considered (Fig. 1.2). The lengths of these sides are $2dx$, $2dy$ and $2dz$. The centre of this element is $O(x, y, z)$, and the temperature at this point is T.

The rate at which heat conduction enters the element through the surface $ABCD$ in the plane $x - dx$ is

$$4\left(\phi_x - \frac{\partial \phi_x}{\partial x}\,dx\right) dy\,dz \tag{1.15}$$

where ϕ_x is the flux of heat conduction along the Ox direction at O.

The rate of heat loss through the surface $A'B'C'D'$ is given by

$$4\left(\phi_x + \frac{\partial \phi_x}{\partial x}\,dx\right) dy\,dz \tag{1.15'}$$

The contribution to the rate of increase of heat in the element from these two surfaces is thus equal to

$$-8\,\frac{\partial \phi_x}{\partial x}\,dx\,dy\,dz \tag{1.16}$$

Similarly, it is obtained through the other surfaces

$$-8\,\frac{\partial \phi_y}{\partial y}\,dy\,dx\,dz \quad \text{and} \quad -8\,\frac{\partial \phi_z}{\partial z}\,dz\,dx\,dy \tag{1.16'}$$

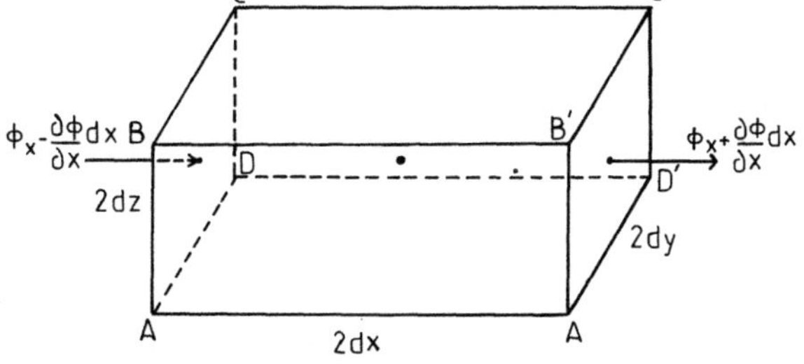

Fig. 1.2. Element of volume, a parallelepiped of sides $2dx$, $2dy$ and $2dz$.

The rate at which the amount of heat in the element increases is also expressed in terms of the rate of increase in temperature:

$$8dx\,dy\,dz\,\frac{\partial T}{\partial t}\,\rho C \tag{1.17}$$

The heat balance within the parallelepiped by considering heat conduction along the three perpendicular axes, leads to

$$\rho C\,\frac{\partial T}{\partial t} = -\frac{\partial \phi_x}{\partial x} - \frac{\partial \phi_y}{\partial y} - \frac{\partial \phi_z}{\partial z} \tag{1.18}$$

As the heat flux is expressed in terms of heat conductivity and the gradient of temperature,

$$\phi_x = -\lambda_1\,\frac{\partial T}{\partial x} \tag{1.2}$$

when the axes of coordinates coincide with the principal axes of heat conducton, Eq. (1.18) is thus rewritten as follows:

$$\rho C\,\frac{\partial T}{\partial t} = \frac{\partial}{\partial x}\left(\lambda_1\,\frac{\partial T}{\partial x}\right) + \frac{\partial}{\partial y}\left(\lambda_2\,\frac{\partial T}{\partial y}\right) + \frac{\partial}{\partial z}\left(\lambda_3\,\frac{\partial T}{\partial z}\right) \tag{1.19}$$

Eq. (1.19) stands for an anisotropic material with the principal axes of heat conduction and the principal thermal conductivities λ_1, λ_2, λ_3.

In the more common case of isotropic material, when thermal conductivity is temperature-dependent, Eq. (1.19) is rewritten with the same value of the thermal conductivity $\lambda = \lambda_1 = \lambda_2 = \lambda_3$.

$$\rho C\,\frac{\partial T}{\partial t} = \frac{\partial}{\partial x}\left(\lambda\,\frac{\partial T}{\partial x}\right) + \frac{\partial T}{\partial y}\left(\lambda\,\frac{\partial T}{\partial y}\right) + \frac{\partial}{\partial z}\left(\lambda\,\frac{\partial T}{\partial z}\right) \tag{1.20}$$

In the very simple case of isotropic material and constant thermal conductivity, the above equation is reduced to

$$\frac{\partial T}{\partial t} = \alpha\left(\frac{\partial^2 T}{\partial x^2} + \frac{\partial^2 T}{\partial y^2} + \frac{\partial^2 T}{\partial z^2}\right) = \alpha\nabla^2 T \tag{1.21}$$

1.3.3 Sphere

Often, in thermoset materials, heat conduction is radial only. The initial and surface conditions are such that the isothermal surfaces are concentric spheres, and the temperature thus depends only upon the radial abscissa r and time t. The equation of heat conduction in rectangular cartesian coordinates (Eq. (1.20)) can then be expressed in terms of the radial abscissa r and time t.

The equation of radial heat convection in a sphere can also be determined by using the following method. The rate of heat leaving the sphere of radius r is

$$\frac{\partial Q}{\partial t} = -4\pi r^2 \lambda\,\frac{\partial T}{\partial r} \tag{1.22}$$

The spherical membrane of thickness dr located between $r - (dr/2)$ and $r + (dr/2)$, the volume of which being $4\pi r^2 dr$, is considered (Fig. 1.3). The rate of heat entering this membrane is $(\partial Q/\partial t)_{r-(dr/2)}$, while the rate of heat leaving

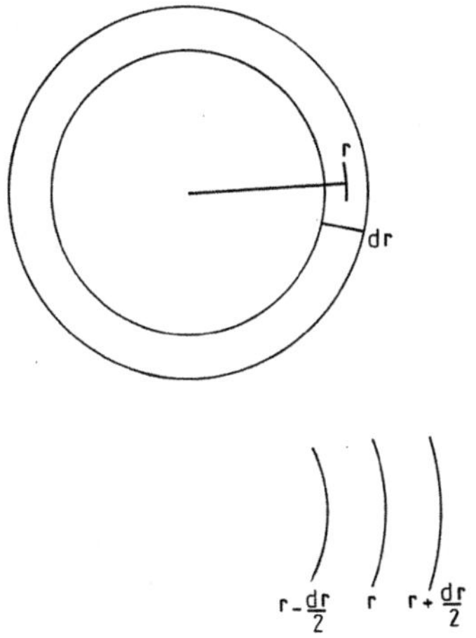

Fig. 1.3. Radial heat conduction from the centre of a sphere.

the membrane is $(\partial Q/\partial t)_{r+(dr/2)}$. The difference between these rates of heat is

$$\left(\frac{\partial Q}{\partial t}\right)_{r-(dr/2)} - \left(\frac{\partial Q}{\partial t}\right)_{r+(dr/2)} = -\frac{\partial}{\partial r}\left(\frac{\partial Q}{\partial t}\right) dr \tag{1.23}$$

But the rate at which the amount of heat varies in the spherical membrane is also given by

$$4\pi r^2 dr \rho C \frac{\partial T}{\partial t} \tag{1.24}$$

Hence, there is immediately the relation

$$4\pi r^2 dr \rho C \frac{\partial T}{\partial t} = -\frac{\partial}{\partial r}\left(-4\pi r^2 \lambda \frac{\partial T}{\partial r}\right) dr \tag{1.25}$$

which can be rewritten as follows, when the thermal conductivity is temperature-dependent:

$$\rho C \frac{\partial T}{\partial t} = \frac{1}{r^2}\frac{\partial}{\partial r}\left(\lambda r^2 \frac{\partial T}{\partial r}\right) \tag{1.26}$$

When the thermal conductivity is constant, this equation reduces to

$$\frac{\partial T}{\partial t} = \frac{\alpha}{r^2}\frac{\partial}{\partial r}\left(r^2 \frac{\partial T}{\partial r}\right) \tag{1.27}$$

which is equivalent to

$$\frac{\partial T}{\partial t} = \alpha\left[\frac{\partial^2 T}{\partial r^2} + \frac{2}{r}\frac{\partial T}{\partial r}\right] \tag{1.28}$$

At the centre ($r = 0$), these equations are not valid. Eq. (1.20) is then used. By remarking that

$$\frac{\partial}{\partial x}\left(\lambda \frac{\partial T}{\partial x}\right) = \frac{\partial}{\partial y}\left(\lambda \frac{\partial T}{\partial y}\right) = \frac{\partial}{\partial z}\left(\lambda \frac{\partial T}{\partial z}\right) = \frac{\partial}{\partial r}\left(\lambda \frac{\partial T}{\partial r}\right) \tag{1.29}$$

we obtain

$$\frac{\partial T}{\partial t} = \frac{3}{\rho C} \frac{\partial}{\partial r}\left(\lambda \frac{\partial T}{\partial r}\right) \tag{1.30}$$

But, because of the symmetry, the temperature gradient at the centre is zero. Hence at the centre, there is

$$\frac{\partial T}{\partial t} = 3 \frac{\lambda}{\rho C} \frac{\partial^2 T}{\partial r^2} \tag{1.31}$$

1.3.4 Cylinder (Infinite or Finite Length)

Infinite Length

When heat conduction is radial, and the initial and surface conditions are such that the isothermal surfaces are concentric cylinders, the temperature depends only upon the radial abscissa r and time t. The equation of heat conduction in cartesian coordinates (if the z axis is parallel to the axis of the cylinder) (Fig. 1.4)

$$\rho C \frac{\partial T}{\partial t} = \frac{\partial}{\partial x}\left(\lambda \frac{\partial T}{\partial x}\right) + \frac{\partial}{\partial y}\left(\lambda \frac{\partial T}{\partial y}\right) \tag{1.32}$$

can be expressed in terms of time t and position r.

The equation of radial heat conduction can also be determined by using the following method. The cylindrical membrane of unit length and of thickness dr, located between the radial positions $r - (dr/2)$ and $r + (dr/2)$ is considered. The rate of heat entering this membrane by conduction is $(dQ/dt)_{r-(dr/2)}$ and the rate of heat leaving this membrane is $(dQ/dt)_{r+(dr/2)}$. The difference between these rates of heat is

$$\left(\frac{dQ}{dt}\right)_{r-(dr/2)} - \left(\frac{dQ}{dt}\right)_{r+(dr/2)} = -\frac{\partial}{\partial r}\left(\frac{dQ}{dt}\right) dr \tag{1.33}$$

The volume of the cylindrical membrane per unit length is $2\pi r\, dr$, and the rate at which the amount of heat varies in this membrane is also

$$2\pi r\, dr \rho C \frac{\partial T}{\partial t} \tag{1.34}$$

As these rates of change of the amount of heat in the cylindrical membrane are identical,

$$2\pi r\, dr \rho C \frac{\partial T}{\partial t} = -\frac{\partial}{\partial r}\left(-2\pi r \lambda \frac{\partial T}{\partial r}\right) dr \tag{1.35}$$

and the final relationship is obtained when the thermal conductivity is concentration-dependent:

$$\rho C \frac{\partial T}{\partial t} = \frac{1}{r} \frac{\partial}{\partial r}\left(r \lambda \frac{\partial T}{\partial r}\right) \tag{1.36}$$

When the thermal conductivity is constant, this equation reduces to

$$\frac{\partial T}{\partial t} = \frac{\alpha}{r} \frac{\partial}{\partial r}\left(r \frac{\partial T}{\partial r}\right) \tag{1.37}$$

Eq. (1.37) can also be written in the following way:

$$\frac{\partial T}{\partial t} = \alpha\left(\frac{\partial^2 T}{\partial r^2} + \frac{1}{r}\frac{\partial T}{\partial r}\right) \tag{1.38}$$

These relations are not valid for $r = 0$. Therefore Eq. (1.32) is used. By remarking that

$$\frac{\partial}{\partial x}\left(\lambda\frac{\partial T}{\partial x}\right) = \frac{\partial}{\partial y}\left(\lambda\frac{\partial T}{\partial y}\right) = \frac{\partial}{\partial r}\left(\lambda\frac{\partial T}{\partial r}\right) \tag{1.39}$$

we obtain

$$\frac{\partial T}{\partial t} = \frac{2}{\rho C}\frac{\partial}{\partial r}\left(\lambda\frac{\partial T}{\partial r}\right) \tag{1.40}$$

Because of the symmetry, the temperature gradient at the axis is zero. Hence at the axis (for $r = 0$), there is

$$\frac{\partial T}{\partial t} = \frac{2\lambda}{\rho C}\frac{\partial^2 T}{\partial r^2} \tag{1.41}$$

Finite Length

When the flow of heat conduction takes place in planes through the axis, in addition to the radial flow, the equation of conduction, when the thermal conductivity is constant, becomes

$$\frac{\partial T}{\partial t} = \alpha\left(\frac{\partial^2 T}{\partial r^2} + \frac{1}{r}\frac{\partial T}{\partial r} + \frac{\partial^2 T}{\partial z^2}\right) \tag{1.42}$$

with the contribution of heat conduction along the longitudinal z axis (Fig. 1.5).

When the thermal conductivity is temperature-dependent, the general equation of heat conduction in the cylinder of finite length is

$$\rho C\frac{\partial T}{\partial t} = \frac{1}{r}\frac{\partial}{\partial r}\left(\lambda r\frac{\partial T}{\partial r}\right) + \frac{\partial}{\partial z}\left(\lambda\frac{\partial T}{\partial z}\right) \tag{1.43}$$

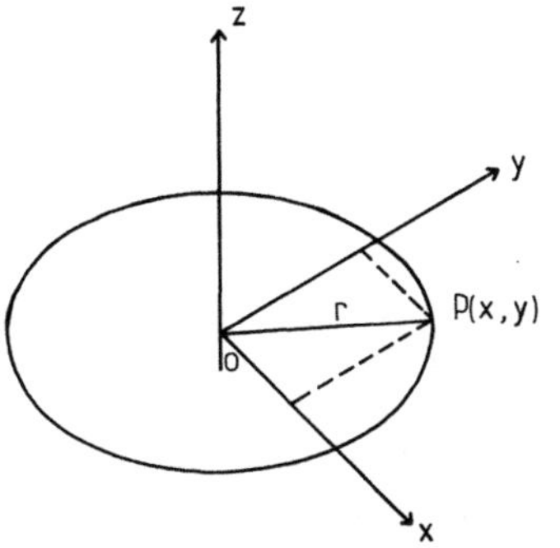

Fig. 1.4. Radial heat conduction in the circular cross-section of a cylinder of infinite length.

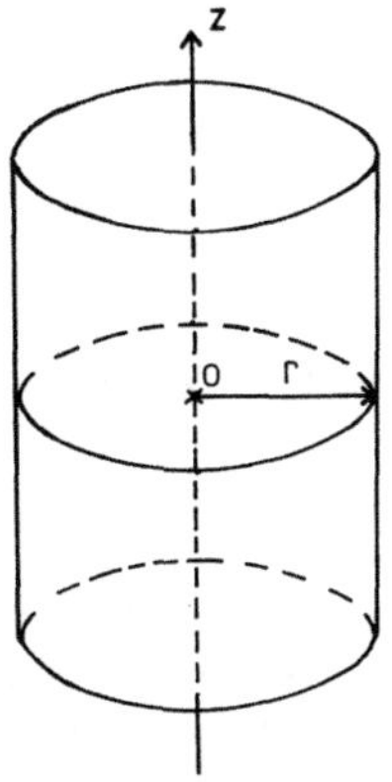

Fig. 1.5. Radial and longitudinal heat conduction in a cylinder of finite length.

1.4 Initial and Boundary Conditions

Before proceeding to the mathematical treatment (or the numerical treatment) of the problems of heat conduction, it is necessary to define the initial and boundary conditions which the temperature satisfies. These conditions result from experiment, and they are expressed by equations.

The temperature is a continuous function of time t and of coordinates x, y, z defining position in the interior of the material. The first derivative of temperature with regard to time t, and second derivative of temperature with regard to x, y and z are also continuous functions.

At the boundary of the material, a discontinuity in the solid appears, and these above assumptions are not valid.

1.4.1 Initial Conditions

The temperature throughout the solid must be known at the instant taken as the origin of the time t. This initial condition of temperature is written as follows:

$$t = 0 \qquad T = f(x, y, z) \qquad \text{within the solid} \tag{1.44}$$

Of course, the solution of the problem must satisfy the initial condition.

The initial conditions are the conditions of temperature at the beginning of the process, or rather of each stage of the process. Very often, two stages are considered, the heating period (in order to heat the material up to a temperature at which the reaction takes place), and the cooling period (after the end of the reaction).

Generally, at the beginning of the stage of heating, the temperature is uniform throughout the solid.

The initial temperature conditions taken for the following stage of cooling are the temperature conditions obtained at the end of the previous stage of heating. Of course, the temperature through the solid is not uniform when equilibrium temperature is not attained at the end of the heating stage.

1.4.2 Boundary or Surface Conditions

Various surface conditions may arise. The following surface conditions are most frequent.

No Heat Flux at the Surface

From a practical point of view, this case is very extreme, as perfect thermal insulators do not really exist. However, from a theoretical point of view, the case is of interest, because mathematical treatment is often possible.

At all points on the surface, the differentiation of temperature in the n direction of the outward normal to the surface is equal to zero.

$$t > 0 \qquad \frac{\partial T}{\partial n} = 0 \qquad \text{(surface)} \tag{1.45}$$

Constant Temperature at the Surface

This is the other extreme case, after the case where there is no heat flux at the surface. In practice, it is difficult to immediately obtain a constant temperature at the surface. However, this is the easiest boundary condition for the mathematical treatment of heat conduction in a solid.

The surrounding atmosphere must be very large and very strongly stirred. Thus the coefficient of heat transfer h is infinite (in Eq. (1.4)), the temperature at the surface of the material considered is equal to the temperature of the surrounding atmosphere, which is constant.

$$t > 0 \qquad T_s = \text{constant} \qquad \text{(surface)} \tag{1.46}$$

Heat Flux at the Surface of Separation of Two Materials of Different Thermal Conductivities

The heat flux has the same value on both sides of the interface. The following equation is obtained over the surface of separation of the two materials denoted 1 and 2:

$$t > 0 \qquad \lambda_1 \frac{\partial T_1}{\partial n} = \lambda_2 \frac{\partial T_2}{\partial n} \qquad \text{(interface)} \tag{1.47}$$

Moreover, the temperature on each side of the surface of separation can be the same or different, according to the problems.

$$\left.\begin{array}{lll} t > 0 & T_1 = T_2 & \text{(interface, perfect contact)} \\ \text{or} \quad t > 0 & T_1 \neq T_2 & \text{(interface)} \end{array}\right\} \tag{1.48}$$

When the two slabs of materials are in intimate contact at the junction, the temperature has the same value on both sides of the interface.

Heat Transfer at the Surface with Forced Convection

When a fluid (gas or liquid) at temperature T_{ext} is forced past the surface of the material considered, the rate of transfer of heat through the surface is proportional to the temperature difference between the surface and the surrounding atmosphere ($T_s - T_{ext}$), the coefficient of proportionality being the coefficient of surface heat transfer h:

$$t > 0 \qquad \phi = h(T_s - T_{ext}) \qquad \text{(surface)} \tag{1.49}$$

The rate of heat evacuated by convection from the surface in the surrounding atmosphere is constantly equal to the rate at which heat is brought to the surface by conduction within the material:

$$t > 0 \qquad -\lambda\left(\frac{\partial T}{\partial n}\right)_s = h(T_s - T_{ext}) \qquad \text{(surface)} \tag{1.50}$$

Depending on the extreme values of the coefficient of surface heat transfer h, the previous simple cases are obtained:

$$h \to 0 \quad \text{no heat flux at the surface} \tag{1.51}$$

$$h \to \infty \quad \text{constant temperature at the surface} \tag{1.52}$$

The value of the coefficient of surface heat transfer h depends on various factors: the velocity and nature of the fluid with its density and viscosity, the shape and dimensions of the surface, the direction of the flow with respect to the surface (parallel, perpendicular). This coefficient can be calculated when the steady state has been attained, by using various dimensionless numbers such as Reynolds's, Prandtl's and Nüsselt's numbers, in many cases and especially for tubes.

Heat Transfer at the Surface with Natural Convection

When a heated surface is exposed to colder air, buoyant forces cause an upward flow of air near the surface, since the density of the air near the heated surface is less than that of the main body of the air. When the surface is colder than the air, because of greater density near the surface, the air flows downward. In both cases, heat is conducted by the gas layers and is carried away by convection or bulk motion. The process is called natural convection or free convection, although both convection and conduction are involved.

Eq. (1.49) is still valid. The value of the coefficient of surface heat transfer can be evaluated by considering the dimensionless numbers such as Grashof's, Prandtl's and Nüsselt's numbers, in many cases. In all cases, this coefficient depends on the solid surface and especially its position, shape, dimensions and temperature, and on the nature of the gas, its density, viscosity, coefficient of heat expansion and temperature. Depending on the value of the product of Grashof's and Prandtl's numbers, the range is turbulent or laminar. The coeffficient of surface heat transfer is proportional to the difference in the temperatures of the heated solid and the bulk air far away from the surface, to the power 0.33 in the turbulent range and to the power 0.25 in the laminar range:

$$\phi = h_0 |T_s - T_{ext}|^m (T_s - T_{ext}) \tag{1.49'}$$

m being 0.33 or 0.25.

The rate of transfer of heat through the solid surface is thus not proportional to the temperature difference between the surface and the surrounding fluid, but to the 1.33th power of it in the turbulent range and to the 1.25th power of it in the laminar range.

1.5 General Equation of Cure Reaction

The cure of thermosets is a process whereby chemical crosslinks are introduced between the chains, resulting in the formation of a three-dimensional network. These curing reactions are highly exothermic, and the process of cure can be described by the enthalpy of cure and the kinetics of reaction, or rather by the rate at which the heat of cure is evolved. Both the enthalpy and the kinetics of cure can be determined by calorimetry. In this technique, enthalpy changes accompanying chemical events in the system are monitored. The result is a profile of the rate of enthalpy change either as a function of temperature in differential scanning calorimetry (DSC) when the sample is heated at a known linear rate, or as a function of time in differential calorimetry (DC) when the calorimeter is held at a constant temperature. Detailed kinetic analysis of the enthalpy profile is only possible in the case of the simplest of reactions and the determination of meaningful kinetic parameters for complex reactions is generally not feasible. However, the kinetics of heat evolved from the overall reaction can be obtained.

1.5.1 Studies of Cure Reaction by Calorimetry

Numerous techniques have been developed to characterise the cure process of thermosets or elastomers, and most of them are time-consuming and tedious. The best approach in such studies at this time are based on DSC and DC.

The principles of these two calorimetry techniques are summarised, as are their advantages and drawbacks.

DC Experiment: Isothermal Conditions

In the DC experiment, the enthalpy of all reactions occurring at a constant applied temperatue up to some given time is determined. The calorimeter is stabilised at the desired temperature, and the sample, previously at room temperature, is introduced. After the sample has been given a complete cure for a known time, it is cooled very quickly to the same initial temperature and thus reintroduced into the calorimeter in order to obtain the reference baseline. The process seems to be very easy, as the kinetics of heat evolved is obtained at a constant temperature; however, many drawbacks appear.

In fact, the applied temperature is the set calorimeter temperature and not the true sample temperature during cure, for the following reasons:

1. It takes some time for the sample to reach the calorimeter temperature, and a heating period exists for the sample to go from room temperature to that of the calorimeter after its introduction.
2. Local heating arises from the exothermic reaction, and the temperature is higher in the sample than in the calorimeter, when heat is emitted from the sample to the calorimeter detector.
3. Gradients of temperature of various sorts are developed within the sample, either during the heating period or during the reaction period.

Moreover, two experimental limitations on the temperature range over which meaningful data can be obtained result from the principle of this technique. The lowest temperature for cure studies is determined by the calorimeter sensitivity. If the cure rate is very low, the rate of enthalpy change is too small to be recorded. The highest temperature at which measurements can be obtained is determined by the induction period for the cure reactions to initiate. If the induction period is shorter than the time required for the calorimeter to establish thermal equilibrium after the sample is introduced, the first part of the cure exotherm is not obtained. As a result, a temperature window of around 30 °C width is found over which meaningful data can be obtained.

DSC Experiment

In the DSC experiment, starting with fresh uncured material, the sample is scanned at a selected heating rate up to a temperature at which the reaction exotherm is complete. After the sample has been given this precure, it is quench-cooled to prevent further reactions and then scanned under conditions identical to the first scan.

The heating rate is an important factor, and the optimal heating rate depends on the size of the sample and the rate of the cure reaction. The following statement can be made: the higher the rate of reaction (or the greater the size of the sample), the lower the heating rate (Chap. 13).

1.5.2 Expression of the Kinetics of Reaction

The assumption in the calorimeter calculation is that the partial enthalpy evolved up to time t, as a fraction of the total enthalpy evolved, represents the conversion up to time t. Such assumptions are the basis for the kinetic analysis of the reaction exotherm. This approach allows simple mathematical reduction of the data to yield apparent kinetic constants which are useful in evaluating the state of cure and the characteristics of the process.

The degree of cure or extent of reaction, at time t is defined by the ratio

$$\frac{Q_t}{Q_\infty} \tag{1.53}$$

in which Q_t is the amount of heat generated up to time t, and Q_∞ is the total amount of heat generated during a complete cure.

The state of cure (SOC) is expressed in percent:

$$\mathrm{SOC} = 100\,\frac{Q_t}{Q_\infty} \tag{1.54}$$

The time and temperature dependence of the degree of cure throughout the curing reaction is represented by a mathematical equation. The following equation is claimed to have a quasi-theoretical justification and has been found to describe satisfactorily experimental data for the cure of both epoxy and polyester resins. According to this equation, the rate of reaction is related to the degree of cure through the expression

$$\frac{1}{Q_\infty}\,\frac{dQ_t}{dt} = \left[k + k_1\left(\frac{Q_t}{Q_\infty}\right)^m\right]\left(1 - \frac{Q_t}{Q_\infty}\right)^n \tag{1.55}$$

in which k and k_1 are coefficients depending on temperature as follows:

$$k = k_0 \exp\left(-\frac{E}{RT}\right) \tag{1.56}$$

and m and n are constants which are independent of temperature. Very often, a simple equation for the kinetics of cure reaction is used with only one rate constant k, this rate constant being expressed in terms of the temperature by the Arrhenius equation

$$\frac{1}{Q_\infty}\frac{dQ_t}{dt} = \left(1 - \frac{Q_t}{Q_\infty}\right)^n k_0 \exp\left(-\frac{E}{RT}\right) \tag{1.57}$$

where

n is the apparent order of the reaction.
E is the apparent energy of activation, which is constant.
k_0 is the pre-exponential factor, which is constant.
R is the ideal gas constant.
T is the absolute temperature.

1.6 General Equation for Heat Transfer with Internal Reaction

The most general equation for heat transferred through a material is obtained by considering not only heat conduction through the solid but also internal heat generated by cure reactions. The rate of change in temperature in the sample is thus given by

$$\rho C \frac{\partial T}{\partial t} = \operatorname{div}(\lambda \operatorname{grad} T) + \rho \frac{\partial Q}{\partial t} \tag{1.58}$$

where Q is the heat generated by the reaction in unit mass, up to time t, at position x, y, z and T is the temperature at position x, y, z and time t. The contribution to this rate of change appears on the right-hand side, the first term due to heat conduction and the second term due to heat evolved from cure reaction. No analytical solution is obtained for this general equation, as the rate of reaction depends on time and temperature. This general problem has thus to be solved by using numerical models with finite differences.

References

1. Fourier JBJ. In: Theorie analytique de la chaleur. Gauthier-Villars, 1822
2. Stefan J. Math Naturw Kl, Sitzer Kais Akad Wiss Wien 1879; 79: 391
3. Boltzmann L. Wied Annal 1884; 22: 291

Chapter 2

Methods of Solution of Equations of Heat Conduction with Constant Thermal Parameters and Without Reaction

2.1 Introduction

Solutions of the heat conduction equation are obtained for a variety of initial and boundary conditions when the thermal parameters – thermal conductivity λ, density ρ, specific heat C, and thus thermal diffusivity α – are constant. Usually solutions have one or two standard forms. One is in the form of a series of error functions or related integrals, which is most suitable for calculation in the early stages of heat conduction, as it converges rather quickly. The other is expressed in terms of a trigonometrical and exponential series, which is of interest for large values of time, because it converges very quickly under these conditions.

In the case of a cylinder of infinite length, and partly for the cylinder of finite length, the trigonometrical series is replaced by a series of Bessel functions to represent the radial heat conduction.

Three methods of solution are described in this chapter:

1. Method of separation of variables, with products of a function of space and a function of time.
2. Method of reflection and superposition, with error functions.
3. Method employing the Laplace transform. It is an operator method which is very powerful in obtaining some types of analytical solutions.

2.2 Separation of Variables

Assuming that the variables are separable is the basic method of obtaining a solution of the partial differential equation of heat conduction.

$$\frac{\partial T}{\partial t} = \alpha \frac{\partial^2 T}{\partial x^2} \tag{2.1}$$

At first, an attempt to find solutions is made by putting the temperature function of space and time:

$$T_{x,t} = T_x T_t \tag{2.2}$$

where T_x and T_t are functions of space x and time t, respectively.

Substitution in Eq. (2.1) gives

$$T_x \frac{dT_t}{dt} = \alpha T_t \frac{d^2 T_x}{dx^2} \tag{2.3}$$

which is rewritten after the separation of variables:

$$\frac{1}{T_t} \frac{dT_t}{dt} = \frac{\alpha}{T_x} \frac{d^2 T_x}{dx^2} \tag{2.3'}$$

Eq. (2.3′) must be verified whatever the time t and space x. The left-hand side depends on time t only, while the right-hand side depends on space x only. The two members are thus equal to the same constant, which is conveniently taken as $-\omega^2 \alpha$.

Therefore, the two ordinary differential equations result in

$$\frac{1}{T_t} \frac{dT_t}{dt} = -\omega^2 \alpha \tag{2.4}$$

$$\frac{1}{T_z} \frac{d^2 T_x}{dx^2} = -\omega^2 \tag{2.5}$$

The solution of these equations being

$$T_t = \exp(-\omega^2 \alpha t) \tag{2.6}$$

$$T_x = A \sin \omega x + B \cos \omega x \quad \text{if } \omega \neq 0 \quad \text{(non-steady state)} \tag{2.7}$$

$$T_x = ax + b \quad \text{if } \omega = 0 \quad \text{(steady state)} \tag{2.7'}$$

A particular solution of Eq. (2.1) is of the form

$$T_{x,t} = (A \sin \omega x + B \cos \omega x) \exp(-\omega^2 \alpha t) \tag{2.8}$$

where ω, A and B are arbitrary constants.

The most general solution of Eq. (2.1) is obtained by summing solutions of the type given by Eq. (2.8), and a linear function $ax + b$:

$$T_{x,t} = ax + b + \sum_{n=0}^{\infty} (A_n \sin \omega_n x + B_n \cos \omega_n x) \exp(-\omega_n^2 \alpha t) \tag{2.9}$$

The values of parameters a, b, A_n, B_n and ω_n have to be determined for each particular problem, by considering that the general solution must satisfy the initial and boundary conditions of the problem.

2.3 Reflection and Superposition

2.3.1 Thin Plane Source

The expression of temperature in terms of space x and time t with the constant A

$$T = At^{-0.5} \exp\left(- \frac{x^2}{4\alpha t} \right) \tag{2.10}$$

is a particular solution of the equation for heat conduction in one dimension when the heat diffusivity α is constant:

$$\frac{\partial T}{\partial t} = \alpha \frac{\partial^2 T}{\partial x^2} \tag{2.1}$$

Eq. (2.1) still holds whatever the scale chosen for temperature. This expression of temperature is symmetrical with respect to $x = 0$, tends to zero when $x \to \pm\infty$, and for $t = 0$ it vanishes everywhere except at $x = 0$, where it becomes infinite.

Consider Q, the amount of heat per unit cross-section, initially located within the place $x = 0$, which is transferred by conduction in one dimension, along the x axis.

The conditions of the problem are

$$t = 0 \quad x = 0 \quad Q \neq 0 \quad x \neq 0 \quad T = 0 \tag{2.11}$$

$$t > 0 \quad \rho C \int_{-\infty}^{+\infty} T\, dx = Q \tag{2.12}$$

By combining Eqs (2.10) and (2.12), the following is found:

$$Q = \frac{\rho C A}{\sqrt{t}} \int_{-\infty}^{+\infty} \exp\left(- \frac{x^2}{4\alpha t} \right) dx \tag{2.13}$$

As it is known that

$$\int_{-\infty}^{+\infty} \exp(-v^2)\, dv = \sqrt{\pi} \tag{2.14}$$

it is obvious that

$$Q = 2A\rho C\sqrt{\pi\alpha} \tag{2.15}$$

and finally the solution is

$$T_{x,t} = \frac{Q}{2\rho C\sqrt{\pi\alpha t}} \exp\left(- \frac{x^2}{4\alpha t} \right) \tag{2.16}$$

This equation describes the transfer by conduction of the amount of heat Q initially located in the plane $x = 0$, per unit area, at time $t = 0$.

2.3.2 Reflection at a Boundary

The concept of reflection at a boundary is very useful for solving other problems in linear heat flow by conduction.

In the previous problem, half the heat is conducted in the direction of positive x and the other half along negative x. If there is an insulated boundary at $x = 0$ protecting the region $x < 0$ from heat, all the heat is conducted in the direction of positive x. The solution for negative x can be considered as reflected at the plane $x = 0$ and superposed on the original distribution in the region $x > 0$. As the original solution is symmetrical with respect to $x = 0$, the temperature distribution for the semi-infinite medium is

$$T_{x,t} = \frac{Q}{\rho C\sqrt{\pi \alpha t}} \exp\left(\frac{-x^2}{4\alpha t}\right) \tag{2.17}$$

2.3.3 Heat Located in a Semi-infinite Medium

For a semi-infinite medium, heat is initially located in the region $x < 0$. The initial and boundary conditions are thus

$$t = 0, \qquad T = T_1 \quad \text{for } x < 0, \qquad T = 0 \quad \text{for } x > 0 \tag{2.18}$$

This is the initial temperature distribution along two identical long metal bars placed in contact end to end, one being hot and the other cold.

The extended distribution is thus composed of an infinite number of line sources of heat. The solution to this problem is thus obtained by superposing the corresponding infinite number of elementary solutions for each line source of heat.

Consider (Fig. 2.1) the heat in the element of width $d\varepsilon$ and unit cross-section, $\rho C T_1 d\varepsilon$. The temperature at point P, distance ε from this element, resulting from conduction of this amount of heat, is at time t

$$dT = \frac{T_1 d\varepsilon}{2\sqrt{\pi \alpha t}} \exp\left(-\frac{\varepsilon^2}{4\alpha t}\right) \tag{2.19}$$

The complete solution due to the initial heat distribution shown in Eq. (2.18), is given by integrating over the space between x and ∞:

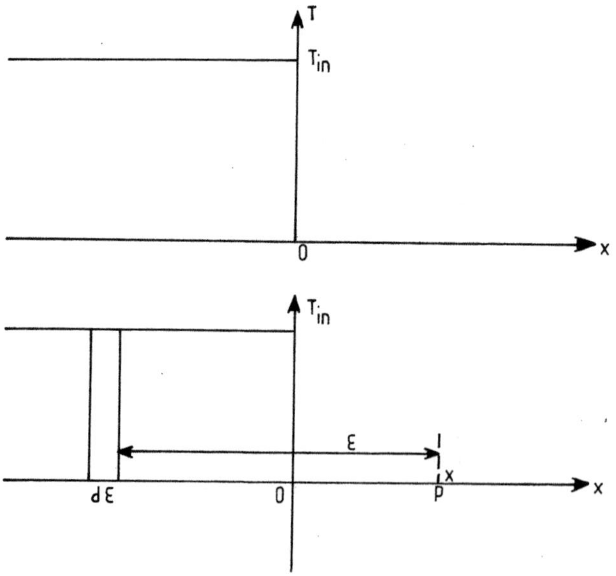

Fig. 2.1. Initial distribution of temperature in two semi-infinite media.

$$T_{x,t} = \frac{T_1}{2\sqrt{\pi \alpha t}} \int_x^{\infty} \exp\left(-\frac{\varepsilon^2}{4\alpha t}\right) d\varepsilon \tag{2.20}$$

or, by putting $v^2 = \varepsilon^2/4\alpha t$:

$$T_{x,t} = \frac{T_1}{\sqrt{\pi}} \int_{x/2\sqrt{\alpha t}}^{\infty} \exp(-v^2)\, dv \tag{2.21}$$

The error function erf and the error function complement erfc are thus introduced:

$$\operatorname{erf}(y) = \frac{2}{\sqrt{\pi}} \int_0^y \exp(-v^2)\, dv \qquad \operatorname{erfc}(y) = \frac{2}{\sqrt{\pi}} \int_y^{\infty} \exp(-v^2)\, dv \tag{2.22}$$

These functions have the following properties:

$$\operatorname{erf}(-y) = -\operatorname{erf}(y), \qquad \operatorname{erf}(\infty) = 1, \quad \operatorname{erf}(0) = 0, \qquad \operatorname{erf}(-\infty) = -1 \tag{2.23}$$

$$\operatorname{erfc}(y) = 1 - \operatorname{erf}(y), \quad \operatorname{erfc}(\infty) = 0, \quad \operatorname{erf}(0) = 1, \quad \operatorname{erfc}(-\infty) = 2 \tag{2.24}$$

Thus Eq. (2.21) becomes

$$T_{x,t} = \frac{T_1}{2} \operatorname{erfc}\left(\frac{x}{2\sqrt{\alpha t}}\right) \tag{2.25}$$

If the initial temperature for $x > 0$ is T_0 instead of 0, this equation becomes

$$T_{x,t} - T_0 = \frac{T_1 - T_0}{2} \operatorname{erfc}\left(\frac{x}{2\sqrt{\alpha t}}\right) \tag{2.26}$$

The initial conditions are verified, for $t = 0$ (Fig. 2.2):

$$\text{if } x < 0, \quad T_{x,0} = T_1, \qquad \text{if } x > 0, \quad T_{x,0} = T_0 \tag{2.27}$$

Moreover the following is obtained, at any time:

$$T_{0,t} = \frac{T_0 + T_1}{2} \tag{2.28}$$

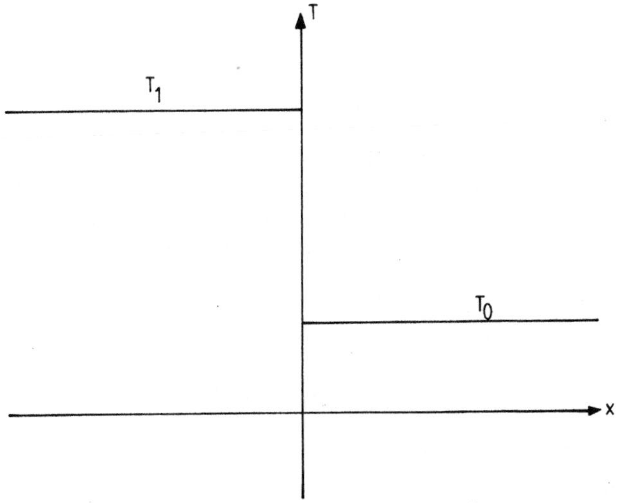

Fig. 2.2. Initial distribution of temperature in two semi-infinite media.

$$\lim_{x \to +\infty} T_{x,t} = T_0 \qquad \lim_{x \to -\infty} T_{x,t} = T_1 \tag{2.29}$$

The above equations play the role of boundary conditions.

2.3.4 Heat Located in a Sheet

In the case of a sheet, the heat Q is initially located in the region $-h < x < h$ as shown in Fig. 2.3. The initial conditions are written as follows:

$$t = 0, \quad T = T_{\text{ext}} \quad \text{for} \quad x < -h \quad \text{and} \quad x > h$$

$$T = T_{\text{in}} \quad \text{for} \quad -h < x < h \tag{2.30}$$

The integration of Eq. (2.20) is thus from $x - h$ to $x + h$ instead of from x to ∞:

$$T_{x,t} - T_{\text{ext}} = \frac{T_{\text{in}} - T_{\text{ext}}}{2\sqrt{\pi \alpha t}} \int_{x-h}^{x+h} \exp\left(-\frac{\varepsilon^2}{4\alpha t}\right) d\varepsilon \tag{2.31}$$

By using the error function shown in Eq. (2.22), the solution of the problem is

$$\frac{T_{x,t} - T_{\text{ext}}}{T_{\text{in}} - T_{\text{ext}}} = \frac{1}{2}\left[\operatorname{erf}\left(\frac{h - x}{2\sqrt{\alpha t}}\right) + \operatorname{erf}\left(\frac{h + x}{2\sqrt{\alpha t}}\right)\right] \tag{2.32}$$

For $x = \pm h$, if $t \to 0$, we obtain

$$T_{-h,0} = T_{h,0} = \frac{T_{\text{in}} + T_{\text{ext}}}{2} \tag{2.33}$$

in the same way as the semi-infinite medium (Eq. (2.28)).

2.4 Laplace Transform

2.4.1 Principle

The application of the Laplace transform to a partial differential equation removes the time variable, leading to an ordinary differential equation. The Laplace transformation is thus a useful mathematical device for the solution of different problems in physical science. In the case of heat conduction, temperature is a function of space and time, intervening in a partial differential equation (Eq. (2.1)). The solution of the ordinary differential equation obtained with the Laplace transform is thus a function of space only and an auxiliary variable p.

If $f(t)$ is a known function of time t for positive values of t, the Laplace transform $\bar{f}(p)$ of the function $f(t)$ is defined by

$$\bar{f}(p) = \int_0^\infty \exp(-pt) f(t)\, dt \tag{2.34}$$

where p is a real positive number which is sufficiently large to make the integral (2.34) converge (p may also be a complex number with a sufficiently large real part). For instance, if $f(t) = \exp(at)$, p must exceed a.

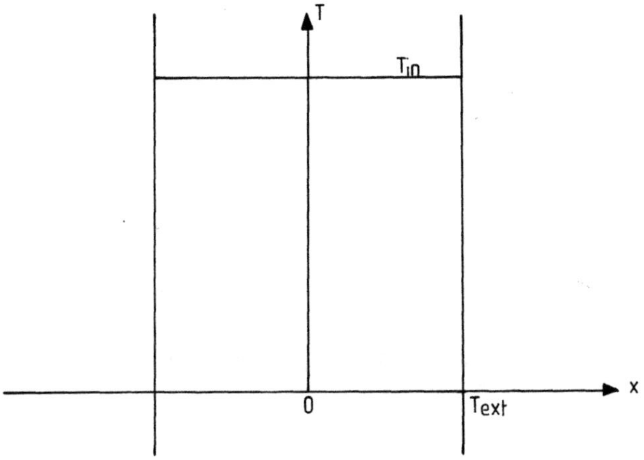

Fig. 2.3. Initial distribution of temperature in a slab of thickness $2h$ located between two semi-infinite media.

Laplace transforms of simple functions are easily obtained by integrating Eq. (2.33), as shown in the following examples:

$$f(t) = a \qquad \bar{f}(p) = \int_0^\infty a \exp(-pt)\, dt = \frac{a}{p} \tag{2.35}$$

a being constant.

$$f(t) = \exp(at) \qquad \bar{f}(p) = \int_0^\infty \exp(-pt)\exp(at)\, dt = \frac{1}{p-a} \tag{2.35'}$$

If f has a derivative, by integrating by part, we obtain

$$\bar{f'}(p) = \int_0^\infty \exp(-pt) f'(t)\, dt = [\exp(-pt) f(t)]_0^\infty - \int_0^\infty -p \exp(-pt) f(t)\, dt$$

If p is positive (and sufficiently large), then $\exp(-pt)f(t) \to 0$ when $t \to \infty$, and the following is obtained:

$$\bar{f'}(p) = -f(0) + p\bar{f}(p) \tag{2.36}$$

Tables of Laplace transforms are available, enabling us to find the function $f(t)$ when the function $\bar{f}(p)$ is known.

2.4.2 Heat Conduction Through a Semi-infinite Medium

The problem of heat conduction through a semi-infinite medium is considered, when the boundary is kept at a constant temperature T_∞, and the initial temperature is T_0 within this medium. The initial and boundary conditions are

$$t = 0 \qquad x > 0 \qquad T = T_0 \tag{2.37}$$

$$t > 0 \qquad x = 0 \qquad T = T_\infty \tag{2.38}$$

Heat conduction is defined by the partial differential equation (2.1), with constant thermal diffusivity α:

$$\frac{\partial T}{\partial t} = \alpha \frac{\partial^2 T}{\partial x^2} \tag{2.1}$$

On multiplying both sides of this equation by $\exp(-pt)$ and integrating with respect to time from 0 to ∞, the following is obtained:

$$\int_0^\infty \exp(-pt) \frac{\partial T}{\partial t} dt = \alpha \int_0^\infty \exp(-pt) \frac{\partial^2 T}{\partial x^2} dt \tag{2.39}$$

According to the principle shown in Eq. (2.36) and to the condition (2.37), the left-hand side, for $x > 0$, becomes

$$p\bar{T}(x, p) - T_0$$

The right-hand side of Eq. (2.39) can be evaluated by interchanging the orders of derivation and integration (which is justified in the present case):

$$\int_0^\infty \frac{\partial^2 T}{\partial x^2} \exp(-pt)\, dt = \frac{\partial^2}{\partial x^2} \int_0^\infty T \exp(-pt)\, dt = \frac{\partial^2 \bar{T}}{\partial x^2} \tag{2.40}$$

Eq. (2.39) is thus reduced to

$$p\bar{T}(x, p) - T_0 = \alpha \frac{\partial^2 \bar{T}}{\partial x^2}(x, p) \tag{2.41}$$

which can be written as

$$\frac{\partial^2 \bar{T}}{\partial x^2}(x, p) = \frac{p}{\alpha}\left(\bar{T}(x, p) - \frac{T_0}{p}\right) \tag{2.41'}$$

The general solution of Eq. (2.41′) is

$$\bar{T}(x, p) - \frac{T_0}{p} = A_p \exp\left(\sqrt{\frac{p}{\alpha}}\, x\right) + B_p \exp\left(-\sqrt{\frac{p}{\alpha}}\, x\right) \tag{2.42}$$

It must satisfy the condition

$$\bar{T} \text{ is bounded when } x \to \infty, \text{ and thus } A_p = 0 \tag{2.43}$$

The boundary condition (Eq. (2.38)) treated in the same way, gives

$$t > 0 \qquad x = 0 \qquad \bar{T} = \int_0^\infty T_\infty \exp(-pt)\, dt = \frac{T_\infty}{p} \tag{2.44}$$

By comparing Eqs (2.42) and (2.44) for $x = 0$, it is obvious that

$$B_p = \frac{T_\infty}{p} - \frac{T_0}{p}$$

Then the relation (2.42) becomes

$$\bar{T} = \frac{T_0}{p} + \frac{T_\infty - T_0}{p} \exp\left(-\sqrt{\frac{p}{\alpha}}\, x\right) \tag{2.45}$$

Reference to the tables of Laplace transforms shows that the following function,

$$T = T_0 + (T_\infty - T_0)\,\mathrm{erfc}\left(\frac{x}{2\sqrt{\alpha t}}\right) \tag{2.46}$$

is the reciprocal Laplace transform of function (2.45). Eq. (2.46) is the same as that obtained in the same case with the error function complement (Eq. (2.26)), if T_∞ is replaced by $(T_1 + T_0)/2$.

Chapter 3

Heat Conduction in a Plane Sheet

3.1 Introduction

Various classical cases of one-dimensional heat conduction in a medium bounded by two parallel planes are described in this chapter. The two parallel planes are located either between $x = 0$ and $x = L$ with a thickness L, or between $x = -R$ and $x = R$ with a thickness of $2R$.

In practice, these cases apply to heat conduction through

1. A plane sheet of material so thin that all the heat passes through the plane faces, and a negligible amount through the edges.
2. A rod of material of length L with no transfer of heat through its surface and heat conduction along the length direction.

Non-steady heat conduction is studied with two various conditions at the two parallel surfaces:

1. A coefficient of surface heat transfer so high that the temperatures at each plane can be assumed to be constant.
2. A finite coefficient of surface heat transfer.

Steady heat conduction is considered, especially through composite slabs. The material is isotropic, and the thermal properties are uniform throughout the sheet. Moreover, these thermal properties do not vary with temperature. The thickness of the sheet is constant, whatever the temperature.

3.2 Non-steady Conduction with Very High Coefficient of Surface Heat Transfer

If the value of the coefficient of surface heat transfer is very high, as well as the heat diffusivity of the surrounding medium (e.g., the case of a metal mould), the temperature of the surrounding medium is uniform and always equal to the temperature at the surface of the sheet. If the surrounding medium is a fluid, the surface temperature of the sheet still remains the same as that of the fluid, because of the forced heat convection, when this fluid is strongly stirred. In both cases, when the volume of the surrounding medium is very large as compared to that of the sheet, the temperature of this medium remains constant.

On the other hand, when the volume of the surrounding medium is not very large, its temperature may vary. This case is studied when the external boundaries of this surrounding medium are perfectly insulated.

3.2.1 Sheet with Uniform Initial Temperature and Constant Surface Temperatures $0 < x < L$

The differential equation to be solved is

$$\frac{\partial T}{\partial t} = \alpha \frac{\partial^2 T}{\partial x^2} \tag{3.1}$$

within the sheet, with $0 < x < L$.

The initial and boundary conditions are the following (Fig. 3.1):

$$t = 0 \qquad 0 < x < L \qquad T = T_{in} \qquad \text{(solid)} \tag{3.2}$$

$$t > 0 \qquad x = 0 \text{ and } x = L \qquad T = T_{ext} \qquad \text{(surfaces)} \tag{3.3}$$

Method of Separation of Variables, with $0 < x < L$

Profiles of Temperature Through the Sheet. The temperature in the sheet is a function of position and time $T_{x,t}$. The following general solution of heat conduction has been obtained:

$$T_{x,t} = ax + b + \sum_{n=0}^{\infty} (A_n \sin \omega_n x + B_n \cos \omega_n x) \exp(-\omega_n^2 \alpha t) \tag{3.4}$$

The values of parameters a, b, A_n, B_n and ω_n have to be determined, by writing that the expression (3.4) must satisfy the initial and boundary conditions of the problem. In this case, when $t \to +\infty$, then $T_{x,t} \to T_{ext}$, and therefore

$$a = 0 \qquad \text{and} \qquad b = T_{ext} \tag{3.5}$$

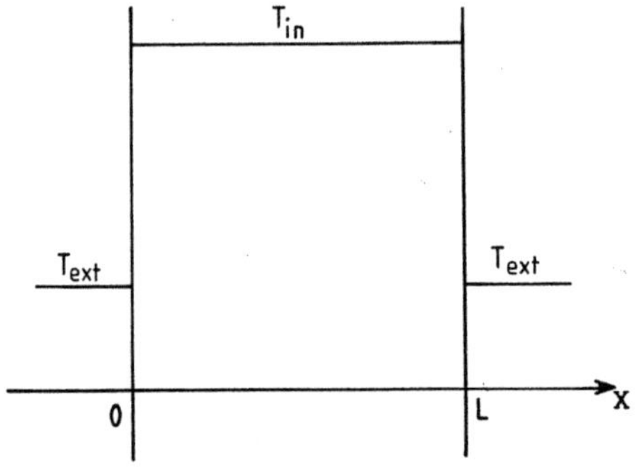

Fig. 3.1. Sheet of thickness L, with $0 < x < L$, and constant temperatures on both surfaces.

Consideration of the boundary condition yields, for $x = 0$ and $x = L$:

$$T_{x,t} - T_{\text{ext}} = 0 = \sum_{n=0}^{\infty} (A_n \sin \omega_n x + B_n \cos \omega_n x) \exp(-\omega_n^2 \alpha t) \tag{3.6}$$

The following is verified:

$$\text{for } x = 0 \quad \text{if } B_n = 0 \tag{3.7}$$

$$\text{for } x = L \quad \text{if } \sin \omega_n L = 0 \quad \text{with} \quad \omega_n = \frac{n\pi}{L} \tag{3.8}$$

Eq. (3.4) is thus reduced to:

$$T_{x,t} - T_{\text{ext}} = \sum_{n=1}^{\infty} A_n \sin\left(\frac{n\pi x}{L}\right) \exp\left(-\frac{n^2\pi^2}{L^2}\alpha t\right) \tag{3.9}$$

This equation must also satisfy the initial condition, with $t = 0$, and $0 < x < L$:

$$T_{\text{in}} - T_{\text{ext}} = \sum_{n=1}^{\infty} A_n \sin \frac{n\pi x}{L} \tag{3.10}$$

or

$$T_{\text{in}} - T_{\text{ext}} = A_1 \sin\left(\frac{\pi x}{L}\right) + \ldots + A_n \sin\left(\frac{n\pi x}{L}\right) + \ldots$$

$$+ A_p \sin\left(\frac{p\pi x}{L}\right) + \ldots \tag{3.10'}$$

The value of coefficient A_n is obtained as follows, by multiplying both sides by $\sin(n\pi x/L)$ and integrating from 0 to L:

$$(T_{\text{in}} - T_{\text{ext}}) \int_0^L \sin\left(\frac{n\pi x}{L}\right) dx = \ldots + A_n \int_0^L \sin^2\left(\frac{n\pi x}{L}\right) dx + \ldots$$

$$+ A_p \int_0^L \sin\left(\frac{n\pi x}{L}\right) \sin\left(\frac{p\pi x}{L}\right) dx + \ldots \tag{3.11}$$

By using the relationship

$$\sin\left(\frac{n\pi x}{L}\right) \sin\left(\frac{p\pi x}{L}\right) = \frac{1}{2}\left[\cos\left((n - p)\frac{\pi x}{L}\right) - \cos\left((n + p)\frac{\pi x}{L}\right)\right]$$

the following is obtained:

$$\int_0^L \sin\left(\frac{n\pi x}{L}\right) \sin\left(\frac{p\pi x}{L}\right) dx = 0 \quad \text{if } p \neq n \qquad \int_0^L \sin^2\left(\frac{n\pi x}{L}\right) dx = \frac{L}{2}$$

Moreover,

$$\int_0^L \sin\left(\frac{n\pi x}{L}\right) dx = \frac{L}{n\pi}(1 - \cos n\pi)$$

Therefore it is clear that Eq. (3.11) becomes

$$(T_{\text{in}} - T_{\text{ext}}) \frac{L}{n\pi}(1 - \cos n\pi) = A_n \frac{L}{2} \tag{3.12}$$

For even values of n, $A_n = 0$. For odd values of n, written in the form $2n + 1$ for convenience, A_{2n+1} is equal to

$$A_{2n+1} = \frac{4(T_{\text{in}} - T_{\text{ext}})}{(2n + 1)\pi} \tag{3.13}$$

and the final solution is

$$\frac{T_{x,t} - T_{\text{ext}}}{T_{\text{in}} - T_{\text{ext}}} = \frac{4}{\pi} \sum_{n=0}^{\infty} \frac{1}{2n + 1} \sin\frac{(2n + 1)\pi x}{L} \exp\left(\frac{-(2n + 1)^2\pi^2}{L^2}\alpha t\right) \tag{3.14}$$

The profile of temperature developed through the sheet is thus expressed in terms of position x and time t, as well as of the initial and external temperatures.

The profiles of temperature developed through half the thickness of the sheet of thickness L are plotted for various values of the dimensionless number $\alpha t/R^2$ or $4\alpha t/L^2$ (Fig. 3.2). Dimensionless values are used for this purpose, either for the temperature with the ratio $(T_{x,t} - T_{\text{ext}})/(T_{\text{in}} - T_{\text{ext}})$ or for the abscissa which is expressed by the ratio x/L.

Of course, for short times, steep gradients of temperature are observed next to the surface (at abscissa 0).

Amount of Heat Entering or Leaving the Sheet $0 < x < L$

The heat flux leaving each surface of the sheet is proportional to the gradient of temperature next to the surface:

$$\phi = \pm\lambda\frac{\partial T}{\partial x} \qquad \text{with sign} + \quad \text{at } x = 0 \qquad \text{sign} - \quad \text{at } x = L \tag{3.15}$$

When heat enters the sheet, the algebraic value of the heat leaving that sheet is negative.

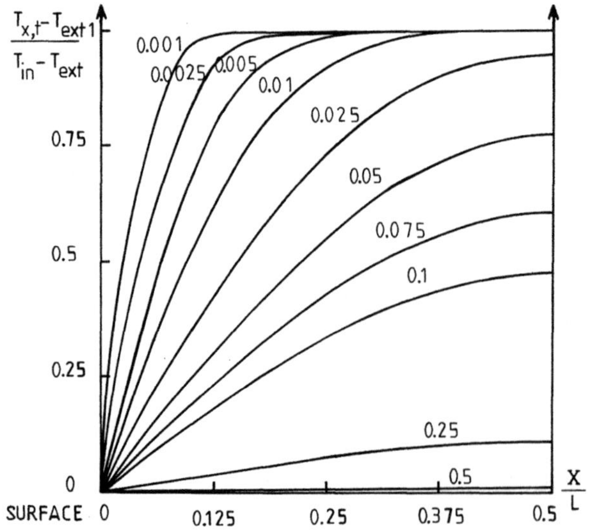

Fig. 3.2. Plane sheet of thickness L with constant temperature on the surfaces. Profiles of temperature for different times (or $4\alpha t/L^2$).

The amount of heat leaving the sheet at time t is obtained by integrating the flux for $x = 0$, with respect to time:

$$Q_t = 2\lambda A \int_0^t \frac{\partial T}{\partial x}(0, \theta)\, d\theta \qquad 0 \leq \theta \leq t \tag{3.16}$$

A being the area of each surface through which heat is transferred.

The gradient of temperature is obtained by deriving the temperature in Eq. (3.14) with respect to x. At position $x = 0$, this gradient of temperature is thus equal to

$$\frac{\partial T}{\partial x}(0, t) = \frac{4(T_{in} - T_{ext})}{L} \sum_{n=0}^{\infty} \exp\left(\frac{-(2n+1)^2\pi^2}{L^2}\alpha t\right) \tag{3.17}$$

and the amount of heat leaving the sheet through the area $2A$ is given by

$$Q_t = \frac{8(T_{in} - T_{ext})A\lambda}{L}\left[\sum_{n=0}^{\infty} \frac{-L^2}{(2n+1)^2\pi^2\alpha} \exp\left(\frac{-(2n+1)^2\pi^2}{L^2}\alpha t\right)\right]_0^t \tag{3.18}$$

or by

$$Q_t = \frac{8A\lambda L}{\pi^2\alpha}(T_{in} - T_{ext}) \sum_{n=0}^{\infty} \frac{1}{(2n+1)^2}\left[1 - \exp\left(\frac{-(2n+1)^2\pi^2}{L^2}\alpha t\right)\right] \tag{3.19}$$

As

$$\sum_{n=0}^{\infty} \frac{1}{(2n+1)^2} = \frac{\pi^2}{8} \tag{3.20}$$

the expression of the amount of heat Q_t can be rewritten as follows:

$$Q_t = \frac{A\lambda L}{\alpha}(T_{in} - T_{ext})\left[1 - \sum_{n=0}^{\infty} \frac{8}{\pi^2(2n+1)^2} \exp\left(\frac{-(2n+1)^2\pi^2}{L^2}\alpha t\right)\right] \tag{3.21}$$

The total amount of heat which can be transferred out of the sheet is

$$Q_\infty = AL\rho C(T_{in} - T_{ext}) \tag{3.22}$$

Heat values Q_t, as well as Q_∞, are negative when $T_{in} < T_{ext}$. Q_∞ is also obtained by using the equation expressing Q_t as a function of time, for infinite value of time.

It is of interest to express the amount of heat Q_t which has entered or left the sheet at time t, as a fraction of the corresponding heat after infinite time Q_∞. When $0 < x < L$ this becomes

$$\frac{Q_\infty - Q_t}{Q_\infty} = \frac{8}{\pi^2} \sum_{n=0}^{\infty} \frac{1}{(2n+1)^2} \exp\left(\frac{-(2n+1)^2\pi^2}{L^2}\alpha t\right) \tag{3.23}$$

The amount of heat leaving (or entering) the sheet up to time t, Q_t, as a fraction of the corresponding amount of heat transferred after infinite time, Q_∞, is expressed either as a function of time (Fig. 3.3, curve 1) or as function of the square root of time (Fig. 3.3, curve 2). Dimensionless numbers are used, with Q_t/Q_∞ and $\alpha t/R^2$ or $4\alpha t/L^2$ (or $\sqrt{\alpha t}/R$ or $2\sqrt{\alpha t}/L$).

A straight line is obtained when the abscissa is $2\sqrt{\alpha t}/L$, showing the linear relationship with square root of time, for values of the ratio $Q_t/Q_\infty < 0.6$.

Case when the Sheet Occupies the Space $-R < x < R$

This condition can be written as follows:

$$0 < x + R < 2R \tag{3.24}$$

It is sufficient, in Eq. (3.14), to replace L by $2R$ and x by $x + R$. Thus $\sin((2n+1)\pi x/L)$ becomes

$$\sin\left(\frac{(2n+1)\pi(x+R)}{2R}\right) = \sin\left(\frac{(2n+1)\pi x}{2R} + n\pi + \frac{\pi}{2}\right)$$

$$= (-1)^n \cos\left(\frac{(2n+1)\pi x}{2R}\right) \tag{3.25}$$

Finally, for the temperature, the following is obtained:

$$\frac{T_{x,t} - T_{\text{ext}}}{T_{\text{in}} - T_{\text{ext}}} = \frac{4}{\pi} \sum_{n=0}^{\infty} \frac{(-1)^n}{2n+1} \cos\frac{(2n+1)\pi x}{2R} \exp\left(\frac{-(2n+1)^2\pi^2}{4R^2}\alpha t\right) \tag{3.26}$$

Amount of Heat Entering or Leaving the Sheet: $-R < x < R$

Eq. (3.23) expressing the heat transferred becomes

$$\frac{Q_\infty - Q_t}{Q_\infty} = \frac{8}{\pi^2} \sum_{n=0}^{\infty} \frac{1}{(2n+1)^2} \exp\left(\frac{-(2n+1)^2\pi^2}{4R^2}\alpha t\right) \tag{3.27}$$

Method of the Laplace Transform with $-R < x < R$

The plane sheet of thickness $2R$, with $-R < x < R$, whose surfaces are maintained at constant temperature T_{ext}, and with uniform temperature T_{in} throughout the sheet initially, is considered (Fig. 3.4). There is a symmetry as regards the plane $x = 0$, and the initial and boundary conditions are

$$t = 0, \quad -R < x < R, \quad T = T_{\text{in}} \quad \text{(solid)} \tag{3.28}$$

$$t > 0, \quad x = \pm R, \quad T = T_{\text{ext}} \quad \text{(surfaces)} \tag{3.29}$$

$$t \geqslant 0, \quad x = 0, \quad \frac{\partial T}{\partial x} = 0 \quad \text{(symmetry)} \tag{3.30}$$

The equations for the Laplace transform are

$$\frac{\partial^2 \bar{T}}{\partial x^2} - q^2\left(\bar{T} - \frac{T_{\text{in}}}{p}\right) = 0, \quad 0 < x < R \quad \text{(see Chap. 2)} \tag{3.31}$$

with

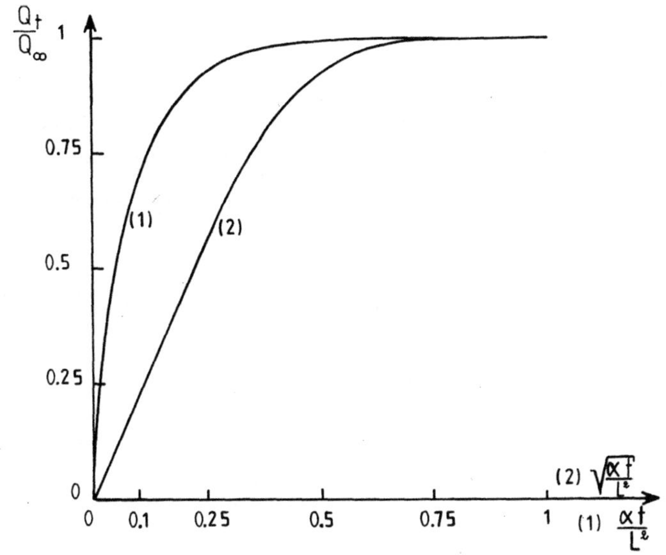

Fig. 3.3. Plane sheet of thickness L with constant temperature on the surfaces. Heat transferred at time t as a fraction of heat transferred after infinite time expressed in terms of time (curve 1) and of the square-root of time (curve 2).

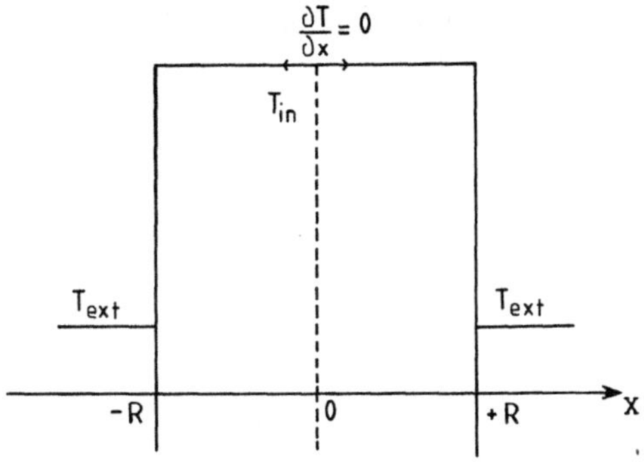

Fig. 3.4. Sheet of thickness $2R$, with $-R < x < R$, and constant temperature on both surfaces.

$$\frac{\partial \bar{T}}{\partial x} = 0, \qquad x = 0 \tag{3.32}$$

and

$$\bar{T} = \frac{T_{ext}}{p}, \qquad x = R \tag{3.33}$$

where q is defined by

$$q^2 = \frac{p}{\alpha} \tag{3.34}$$

The solution of these equations is

$$\bar{T} - \frac{T_{in}}{p} = \frac{T_{ext} - T_{in}}{p} \frac{\cosh qx}{\cosh qR} \tag{3.35}$$

The hyperbolic functions in the above equation can be expressed in terms of exponentials, and expanded in a series with the help of the binomial expression

$$\bar{T} - \frac{T_{in}}{p} = \frac{T_{ext} - T_{in}}{p} \frac{\exp(qx) + \exp(-qx)}{\exp(qR)[1 + \exp(-2qR)]} \tag{3.36}$$

and

$$\bar{T} - \frac{T_{in}}{p} = \frac{T_{ext} - T_{in}}{p} [\exp(-q(R - x)) + \exp(-q(R + x))]$$
$$\times \sum_{n=0}^{\infty} (-1)^n \exp(-2nqR) \tag{3.37}$$

From a table of transforms, for $T_{x,t}$, the following is obtained:

$$\frac{T_{x,t} - T_{in}}{T_{ext} - T_{in}} = \sum_{n=0}^{\infty} (-1)^n \operatorname{erfc} \frac{(2n + 1)R - x}{2\sqrt{\alpha t}}$$
$$+ \sum_{n=0}^{\infty} (-1)^n \operatorname{erfc} \frac{(2n + 1)R + x}{2\sqrt{\alpha t}} \tag{3.38}$$

The amount of heat entering or leaving the sheet of thickness $2R$ is obtained by integrating the above equation. This amount is expressed in terms of time t, as a fraction of the heat transferred after infinite time Q_∞:

$$\frac{Q_t}{Q_\infty} = \frac{2}{R} \sqrt{\alpha t} \left\{ \frac{1}{\sqrt{\pi}} + 2 \sum_{n=1}^{\infty} (-1)^n \operatorname{ierfc} \frac{nR}{\sqrt{\alpha t}} \right\} \tag{3.39}$$

The expression ierfc represents the integral of the error function complement, defined by

$$\operatorname{ierfc} x = \int_x^\infty \operatorname{erfc} y \, dy \tag{3.40}$$

Remarks. The following remarks are of interest.

1. The solutions for the temperature profile and heat transferred are obtained either by way of trigonometrical series or by way of error function.
2. For short durations of time, the series with error function converges very quickly, while the trigonometrical series needs a high number of terms.
3. For short durations of time, when $Q_t/Q_\infty < 0.5$, the series with the integral of the error function complement is negligible, and Eq. (3.39) can be reduced to

$$\frac{Q_t}{Q_\infty} = \frac{2}{R} \sqrt{\frac{\alpha t}{\pi}} \quad \text{for a thickness of } 2R \tag{3.41}$$

and of course to

$$\frac{Q_t}{Q_\infty} = \frac{4}{L} \sqrt{\frac{\alpha t}{\pi}} \quad \text{for a thickness of } L \tag{3.41'}$$

4. For long durations of time, when $Q_t/Q_\infty > 0.6$, the trigonometrical series

converges very quickly, and Eq. (3.27) is reduced by keeping the first term:

$$\ln\left(\frac{Q_\infty - Q_t}{Q_\infty}\right) = \ln\left(\frac{8}{\pi^2}\right) - \frac{\pi^2}{4R^2}\alpha t \quad \text{with a thickness of } 2R \qquad (3.42)$$

$$\ln\left(\frac{Q_\infty - Q_t}{Q_\infty}\right) - \ln\left(\frac{8}{\pi^2}\right) - \frac{\pi^2}{L^2}\alpha t \quad \text{with a thickness of } L \qquad (3.42')$$

3.2.2 Sheet with Initial Temperature $f(x)$ and Different Constant Surface Temperatures

This is the case of heat flow through a wall, when the surface temperatures are kept constant, and the initial temperature throughout the solid is not uniform. Various examples are of interest, when the constant surface temperatures are the same or different, and when the initial temperature throughout the solid is uniform or not.

The differential equation of heat conduction throughout the solid of thickness L is

$$\frac{\partial T}{\partial t} = \alpha \frac{\partial^2 T}{\partial x^2} \quad \text{for} \quad 0 < x < L \qquad (3.1)$$

The initial and boundary conditions are

$$t = 0, \quad 0 < x < L, \quad T = f(x) \quad \text{(solid)} \qquad (3.43)$$

$$t > 0, \quad x = 0, \quad T = T_{\text{ext1}} = T_1 \quad \text{(surface)}$$

$$x = L, \quad T = T_{\text{ext2}} = T_2 \quad \text{(surface)} \qquad (3.44)$$

The general method is to reduce this problem, with constant and different surface temperatures, to a case of a steady temperature, and a case where the surfaces are kept at a zero temperature. The function T is written as follows:

$$T = V + W \qquad (3.45)$$

where V is a function of x only which satisfies

$$\frac{d^2V}{dx^2} = 0, \quad 0 < x < L \qquad (3.46)$$

throughout the solid, and

$$V = T_1, \quad x = 0$$

$$V = T_2, \quad x = L \qquad (3.47)$$

at the surfaces of the sheet. W is a function of x and t, which satisfies

$$\frac{\partial W}{\partial t} = \alpha \frac{\partial^2 W}{\partial x^2}, \quad 0 < x < L \qquad (3.48)$$

throughout the solid, and

$$W = 0, \quad \text{for} \quad x = 0 \quad \text{and} \quad x = L \qquad (3.49)$$

at the surfaces of the sheet, and is initially

$$t = 0, \quad W = f(x) - V, \quad 0 < x < L \qquad (3.50)$$

Clearly, the function (3.45) satisfies all the conditions of the problem, so that the solution is reduced to that of two problems, one of a steady temperature, and one of a variable temperature with a given initial temperature and constant surface temperature.

It is found at once that

$$V = T_1 + (T_2 - T_1)\frac{x}{L} \tag{3.51}$$

It follows from Sect. 3.2.1 that

$$W = \sum_1^\infty A_n \sin\frac{n\pi x}{L}\exp\left(-\frac{n^2\pi^2}{L^2}\alpha t\right) \tag{3.52}$$

where the coefficient A_n is obtained by

$$A_n\frac{L}{2} = \int_0^L\left[f(x') - \left\{T_1 + (T_2 - T_1)\frac{x'}{L}\right\}\right]\sin\frac{n\pi x'}{L}\,dx' \tag{3.53}$$

The profile of temperature is thus given by

$$T_{x,t} = T_1 + (T_2 - T_1)\frac{x}{L} + \frac{2}{\pi}\sum_1^\infty \frac{T_2\cos n\pi - T_1}{n}\sin\frac{n\pi x}{L}\exp\left(\frac{n^2\pi^2}{-L^2}\alpha t\right)$$
$$+ \frac{2}{L}\sum_1^\infty \sin\frac{n\pi x}{L}\exp\left(\frac{n^2\pi^2}{-L^2}\alpha t\right)\int_0^L f(y)\sin\frac{n\pi y}{L}\,dy \tag{3.54}$$

Case of Constant Initial Temperature: $f(x) = T_{in}$ $\quad 0 < x < L$

This is the case of heat flow through a wall of thickness L, when one side is kept at a constant temperature T_1 and the other $x = L$ at T_2, and the wall is initially at a uniform temperature T_{in} (Fig. 3.5). In this case there is a finite interval of time during which the steady-state condition is set up. During this time, the temperature within the wall changes according to

$$T_{x,t} = T_1 + (T_2 - T_1)\frac{x}{L} + \frac{2}{\pi}\sum_1^\infty \frac{T_2\cos n\pi - T_1}{n}\sin\frac{n\pi x}{L}\exp\left(-\frac{n^2\pi^2}{L^2}\alpha t\right)$$
$$+ \frac{4T_{in}}{\pi}\sum_{m=0}^\infty \frac{1}{2m+1}\sin\frac{(2m+1)\pi x}{L}\exp\left(-\frac{(2m+1)^2\pi^2}{L^2}\alpha t\right) \tag{3.55}$$

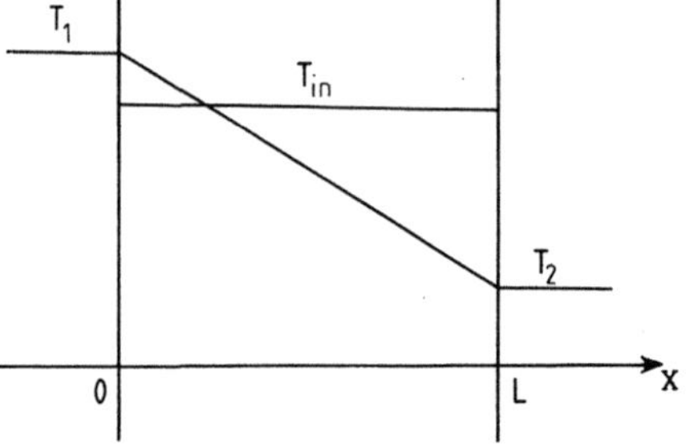

Fig. 3.5. Sheet of thickness L, with $0 < x < L$, initially at uniform temperature T_{in}, with constant temperature at each surface: T_1 for $x = 0$, T_2 for $x = L$.

As time approaches infinity, the terms involving the exponentials vanish, and a linear temperature distribution within the wall is obtained:

$$T_x = T_1 + (T_2 - T_1)\frac{x}{L} \tag{3.56}$$

The rate at which heat emerges from unit area of side $x = L$ of the wall is given by

$$\phi = -\lambda \frac{\partial T}{\partial x}(L, t) \tag{3.15}$$

The amount of heat which has left the unit area of the outer surface of the wall at time t is thus obtained by integrating the flux with respect to time:

$$Q'_t = -\lambda \int_0^t \frac{\partial T}{\partial x}(L, \theta)\, d\theta, \qquad 0 \leqslant \theta \leqslant t \tag{3.57}$$

The gradient of temperature at the plane $x = L$ is

$$\frac{\partial T}{\partial x}(L, t) = (T_2 - T_1)\frac{1}{L} + \frac{2}{L}\sum_0^{\infty}(T_2 \cos n\pi - T_1)(-1)^n \exp\left(-\frac{n^2\pi^2}{L^2}\alpha t\right)$$
$$- 4\frac{T_{\text{in}}}{L}\sum_{m=0}^{\infty}\exp\left(\frac{-(2m+1)^2\pi^2}{L^2}\alpha t\right) \tag{3.58}$$

and the amount of heat passing the unit area of the wall is thus

$$Q'_t = -\frac{\lambda t}{L}(T_2 - T_1) + \frac{2\lambda L}{\alpha\pi^2}\sum_1^{\infty}\frac{(-1)^n T_1 - T_2}{n^2}\left\{1 - \exp\left(\frac{-n^2\pi^2}{L^2}\alpha t\right)\right\}$$
$$+ \frac{4\lambda L T_{\text{in}}}{\pi^2\alpha}\sum\frac{1}{(2m+1)^2}\left\{1 - \exp\left(\frac{-(2m+1)^2\pi^2}{L^2}\alpha t\right)\right\} \tag{3.59}$$

The amount of heat which has left the sheet through the side $x = 0$ per unit area, Q''_t, is obtained by permuting T_1 and T_2 in the expression of Q'_t.

The amount of heat which has been accumulated in the wall at time t, Q_t, is given by

$$Q_t = -Q'_t - Q''_t$$
$$= \frac{2\lambda L}{\alpha\pi^2}\sum_1^{\infty}\frac{(-(-1)^n + 1)(T_1 + T_2)}{n^2}\left\{1 - \exp\left(\frac{-n^2\pi^2}{L^2}\alpha t\right)\right\}$$
$$- \frac{8\lambda L T_{\text{in}}}{\pi^2\alpha}\sum_0^{\infty}\frac{1}{(2m+1)^2}\left\{1 - \exp\left(\frac{-(2m+1)^2\pi^2}{L^2}\alpha t\right)\right\} \tag{3.59'}$$

The corresponding amount for infinite time, Q_∞, is given by

$$Q_\infty = \rho C L\left(\frac{T_1 + T_2}{2} - T_{\text{in}}\right) \quad \text{with} \quad \rho C = \frac{\lambda}{\alpha}$$

Moreover

$$\sum_{m=0}^{\infty}\frac{1}{(2m+1)^2} = \frac{\pi^2}{8}$$

Therefore the amount of heat which has been accumulated in the wall at time t, Q_t, as a fraction of the corresponding amount accumulated during infinite time, is expressed in terms of time by

$$\frac{Q_\infty - Q_t}{Q_\infty} = \frac{8}{\pi^2} \sum_{n=0}^{\infty} \frac{1}{(2n+1)^2} \exp\left(-\frac{(2n+1)^2\pi^2}{L^2}\alpha t\right) \tag{3.60}$$

where L is the thickness of the wall. This expression is identical to Eq. (3.23).

In the simple case where the wall is initially at zero temperature and the temperature at the side through which heat emerges ($x = L$) is maintained at zero temperature, Eq. (3.60) becomes

$$Q'_t = \frac{T_1 t\lambda}{L} + \frac{2\lambda L}{\alpha\pi^2} \sum_{n=1}^{\infty} \frac{(-1)^n}{n^2} T_1 - \frac{2\lambda L T_1}{\alpha\pi^2} \sum_{n=1}^{\infty} \frac{(-1)^n}{n^2} \exp\left(\frac{-n^2\pi^2}{L^2}\alpha t\right) \tag{3.61}$$

When $t \to \infty$, this curve $Q'_t = f(t)$ approaches the line

$$Q_t = \frac{T_1\lambda}{L}\left(t - \frac{L^2}{6\alpha}\right) \tag{3.62}$$

because

$$\sum_{n=1}^{\infty} \frac{(-1)^n}{n^2} = -\frac{\pi^2}{12} \quad \text{and} \quad \exp\left(-\frac{n^2\pi^2}{L^2}\alpha t\right) \to 0 \tag{3.63}$$

The intersection of this line with the time-axis has an abscissa t given by

$$t = \frac{L^2}{6\alpha} \tag{3.64}$$

The approach to steady-state heat flow through the wall is illustrated in Fig. 3.6.

3.2.3 Sheet Immersed in a Well-Stirred Fluid of Finite Volume

The plane sheet initially at temperature T_{is} is immersed in a limited volume of fluid initially at temperature T_{ie}. It is assumed that there is no transfer of heat between this fluid and the external medium. As there is a limited volume of fluid, the temperature in the fluid falls as heat enters the sheet. When the fluid is well-stirred the temperature of the fluid depends only on time. Moreover, the external surface of the fluid is perfectly insulated. This temperature is determined by the condition that heat balance is fulfilled.

The infinite sheet of thickness $2R$ is immersed in the fluid of thickness $2L'$. (Fig. 3.7). The sheet occupies the space

$$-R \leqslant x \leqslant R \tag{3.65}$$

and the fluid of limited volume occupies the space

$$-L' - R \leqslant x \leqslant -R \quad \text{and} \quad R \leqslant x \leqslant R + L' \tag{3.66}$$

The temperature in the fluid is always uniform, while the temperature in the sheet varies with time and space.

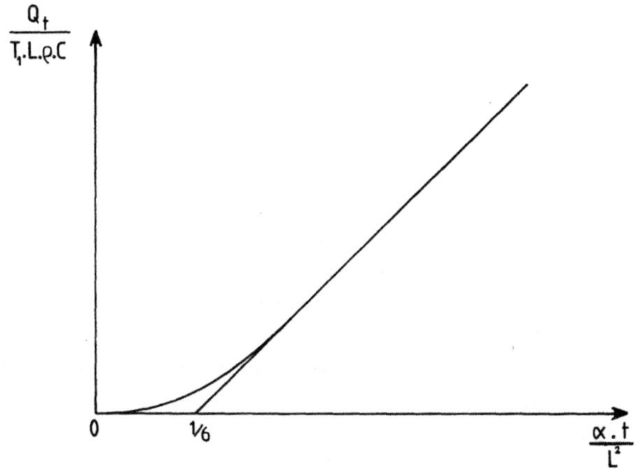

Fig. 3.6. Curve showing the approach to steady-state heat flow through a plane wall.

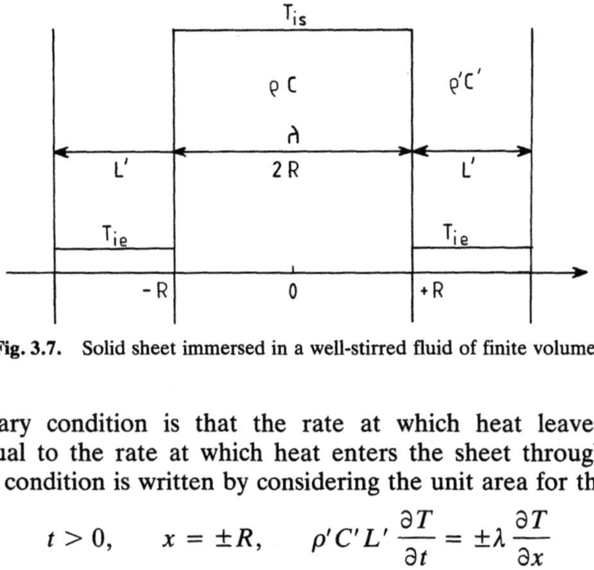

Fig. 3.7. Solid sheet immersed in a well-stirred fluid of finite volume.

The boundary condition is that the rate at which heat leaves the fluid is constantly equal to the rate at which heat enters the sheet through the surfaces $x = \pm R$. This condition is written by considering the unit area for these surfaces:

$$t > 0, \qquad x = \pm R, \qquad \rho' C' L' \frac{\partial T}{\partial t} = \pm \lambda \frac{\partial T}{\partial x} \tag{3.67}$$

The equation of heat conduction through the solid sheet is

$$\frac{\partial T}{\partial t} = \alpha \frac{\partial^2 T}{\partial x^2} \tag{3.1}$$

with the initial conditions

$$t = 0, \qquad -R < x < R, \qquad T = T_{\text{is}} \tag{3.68}$$

The solution of this problem is given in a form expressing the amount of heat transferred at time t, Q_t, as a fraction of the amount of heat transferred after infinite time, Q_∞:

$$\frac{Q_\infty - Q_t}{Q_\infty} = \sum_{n=1}^{\infty} \frac{2\beta(1+\beta)}{1 + \beta + \beta^2 q_n^2} \exp\left(-\frac{q_n^2}{R^2}\alpha t\right) \tag{3.69}$$

where the q_ns are the non-zero positive roots of

$$\tan q_n = -\beta q_n \tag{3.70}$$

and β is the ratio of the values of parameters in the fluid and in the sheet:

$$\beta = \frac{\rho' C' L'}{\rho C R} = \frac{T_\infty - T_{\text{is}}}{T_{\text{ie}} - T_\infty} \tag{3.71}$$

The positive roots of Eq. (3.70) are given in Table 3.1.

The amount of heat which has entered (or left) the sheet at time t, as a fraction of the corresponding quantity after infinite time, is expressed in terms of time, or rather of the dimensionless number $\alpha t/R^2$, for various values of the dimensionless number β (Fig. 3.8). Of course, for a very high value of β, the kinetics of heat transferred are the same as that obtained with an infinite volume of the surrounding atmosphere (Sect. 3.2.1).

Another result of interest is shown by drawing the value of the amount of heat transferred as a function of time, or rather of the dimensionless number $\alpha t/R^2$, for various values of the dimensionless number β (Fig. 3.9). The ratio between the amount of heat transferred up to time t and the amount transferred up to infinite time, is also equal to the ratio $(T_{\text{mean},t} - T_{\text{is}})/(T_{\text{ie}} - T_{\text{is}})$, where

- $T_{\text{mean},t}$ is the average temperature in the sheet at time t.
- T_{is} is the initial temperature in the sheet.
- T_{ie} is the initial temperature of the well-stirred fluid.

The profile of temperature within the solid sheet is given by the expression

$$\frac{T_{x,t} - T_{\text{is}}}{T_\infty - T_{\text{is}}} = 1 + \sum \frac{2(1+\beta)\cos q_n x/R}{(1 + \beta + \beta^2 q_n^2)\cos q_n} \exp\left(-\frac{q_n^2}{R^2}\alpha t\right) \tag{3.72}$$

It is often interesting to have a good knowledge of the profiles of temperature developed within the sheet during the process, for various values of time, or rather of the dimensionless number $\alpha t/R^2$. The profiles of temperature growing within the sheet are calculated by considering the dimensionless number $(T_{x,t} - T_{\text{is}})/(T_\infty - T_{\text{is}})$ and the relative position x/R, for various values of $\alpha t/R^2$, R representing half the thickness of the sheet. These profiles of temperature are drawn for various values of the dimensionless number β. Three particular values of β have been selected:

$\beta = 1$	in Fig. 3.10
$\beta = 5$	in Fig. 3.11
$\beta = 200$	in Fig. 3.12

Of course, for low values of β, or for low volumes of the surrounding atmosphere compared with the volume of the sheet, a variation of temperature in the surrounding atmosphere is observed, with the obvious following statement: the lower the value of β, the higher the variation of temperature in the surrounding atmosphere. For a high value of β, around 200, there is no significant change of temperature in the surrounding atmosphere.

When the volume of fluid is infinite, $\beta \to \infty$, the roots of $\tan q_n = -\beta q_n$ are given by

$$q_n = (n + \tfrac{1}{2})\pi \tag{3.73}$$

Table 3.1. Roots of $\tan q = -bq$

b	q_1	q_2	q_3	q_4	q_5	q_6	q_7	q_8	q_9	q_{10}
0.00	3.1416	6.2832	9.4248	12.566	15.708	18.850	21.991	25.133	28.274	31.416
0.10	2.8628	5.7606	8.7083	11.703	14.733	17.791	20.867	23.957	27.058	30.165
0.25	2.5704	5.3540	8.3029	11.335	14.408	17.503	20.612	23.729	26.851	29.978
0.50	2.2889	5.0870	8.0962	11.173	14.276	17.393	20.518	23.646	26.778	29.912
0.75	2.1301	4.9743	8.0188	11.115	14.231	17.355	20.485	23.618	26.753	29.890
1.00	2.0288	4.9132	7.9787	11.086	14.207	17.336	20.469	23.604	26.741	29.879
1.50	1.9071	4.8490	7.9378	11.056	14.184	17.317	20.453	23.590	26.728	29.867
2.50	1.7906	4.7956	7.9045	11.032	14.165	17.302	20.440	23.579	26.719	29.859
4.00	1.7155	4.7648	7.8857	11.018	14.155	17.293	20.433	23.573	26.713	29.854
9.00	1.6385	4.7358	7.8681	11.006	14.145	17.285	20.426	23.567	26.708	29.849
1000	1.5714	4.7126	7.8541	10.996	14.137	17.279	20.420	23.562	26.704	29.845
∞	1.5708	4.7124	7.8540	10.996	14.137	17.279	20.420	23.562	26.704	29.845

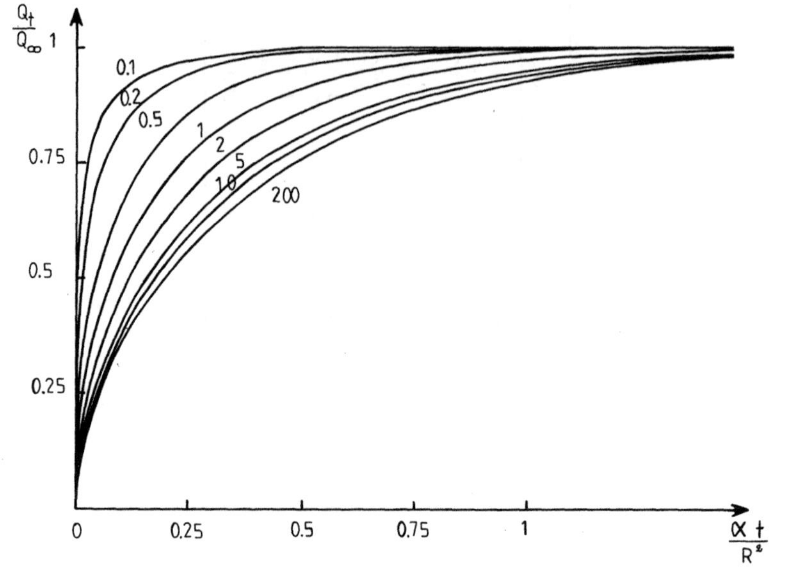

Fig. 3.8. Sheet immersed in a well-stirred fluid of finite volume. Heat which has left (or entered) the sheet at time t as a fraction of the corresponding amount after infinite time, expressed in function of time, or rather of the dimensionless number $\alpha t/R^2$, for various values of β.

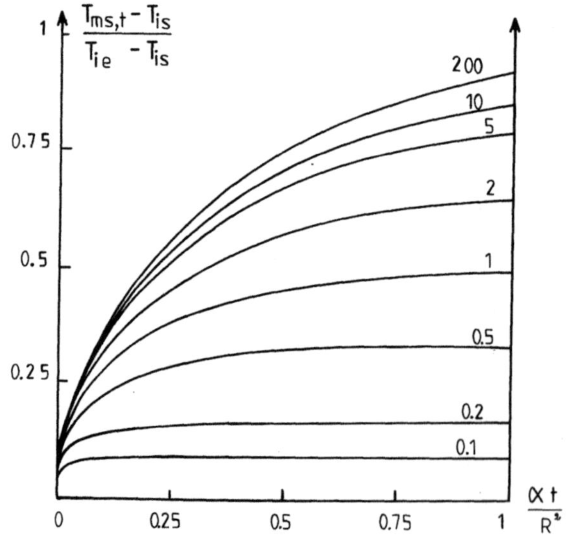

Fig. 3.9. Sheet immersed in a well-stirred fluid of finite volume. Amount of heat transferred (or $(T_{\text{mean},t} - T_{\text{is}})/(T_{\text{ie}} - T_{\text{is}})$) as a function of time, or rather of the dimensionless number $\alpha t/R^2$, for various values of the dimensionless number β.

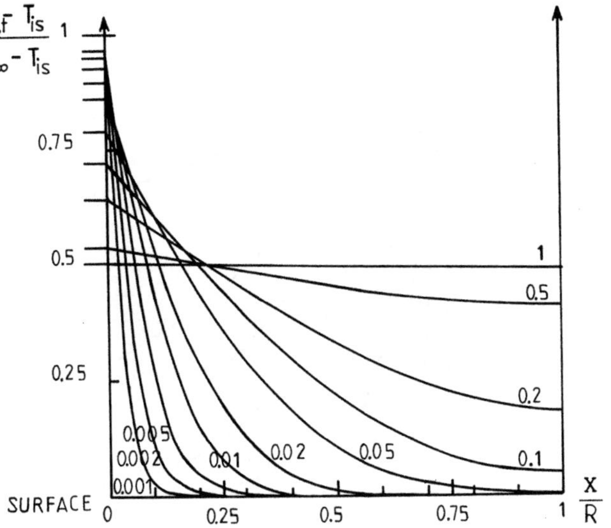

Fig. 3.10. Sheet immersed in a well-stirred fluid of finite volume. Profiles of temperature developed within the sheet and the surrounding atmosphere, for various values of time (or $\alpha t/R^2$), with $\beta = 1$.

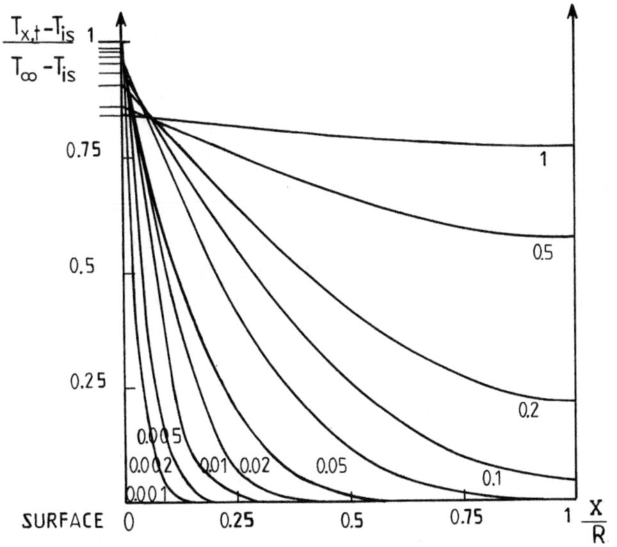

Fig. 3.11. Sheet immersed in a well-stirred fluid of finite volume. Profiles of temperature developed within the sheet and the surrounding atmosphere, for various values of time (or $\alpha t/R^2$), with $\beta = 5$.

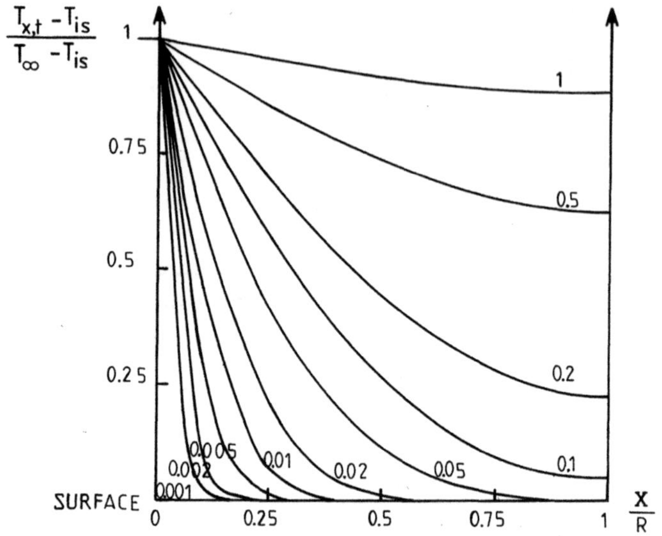

Fig. 3.12. Sheet immersed in a well-stirred fluid of finite volume. Profiles of temperature developed within the sheet and the surrounding atmosphere, for various values of time (or $\alpha t/R^2$), with $\beta = 200$.

and the amount of heat transferred in the sheet at time t is thus expressed by

$$\frac{Q_\infty - Q_t}{Q_\infty} = \frac{8}{\pi^2} \sum_0^\infty \frac{1}{(2n+1)^2} \exp\left(\frac{-(2n+1)^2\pi^2}{4R^2} \alpha t\right) \tag{3.23}$$

3.2.4 Sandwich Material Immersed in a Well-Stirred Fluid of Infinite Volume

Sandwich material can be of interest sometimes. The following case is simple, when the components of the sandwich are made of the same material, and when the sandwich is immersed in a well-stirred fluid of infinite volume. Thus the temperature in the fluid is uniform and constant, and equal to the temperature on the external surfaces of the sandwich if the coefficient of heat transfer is very high. Moreover, the temperature is not the same in all components of the sandwich. The sandwich is thus made by inserting a sheet at uniform temperature T_{is} between two other sheets of the same material at uniform temperature T_{ext}.

The thickness of the sandwich is L, the thickness of the central sheet is $2H < L$. The function $Y = T_{x,t} - T_{ext}$ verifies the following initial and boundary conditions (Fig. 3.13):

$$t = 0, \quad 0 \leqslant x < \frac{L}{2} - H \quad \text{or} \quad \frac{L}{2} + H < x \leqslant L, \quad T - T_{ext} = 0$$

$$\frac{L}{2} - H < x < \frac{L}{2} + H, \qquad T - T_{ext} = T_{is} - T_{ext} \tag{3.74}$$

$$t > 0, \quad x = 0 \quad \text{or} \quad x = L, \qquad T - T_{ext} = 0 \tag{3.75}$$

It is the same problem as in Sect. 3.2.2. The solution is given by Eq. (3.54), by

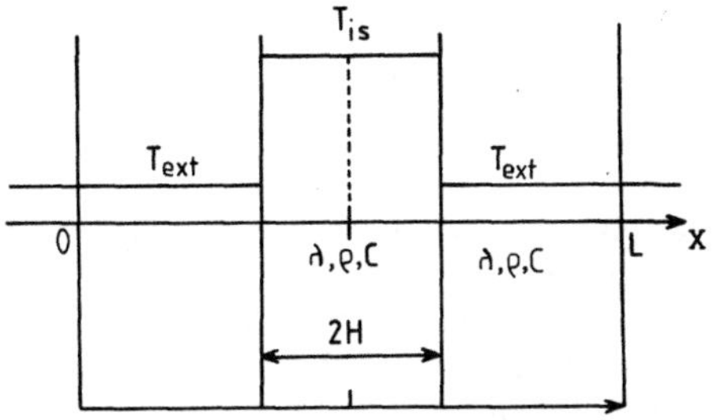

Fig. 3.13. Sandwich with the same material of thickness L, with $0 < x < L$, immersed in a well-stirred fluid of infinite volume, with constant temperature on the surfaces. Heated sheet of thickness $2H$ at the centre of the sheet.

replacing $T_{x,t}$ by $T_{x,t} - T_{\text{ext}}$, as well as T_1 and T_2 by 0. Moreover in this case,

$$\int_0^L f(y) \sin\frac{n\pi y}{L}\, dy = \int_{L/2-H}^{L/2+H} (T_{\text{is}} - T_{\text{ext}}) \sin\frac{n\pi y}{L}\, dy$$

If this integral is called b_n, then

$$b_n = (T_{\text{is}} - T_{\text{ext}}) \frac{L}{n\pi} \left[\cos n\pi\left(\frac{1}{2} - \frac{H}{L}\right) - \cos n\pi\left(\frac{1}{2} + \frac{H}{L}\right)\right]$$

$$b_n = (T_{\text{is}} - T_{\text{ext}}) \frac{2L}{n\pi} \sin\frac{n\pi H}{L} \sin\frac{n\pi}{2} \tag{3.76}$$

If n is even, $b_n = 0$. If n is odd, it is replaced by $2n - 1$. The solution for the profile of temperature is thus given by

$$\frac{T_{x,t} - T_{\text{ext}}}{T_{\text{is}} - T_{\text{ext}}} = \frac{4}{\pi} \sum_{n=1}^{\infty} \frac{(-1)^{n-1}}{(2n-1)} \sin\left(\frac{(2n-1)\pi H}{L}\right) \sin\left(\frac{(2n-1)\pi x}{L}\right)$$

$$\exp\left(\frac{-(2n-1)^2\pi^2}{L^2}\alpha t\right) \tag{3.77}$$

The amount of heat transferred out of the sandwich up to infinite time, per unit area, is obviously

$$Q_\infty = 2H\rho C(T_{\text{is}} - T_{\text{ext}}) \tag{3.78}$$

The amount of heat located in the sheet of thickness L at time t, per unit area, can be evaluated by

$$Q_\infty - Q_t = \int_0^L (T_{x,t} - T_{\text{ext}})\rho C\, dx \tag{3.79}$$

The amount of heat transferred out of the sandwich at time t, Q_t, can thus be expressed in terms of time by the relation:

$$\frac{Q_\infty - Q_t}{Q_\infty} = \frac{4L}{\pi^2 H} \sum_{n=1}^{\infty} \frac{(-1)^{n-1}}{(2n-1)^2} \sin\left(\frac{(2n-1)\pi H}{L}\right) \exp\left(\frac{-(2n-1)^2\pi^2}{L^2}\alpha t\right) \tag{3.80}$$

When $2H = L$, these relations (3.77) and (3.80) are the same as those obtained for the simple sheet (Eqs (3.14) and (3.23)) of thickness L.

3.3 Non-steady Conduction with a Finite Coefficient of Surface Heat Transfer

The sheet, initially at a temperature which is a function of space, is immersed in a medium of infinite volume, and the coefficient of surface heat transfer is finite. The temperature in the external medium is constant at a position far from the sheet. Two cases are considered, when the sheet is located either between the abscissae 0 and L or $-R$ and $+R$.

Sheet Located Between the Abscissae 0 *and* L*, and Initial Temperature* $f(x)$ *in the Sheet*

The equation for heat conduction within the sheet is

$$\frac{\partial T}{\partial t} = \alpha \frac{\partial^2 T}{\partial x^2}, \qquad 0 < x < L \tag{3.1}$$

The initial and boundary conditions are

$$t = 0, \qquad T_{x,0} - T_{\text{ext}} = f(x), \qquad \text{(sheet)} \tag{3.81}$$

$$t > 0, \qquad \lambda \frac{\partial T}{\partial x} = h(T_{0,t} - T_{\text{ext}}), \qquad x = 0 \tag{3.82}$$

$$t > 0, \qquad -\lambda \frac{\partial T}{\partial x} = h(T_{L,t} - T_{\text{ext}}), \qquad x = L \tag{3.82'}$$

The expression

$$T_{x,t} - T_{\text{ext}} = (A \cos \omega x + B \sin \omega x) \exp(-\omega^2 \alpha t) \tag{3.83}$$

is a solution of the problem, by considering the method of separation of variables.

The boundary conditions need

$$\frac{B}{A} = \frac{h}{\omega \lambda} \tag{3.84}$$

$$\lambda\omega(A \sin \omega L - B \cos \omega L) = h(A \cos \omega L + B \sin \omega L) \tag{3.85}$$

By putting $\bar{h} = h/\lambda$, the previous two equations give

$$\tan \omega L = \frac{2\omega \bar{h}}{\omega^2 - \bar{h}^2} \tag{3.86}$$

The following expression,

$$T_{x,t} - T_{\text{ext}} = \sum_{n=0}^{\infty} A_n \left(\cos \omega_n x + \frac{\bar{h}}{\omega_n} \sin \omega_n x \right) \exp(-\omega_n^2 \alpha t) \tag{3.87}$$

where ω_ns are the positive roots of Eq. (3.86), and A_ns are constants to be determined, is a solution which satisfies the equation of heat conduction within the sheet and the boundary conditions.

The initial condition enables us to obtain A_n:

$$A_n = \frac{2\omega_n^2}{(\omega_n^2 + \bar{h}^2)L + 2\bar{h}} \int_0^L f(x)\left[\cos \omega_n x + \frac{\bar{h}}{\omega_n}\sin \omega_n x\right] dx \tag{3.88}$$

The temperature is thus given by the equation

$$T_{x,t} - T_{\text{ext}} = 2 \sum_{n=1}^{\infty} \frac{\omega_n \cos \omega_n x + \bar{h} \sin \omega_n x}{(\omega_n^2 + \bar{h}^2)L + 2\bar{h}}$$

$$\times \exp(-\omega_n^2 \alpha t) \int_0^L f(y)(\omega_n \cos \omega_n y + \bar{h} \sin \omega_n y)\, dy \tag{3.89}$$

Sheet Located Between the Abscissae $-R$ and R, and the Initial Temperature $f(x)$

The Eq. (3.89) obtained previously is rewritten with the sheet of thickness $2R$:

$$T_{x,t} - T_{\text{ext}} = \sum \frac{C_n \cos \omega_n x + D_n \sin \omega_n x}{(\omega_n^2 + \bar{h}^2)R + \bar{h}} \exp(-\omega_n^2 \alpha t)$$

$$\times \int_{-R}^{+R} f(x)[C_n \cos \omega_n x + D_n \sin \omega_n x]\, dx \tag{3.90}$$

The coefficients C_n, D_n and ω_n, obtained by considering the boundary and initial conditions, are

$$C_n = \bar{h} \sin \omega_n R + \omega_n \cos \omega_n R \tag{3.91}$$

$$D_n = \bar{h} \cos \omega_n R - \omega_n \sin \omega_n R \tag{3.92}$$

and ω_n are the positive roots of

$$\tan 2\omega R = \frac{2\omega\bar{h}}{\omega^2 - \bar{h}^2} \tag{3.93}$$

As Eq. (3.93) can be written as follows,

$$(\bar{h} \sin \omega R + \omega \cos \omega R)(\bar{h} \cos \omega R - \omega \sin \omega R) = 0 \tag{3.93'}$$

ω_n are also the positive roots of the following two equations:

$$\omega \tan \omega R = \bar{h} \tag{3.94}$$

$$\omega \cot \omega R = -\bar{h} \tag{3.95}$$

When ω_n is a root of Eq. (3.94) there is

$$D_n = 0 \quad \text{and} \quad C_n^2 = \bar{h}^2 + \omega_n^2 \tag{3.96}$$

If the initial temperature f is an even function, Eq. (3.90) can be written as follows:

$$T_{x,t} - T_{\text{ext}} = \sum_{n=1}^{\infty} \frac{2(\bar{h}^2 + \omega_n^2)\cos \omega_n x}{(\omega_n^2 + \bar{h}^2)R + \bar{h}} \exp(-\omega_n^2 \alpha t) \int_0^R f(y) \cos \omega_n y\, dy \tag{3.97}$$

If the temperature is an odd function of x, the following equation is obtained:

$$T_{x,t} - T_{\text{ext}} = \sum_{n=1}^{\infty} \frac{2(\bar{h}^2 + \omega_n^2) \sin \omega_n x}{(\omega_n^2 + \bar{h}^2)R + \bar{h}} \exp(-\omega_n^2 \alpha t) \int_0^R f(y) \sin \omega_n y \, dy \tag{3.97'}$$

where the ω_ns are the positive roots of Eq. (3.95).

This equation is also the solution of the problem of heat conduction in the slab, $0 < x < R$, with the surface $x = 0$ maintained at the external temperature resulting from an infinite coefficient of surface heat transfer, and the other surface $x = R$ in contact with the external atmosphere at T_{ext} and a finite coefficient of surface heat transfer.

Remark. Eq. (3.94) can be solved by using the data in Table 3.2.

Eq. (3.95) can be written as follows:

$$\tan \omega R = -\frac{\omega R}{\bar{h} R}$$

On putting $q = \omega R$ and $M = \bar{h} R$, this equation becomes

$$\tan q = -\frac{q}{M}$$

which is the same as Eq. (3.70), with $\beta = 1/M$.

Sheet Located Between the Abscissae −R and R, and a Constant Initial Temperature

The profile of temperature developed within the sheet is thus given by the relation

$$\frac{T_{x,t} - T_{\text{ext}}}{T_{\text{in}} - T_{\text{ext}}} = \sum_{n=1}^{\infty} \frac{2M \cos \omega'_n \dfrac{x}{R}}{(\omega'^2_n + M^2 + M) \cos \omega'_n} \exp\left(\frac{-\omega'^2_n \alpha t}{R^2}\right) \tag{3.98}$$

where ω'_n are the positive roots of

$$\omega' \tan \omega' = M \tag{3.99}$$

and the dimensionless number M is given by

$$M = R\bar{h} = \frac{Rh}{\lambda} \tag{3.100}$$

The profiles of the temperature developed within half the thickness of the sheet are drawn for various times by considering the following dimensionless numbers:

$$\frac{T_{x,t} - T_{\text{ext}}}{T_{\text{in}} - T_{\text{ext}}}, \frac{x}{R} \text{ and } \frac{\alpha t}{R^2}$$

Graphs are plotted for two different values of the dimensionless number M: $M = 0.2$ (Fig. 3.14) and $M = 1$ (Fig. 3.15). The effect of this number, and especially of the value of the coefficient of heat convection at the surface is thus clearly shown. The higher the value of M and h, the steeper the gradients of

Table 3.2. Roots of $\omega \tan \omega = M$

M	ω_1	ω_2	ω_3	ω_4	ω_5	ω_6	ω_7	ω_8	ω_9	ω_{10}
0.00	0.0000	3.1416	6.2832	9.4248	12.566	15.708	18.850	21.991	25.133	28.274
0.01	0.0998	3.1448	6.2848	9.4258	12.567	15.709	18.850	21.992	25.133	28.275
0.10	0.3111	3.1731	6.2991	9.4354	12.574	15.714	18.855	21.996	25.137	28.278
0.20	0.4328	3.2039	6.3148	9.4459	12.582	15.721	18.860	22.000	25.141	28.281
0.50	0.6533	3.2923	6.3616	9.4775	12.606	15.740	18.876	22.014	25.153	28.292
1.00	0.8603	3.4256	6.4373	9.5293	12.645	15.771	18.902	22.036	25.172	28.310
2.00	1.0769	3.6436	6.5783	9.6296	12.722	15.834	18.955	22.081	25.212	28.345
5.00	1.3138	4.0336	6.9096	9.8928	12.935	16.011	19.106	22.213	25.328	28.448
10.0	1.4289	4.3058	7.2281	10.200	13.214	16.259	19.327	22.411	25.506	28.611
100	1.5552	4.6658	7.7764	10.887	13.998	17.109	20.221	23.333	26.445	29.558
1000	1.5692	4.7077	7.8461	10.985	14.123	17.261	20.400	23.538	26.677	29.815
∞	1.5708	4.7124	7.8540	10.996	14.137	17.279	20.420	23.562	26.704	29.845

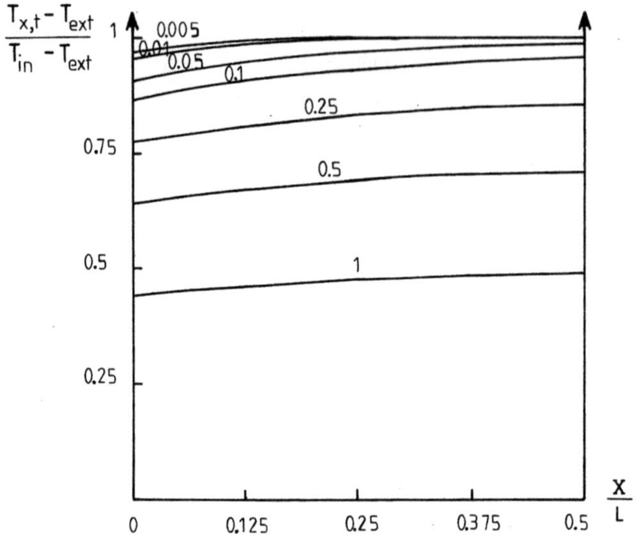

Fig. 3.14. Plane sheet of thickness $2R$, $-R < x < R$, with a finite coefficient of surface heat convection. Profiles of temperature developed within the sheet for various times (or $\alpha t/R^2$), with $M = 0.2$.

temperature. These two figures can be compared with Fig. 3.2, plotted with a very high coefficient of heat convection at the surface.

The amount of heat leaving the sheet up to time t, through the unit area of the surface, at abscissa R, is expressed by the following equation:

$$Q_t = -\lambda \int_0^t \frac{\partial T}{\partial x}(R, \theta)\, d\theta \tag{3.16'}$$

Therefore the amount of heat leaving (or entering) the sheet up to time t (through both surfaces), Q_t, is expressed as a fraction of Q_∞, the corresponding quantity after infinite time, by

$$\frac{Q_\infty - Q_t}{Q_\infty} = \sum_{n=1}^{\infty} \frac{2M^2}{\omega'^2_n[\omega'^2_n + M^2 + M]} \exp\left(-\frac{\omega'^2_n}{R^2}\alpha t\right) \tag{3.101}$$

In Fig. 3.16, the amount of heat entering (or leaving) the sheet up to time t, Q_t, as a fraction of Q_∞, the corresponding amount after infinite time, is expressed as a function of time (or rather of the dimensionless number $\alpha t/L^2$), for various values of the dimensionless number M. For a high value of M, around 10000, the curve obtained is the same as that shown in Fig. 3.3 with the constant temperature kept at the surfaces of the sheet.

3.4 Steady Conduction

The mechanism of heat transfer by steady unidirectional conduction is given by the Fourier relation expressing the heat flux in terms of the gradient of

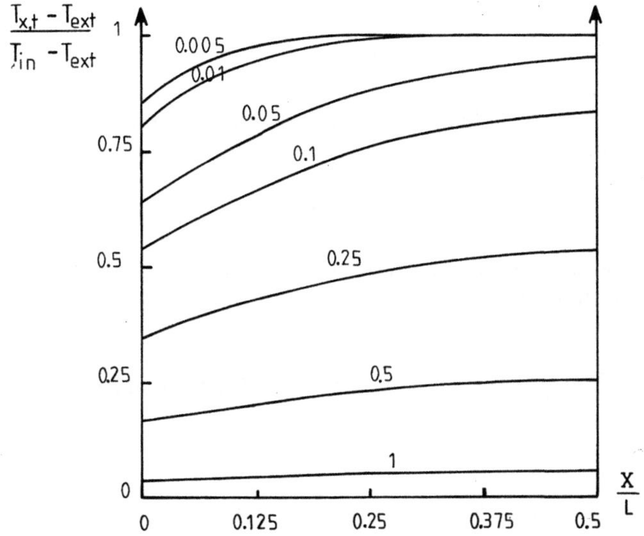

Fig. 3.15. Plane sheet of thickness $2R$, with a finite coefficient of surface heat convection. Profiles of temperature developed within the sheet for various times (or $\alpha t/R^2$), with $M = 1$.

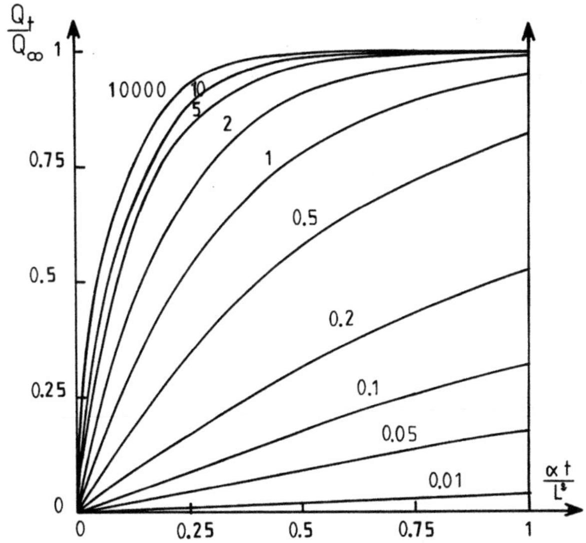

Fig. 3.16. Plane sheet of thickness $2R$, with a finite coefficient of surface heat convection. Amount of heat transferred at time t as a fraction of the corresponding amount after infinte time, expressed as a function of time (or $\alpha t/R^2$), for various values of the dimensionless number M.

temperature:

$$\phi = -\lambda \frac{\partial T}{\partial x} \tag{3.15}$$

On the other hand, the equation of heat conduction in the case of steady flow becomes

$$\frac{\partial}{\partial x}\left(\lambda \frac{\partial}{\partial x}\right) = \rho C \frac{\partial T}{\partial t} = 0 \tag{3.102}$$

as temperature does not vary with time.

Thus, if the thermal conductivity is constant, the gradient of temperature within the sheet is constant at any place within the sheet:

$$\frac{\partial T}{\partial x} = \text{constant} \tag{3.103}$$

Two cases are considered, according to the values of the coefficient of surface heat transfer: infinite or finite.

3.4.1 Given Temperatures at the Two External Surfaces

The temperatures at the surfaces of the sheet are constant and equal to (Fig. 3.17)

$$T = T_0 \qquad x = 0 \qquad \text{(first surface)}$$
$$T = T_1 \qquad x = L \qquad \text{(second surface)} \tag{3.104}$$

The heat flux by conduction through the sheet of thickness L is thus given by

$$\phi = \lambda \frac{T_0 - T_1}{L} = \frac{T_0 - T_1}{L/\lambda} \qquad \text{per unit area} \tag{3.105}$$

where L/λ represents the thermal resistance of the sheet to heat conduction.

When the composite wall is made of n sheets of thickness $L_1, L_2, \ldots, L_n$ with thermal conductivities $\lambda_1, \lambda_2, \ldots, \lambda_n$, and the sheets are in perfect contact at their surfaces of separation, the fall of temperature over the whole composite is the sum of the falls of temperature over the component sheets, since the heat flux is the same throughout the wall (Fig. 3.18).

$$\phi = \frac{T_0 - T_1}{L_1/\lambda_1} = \frac{T_1 - T_2}{L_2/\lambda_2} = \frac{T_2 - T_3}{L_3/\lambda_3} = \cdots = \frac{T_{n-1} - T_n}{L_n/\lambda_n} = \frac{T_0 - T_n}{\sum_{i=1}^{n} L_i/\lambda_i} \tag{3.106}$$

Of course the thermal resistance of the composite wall is the sum of the thermal resistance of the separate layers.

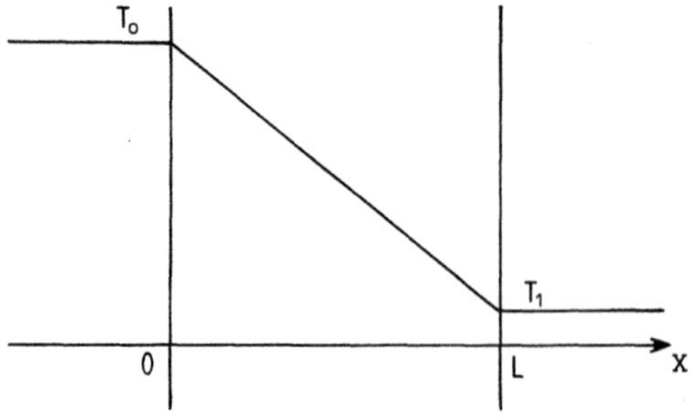

Fig. 3.17. Steady heat conduction within a sheet with a very high coefficient of surface heat transfer.

In the more general case with contact resistances between the sheets located in the composite wall, the total resistance of the composite wall is the sum of the thermal resistances of the separate sheets plus the sum of the thermal resistances due to the contacts between them.

3.4.2 Finite Coefficient of Surface Heat Transfer

This case occurs much more frequently than the previous case with an infinite coefficient of surface heat transfer. The sheet of thickness L separates two media kept at given constant temperatures T_{ext0} and T_{ext1}, and the temperature at each position is constant.

In the first case, the coefficient of surface heat transfer is infinite on the left of the sheet, and finite on the right (Fig. 3.19). The result for the constant heat flux still holds, and is written as follows:

$$\phi = \lambda \frac{T_0 - T_x}{x} = \lambda \frac{T_1 - T_x}{L} = h(T_0 - T_{ext1}) = \frac{T_0 - T_{ext1}}{(L/\lambda) + (1/h)} \tag{3.107}$$

The temperature at position x within the sheet is thus given by

$$\frac{T_0 - T_x}{T_0 - T_{ext1}} = \frac{x}{L + \lambda/h} \tag{3.108}$$

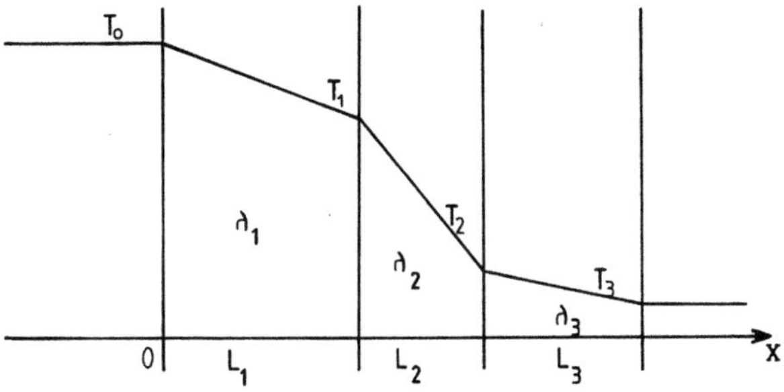

Fig. 3.18. Steady heat conduction through a composite wall made of various sheets.

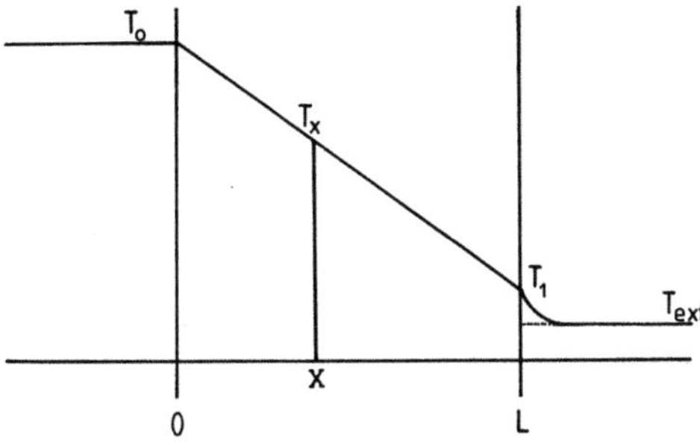

Fig. 3.19. Steady heat conduction through a wall with an infinite coefficient of surface heat transfer on one surface and a finite coefficient on the other surface.

and the value of heat flux through the wall can be expressed in terms of the temperatures:

$$\phi = (T_0 - T_{ext1})\frac{h\lambda}{Lh + \lambda} = (T_0 - T_1)\frac{\lambda}{L} \tag{3.109}$$

In the following case, the sheet of thickness L is located between two media kept at constant temperatures T_{ext0} and T_{ext1}, while the constant temperatures of the sheet surfaces are T_0 and T_1 (Fig. 3.20). The constant heat flux is given by

$$\phi = h_0(T_{ext0} - T_0) = \frac{\lambda}{L}(T_0 - T_1) = h_1(T_1 - T_{ext1}) = \frac{\lambda}{x}(T_0 - T_x) \tag{3.110}$$

The flux can be expressed in terms of the thermal resistances:

$$\phi = \frac{T_{ext0} - T_0}{1/h_0} = \frac{T_0 - T_1}{L/\lambda} = \frac{T_1 - T_{ext1}}{1/h_1} = \frac{T_{ext0} - T_{ext1}}{(1/h_0) + (L/\lambda) + (1/h_1)}$$

$$= \frac{T_0 - T_x}{(x/\lambda)} \tag{3.111}$$

the total resistance being the sum of the three thermal resistances.

The temperature at any position within the sheet is easily determined as a function of the temperatures of the two media:

$$\frac{T_{ext0} - T_x}{T_{ext0} - T_{ext1}} = \frac{(1/h_0) + (x/\lambda)}{(1/h_0) + (L/\lambda) + (1/h_1)} \tag{3.112}$$

3.4.3 Thermal Conductivity a Function of Temperature

When the traditional conductivity is a function of temperature, the differential equation of heat conduction under steady state, becomes

$$\frac{d}{dx}\left(\lambda\frac{dT}{dx}\right) = 0 \tag{3.113}$$

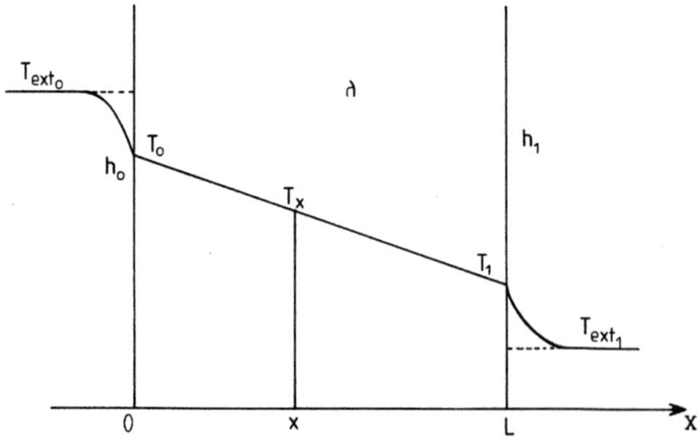

Fig. 3.20. Steady heat conduction through a wall with a finite coefficient of surface heat transfer on both surfaces.

and the heat flux by conduction through the solid is constant:

$$\phi = -\lambda \frac{dT}{dx} \tag{3.15}$$

with variable thermal conductivity.

This heat flux through the sheet can be obtained by integrating between the surface temperatures T_0 and T_1 of the sheet of thickness L:

$$\phi = \frac{\lambda_{av}}{L}(T_0 - T_1) = -\frac{1}{L}\int_{T_0}^{T_1} \lambda \, dT \tag{3.114}$$

and the average thermal conductivity over the temperature range $[T_0, T_1]$ in the sheet given by

$$\lambda_{av} = \frac{1}{T_1 - T_0}\int_{T_0}^{T_1} \lambda \, dT \tag{3.115}$$

The profile of the temperature is obtained by integrating the differential equation (3.15):

$$\phi \, dx = -\lambda_T \, dT \tag{3.15'}$$

$$x - x_0 = \frac{-1}{\phi}\int_{T_0}^{T} \lambda_\theta \, d\theta \tag{3.116}$$

where θ varies from T_0 to T.

Chapter 4

Heat Conduction in a Sphere

4.1 Introduction

4.1.1 General Equation

If the thermal conductivity λ is constant, the general equation, when expressed in spherical polar coordinates, is

$$\frac{\partial T}{\partial t} = \alpha\left[\frac{1}{r^2}\frac{\partial}{\partial r}\left(r^2\frac{\partial T}{\partial r}\right) + \frac{1}{r^2 \sin\theta}\frac{\partial}{\partial \theta}\left(\sin\theta\frac{\partial T}{\partial \theta}\right) + \frac{1}{r^2 \sin^2\theta}\frac{\partial^2 T}{\partial \phi^2}\right] \tag{4.1}$$

where θ is colatitude and ϕ longitude.

4.1.2 Case of Radial Conduction

This case is very simple. When heat conduction is strictly radial, the heat conduction equation for a constant thermal conductivity and without reaction, takes the form of

$$\frac{\partial T}{\partial t} = \alpha\left[\frac{\partial^2 T}{\partial r^2} + \frac{2}{r}\frac{\partial T}{\partial r}\right] \tag{4.2}$$

The isothermal surfaces are concentric spheres and the temperature thus depends only on the position r and time t.

On putting

$$U = Tr \tag{4.3}$$

the equation for radial heat conduction becomes

$$\frac{\partial U}{\partial t} = \alpha\frac{\partial^2 U}{\partial r^2} \qquad 0 < r < R \tag{4.4}$$

Since this is the equation for linear heat flow in one dimension, the solutions of many problems for radial heat conduction in a spherical medium can be deduced from those of the corresponding linear problems.

4.2 Non-steady Conduction with Very High Coefficient of Surface Heat Transfer

4.2.1 Sphere with Uniform Initial Temperature and Constant Surface Temperature

The problem can be solved either by using the method of separation of variables or the error function with the method of reflection and superposition.

Method of Separation of Variables

The initial and boundary conditions are written as follows (Fig. 4.1):

$$t = 0, \quad 0 \leq r < R, \quad T = T_{in}, \quad U = rT_{in} \tag{4.5}$$

$$t > 0, \quad r = 0, \quad U = 0$$

$$t > 0, \quad r = R, \quad U = RT_{ext} \tag{4.6}$$

These are the equations of heat conduction in a plane sheet of thickness R, with its ends, $r = 0$ and $r = R$, kept at constant temperature, 0 and RT_{ext} respectively, and with initial temperatures $f(r) = rT_{in}$ within the sheet. The problem was considered in the previous chapter for heat conduction through a sheet (Sect. 3.2.2).

The general equation for the profile of temperature within the sphere is thus given by

$$\frac{T_{r,t} - T_{in}}{T_{ext} - T_{in}} = 1 + \frac{2R}{\pi r} \sum_{n=1}^{\infty} \frac{(-1)^n}{n} \sin\frac{n\pi r}{R} \exp\left(-\frac{n^2\pi^2}{R^2}\alpha t\right) \tag{4.7}$$

The temperature at the centre of the sphere is given by the limit when $r \to 0$. Since $(\sin x)/x \to 1$ when $x \to 0$,

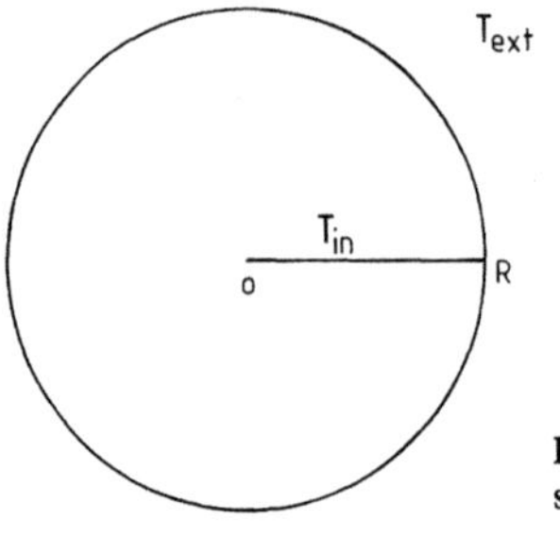

Fig. 4.1. Sphere with initial uniform temperature T_{in} immersed in surrounding atmosphere kept at T_{ext}.

$$\frac{T_{0,t} - T_{\text{in}}}{T_{\text{ext}} - T_{\text{in}}} = 1 + 2 \sum (-1)^n \exp\left(-\frac{n^2\pi^2}{R^2}\alpha t\right) \tag{4.8}$$

Temperature distributions at various times (or rather at various values of the dimensionless number $\alpha t/R^2$), in a sphere of radius R, with uniform initial temperature T_{in} and constant surface temperature T_{ext}, are given in Fig. 4.2. The dimensionless numbers $(T_{r,t} - T_{\text{in}})/(T_{\text{ext}} - T_{\text{in}})$ and $\alpha t/R^2$ are used.

Eq. (4.7) is used for calculating temperature distributions at various places within the sphere, while temperatures at the middle of the sphere ($r = 0$) are calculated with the help of Eq. (4.8).

Of course, steep gradients of temperature next to the surface are observed at the beginning of the process.

The amount of heat which has left or entered the sphere at time t, Q_t, is expressed as a fraction of the corresponding quantity after infinite time Q_∞, by the relation

$$\frac{Q_\infty - Q_t}{Q_\infty} = \frac{6}{\pi^2} \sum \frac{1}{n^2} \exp\left(-\frac{n^2\pi^2}{R^2}\alpha t\right) \tag{4.9}$$

where heat transmitted after infinite time is given by

$$Q_\infty = \tfrac{4}{3}\pi R^3 \rho C |T_{\text{in}} - T_{\text{ext}}| \tag{4.10}$$

The heat transferred after time t through the surface of the sphere, Q_t, as a fraction of the corresponding quantity after infinite time Q_∞, is expressed either in terms of time (or $\alpha t/R^2$) in curve 1, or of square root of time (or $\sqrt{\alpha t}/R$) in curve 2, in Fig. 4.3.

A vertical tangent is observed in curve 1 at the beginning of the process. The

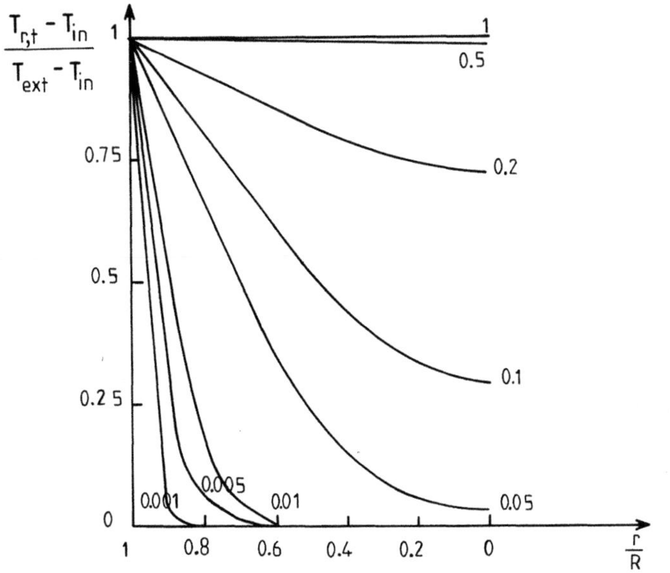

Fig. 4.2. Temperature distribution developed within a sphere at various times (or rather $\alpha t/R^2$). The initial temperature within the sphere is uniform, and the temperature on the surface of the sphere is constant.

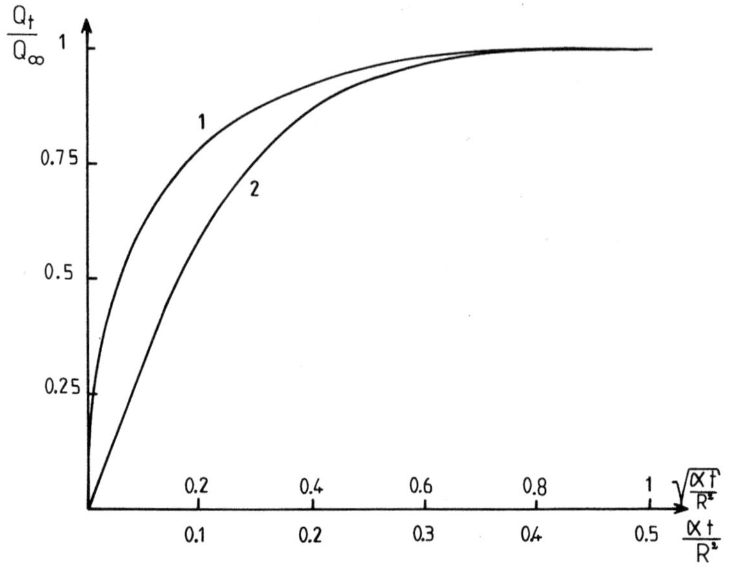

Fig. 4.3. Amount of heat transferred through the surface of the sphere at time t, as a fraction of the corresponding quantity after infinite time, expressed in terms of time (or $\alpha t/R^2$) in curve 1, and in terms of the square root of time (or $\sqrt{\alpha t}/R$) in curve 2.

series in Eq. (4.9) converges rather slowly for short times: in order to obtain a relative error of 10^{-5}, 23 terms must be used when $Q_t/Q_\infty = 0.1$, 12 terms for $Q_t/Q_\infty = 0.2$, seven terms for $Q_t/Q_\infty = 0.4$, four terms for $Q_t/Q_\infty = 0.75$, and two terms only for $Q_t/Q_\infty = 0.95$.

In curve 2, the square root of time dependency of heat transferred can be observed for $Q_t/Q_\infty < 0.35$.

Method of Reflection and Superposition with the Error Function

As shown for the sheet, the profile of concentration is thus given for the sphere by the relation

$$\frac{T_{r,t} - T_{\text{in}}}{T_{\text{ext}} - T_{\text{in}}} = \frac{R}{r} \sum_{n=0}^{\infty} \left\{ \operatorname{erfc} \frac{(2n+1)R - r}{2\sqrt{\alpha t}} - \operatorname{erfc} \frac{(2n+1)R + r}{2\sqrt{\alpha t}} \right\} \tag{4.11}$$

The temperature at the centre of the sphere $T_{0,t}$ is given by the limit when $r \rightarrow 0$. The derivative of the error-function complement and the formula of finite increments are used:

$$\operatorname{erfc}'(y) = \frac{-2}{\sqrt{\pi}} \exp(-y^2)$$

$$\operatorname{erfc}(y-h) - \operatorname{erfc}(y+h) \sim -2h \frac{-2}{\sqrt{\pi}} \exp(-y^2)$$

giving

$$\frac{T_{0,t} - T_{\text{in}}}{T_{\text{ext}} - T_{\text{in}}} = \frac{2R}{\sqrt{\pi \alpha t}} \sum_{n=0}^{\infty} \exp\left(-(2n+1)^2 \frac{R^2}{4\alpha t}\right) \tag{4.12}$$

(This value is twice the value given in Eq. 7, Sect. 9.3 of Carslaw and Jaeger [1].)

The amount of heat entering or leaving the sphere is expressed as a fraction of the corresponding heat amount transmitted after infinite time by

$$\frac{Q_t}{Q_\infty} = \frac{6}{R}\sqrt{\alpha t}\left\{\frac{1}{\sqrt{\pi}} + 2\sum_{n=1}^{\infty} \operatorname{ierfc}\frac{nR}{\sqrt{\alpha t}}\right\} - \frac{3\alpha}{R^2}t \tag{4.13}$$

Remarks. Some remarks are of interest for practical use.

1. The expressions for the temperature profiles and heat transferred are given either by way of trigonometrical series or by way of error functions.
2. For small times, the series with error function converges more quickly than the trigonometrical series.
3. For small times, when

$$\frac{Q_t}{Q_\infty} < 0.5 \tag{4.14}$$

Eq. (4.13) can be reduced to the relationship

$$\frac{Q_t}{Q_\infty} = \frac{6}{\sqrt{\pi}}\sqrt{\frac{\alpha t}{R^2}} - 3\frac{\alpha t}{R^2} \tag{4.15}$$

which can be rewritten as follows:

$$\frac{1}{\sqrt{\pi}} - \sqrt{\frac{1}{\pi} - \frac{1}{3}\frac{Q_t}{Q_\infty}} = \sqrt{\frac{\alpha t}{R^2}} \tag{4.16}$$

4. For long times, when

$$\frac{Q_t}{Q_\infty} > 0.8 \tag{4.17}$$

the trigonometrical series converges very quickly, and the first term becomes preponderent:

$$\ln\left(\frac{Q_\infty - Q_t}{Q_\infty}\right) = \ln\frac{6}{\pi^2} - \frac{\pi^2}{R^2}\alpha t \tag{4.18}$$

4.2.2 Sphere in Contact with a Well-Stirred Fluid of Finite Volume

The problem and method of solution are similar to those of the plane sheets and the results are given without calculation. The sphere of radius R is in contact with the volume V' of well-stirred fluid of density ρ' and specific heat C'. There is no transfer of heat through the external surface of the fluid. The boundary condition at the surface of the sphere, $r = R$, for $t > 0$, is

$$-4\pi R^2 \lambda \frac{\partial T}{\partial r} = V'\rho' C' \frac{\partial T}{\partial t}, \qquad r = R \tag{4.19}$$

expressing the fact that the rate at which heat leaves the sphere is always equal to that at which heat enters the volume V' of fluid, with the assumption that the temperature of the fluid is equal to the surface temperature of the sphere.

The initial conditions are as follows:

$$t = 0, \quad 0 \leq r < R, \quad T = T_{\text{is}} = ct \quad \text{(sphere)}$$
$$r \geq R, \quad T - T_{\text{ia}} = ct \quad \text{(fluid)} \tag{4.20}$$

The solution of the equation of heat conduction when the temperature is initially uniform within the sphere, is given as follows. Within the sphere, $0 < r < R$:

$$\frac{T_{r,t} - T_{\text{ia}}}{T_{\text{is}} - T_{\text{ia}}} = \frac{1}{k+1} - \frac{kR}{r} \sum_{n=1}^{\infty} \frac{6}{k^2 q_n^2 + 9k + 9} \frac{1}{\sin q_n} \sin \frac{q_n r}{R} \exp\left(-\frac{q_n^2}{R^2} \alpha t\right) \tag{4.21}$$

where k is the ratio of the heat capacity of the fluid to that of the sphere,

$$k = \frac{3V'\rho'C'}{4\pi R^3 \rho C} \tag{4.22}$$

and the q_ns are the roots of

$$\tan q_n = \frac{3q_n}{3 + kq_n^2} \tag{4.23}$$

The roots of Eq. (4.23) are given in Table 4.1.

For the centre ($r \to 0$), $(R/r)\sin(q_n r/R)$ can be replaced by q_n, and Eq. (3.21) becomes

$$\frac{T_{0,t} - T_{\text{ia}}}{T_{\text{is}} - T_{\text{ia}}} = \frac{1}{k+1} - 6k \sum_{n=1}^{\infty} \frac{q_n}{(k^2 q_n^2 + 9k + 9)\sin q_n} \exp\left(-\frac{q_n^2}{R^2} \alpha t\right) \tag{4.21'}$$

It diverges for $t = 0$. The following terms of this series are positive or negative, alternately.

The temperature of the fluid which is the same as that of the surface ($r = R$) is given by

Table 4.1. Roots of $\tan q = 3q/(3 + Kq^2)$

K	q_1	q_2	q_3	q_4	q_5	q_6	q_7	q_8	q_9	q_{10}
0.00	4.4934	7.7253	10.904	14.066	17.221	20.371	23.519	26.666	29.812	32.956
0.10	4.3538	7.4881	10.575	13.651	16.726	19.803	22.883	25.968	29.058	32.152
0.25	4.1811	7.2169	10.235	13.269	16.321	19.390	22.472	25.565	28.666	31.773
0.50	3.9720	6.9387	9.9421	12.986	16.058	19.149	22.252	25.363	28.480	31.602
0.75	3.8289	6.7804	9.7983	12.861	15.950	19.054	22.168	25.289	28.414	31.542
1.00	3.7264	6.6814	9.7156	12.793	15.892	19.005	22.125	25.250	28.379	31.511
1.50	3.5909	6.5664	9.6254	12.720	15.833	18.954	22.081	25.212	28.345	31.479
2.50	3.4478	6.4618	9.5482	12.660	15.783	18.913	22.045	25.180	28.317	31.454
4.00	3.3485	6.3978	9.5029	12.625	15.755	18.889	22.025	25.163	28.301	31.440
9.00	3.2410	6.3353	9.4599	12.593	15.729	18.867	22.006	25.146	28.286	31.427
1000	3.1425	6.2837	9.4251	12.567	15.708	18.850	21.991	25.133	28.274	31.416
∞	3.1416	6.2832	9.4248	12.566	15.708	18.850	21.991	25.133	28.274	31.416

$$\frac{T_{R,t} - T_{ia}}{T_{is} - T_{ia}} = \frac{1}{k+1} - \sum_{n=1}^{\infty} \frac{6k}{k^2 q_n^2 + 9k + 9} \exp\left(-\frac{q_n^2}{R^2}\alpha t\right) \tag{4.24}$$

The temperature at equilibrium, when $t \to \infty$, is as follows:

$$\frac{T_{r,\infty} - T_{ia}}{T_{is} - T_{ia}} = \frac{T_{R,\infty} - T_{ia}}{T_{is} - T_{ia}} = \frac{1}{k+1} \tag{4.24'}$$

Some profiles of temperature developed within the sphere are plotted for various values of time (or rather of $\alpha t/R^2$),

for $k = 1$ (Fig. 4.4)
for $k = 5$ (Fig. 4.5)
for $k = 200$ (Fig. 4.6)

The effect of the value of the ratio k on the profiles of temperature is especially sensitive next to the surface of the sphere.

The total amount of heat Q_t which has entered (or left) the sphere at time t is expressed as a fraction of the corresponding quantity after infinite time Q_∞, by the relation

$$\frac{Q_\infty - Q_t}{Q_\infty} = \sum_{n=1}^{\infty} \frac{6k(k+1)}{9 + 9k + q_n^2 k^2} \exp\left(-\frac{q_n^2}{R^2} Dt\right) \tag{4.25}$$

The amount of heat Q_t transferred through the surface of the sphere after time t, as a fraction of the corresponding quantity after infinite time Q_∞, is expressed in terms of time (or of $\alpha t/R^2$), for various values of the ratio k (Fig. 4.7). For a high value of k, around 200, the kinetics of heat transferred are about the same as that obtained with an infinite volume of the surrounding atmosphere.

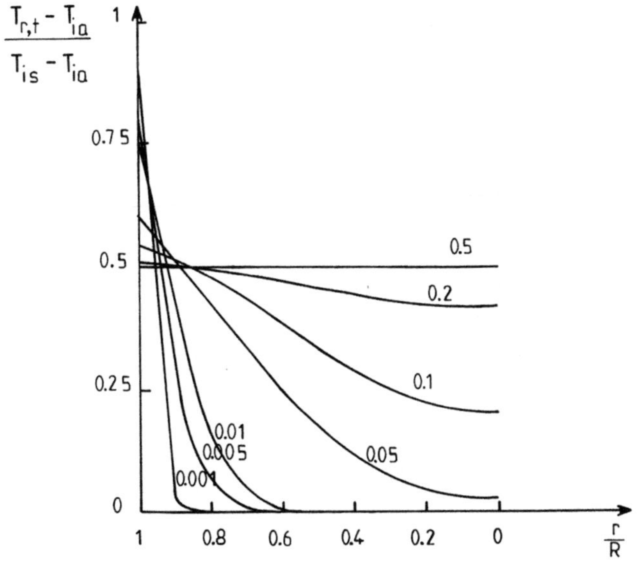

Fig. 4.4. Temperature distribution within the sphere at various times (or $\alpha t/R^2$), for $k = 1$.

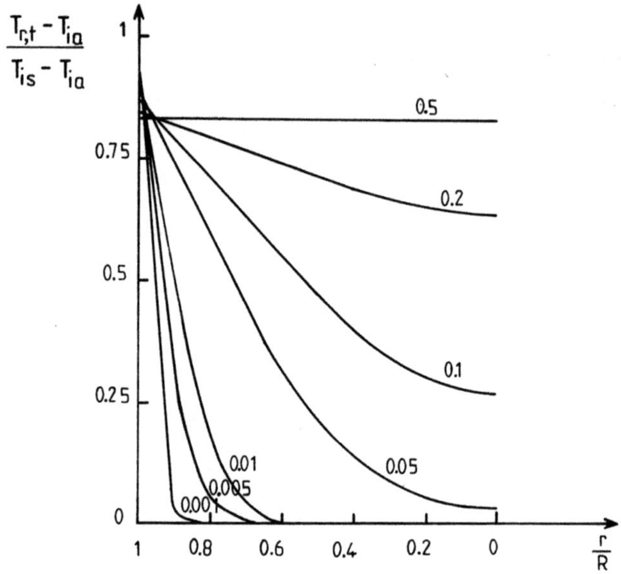

Fig. 4.5. Temperature distribution within the sphere at various times (or $\alpha t/R^2$), for $k = 5$.

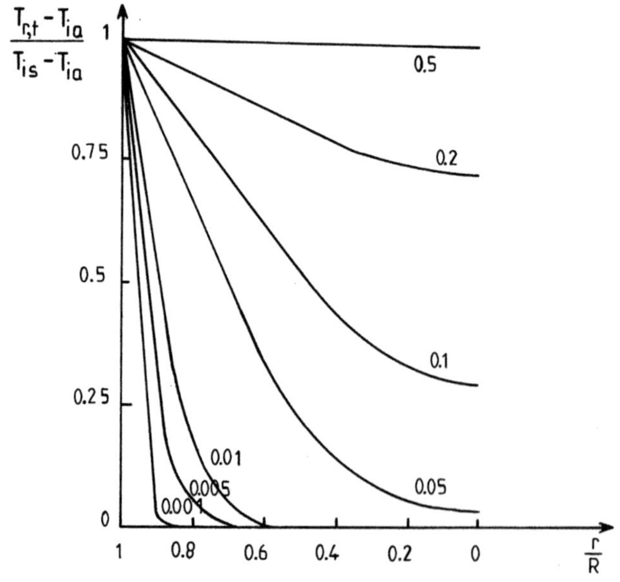

Fig. 4.6. Temperature distribution within the sphere at various times (or $\alpha t/R^2$), for $k = 200$.

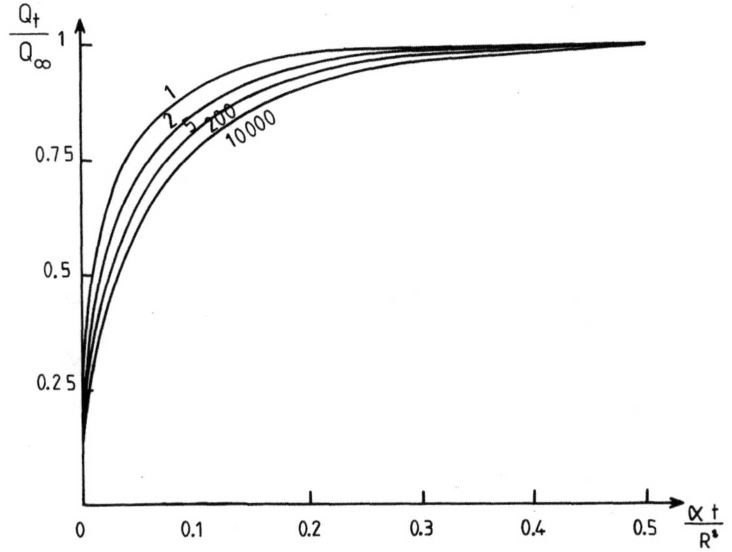

Fig. 4.7. Heat transferred at time t as a fraction of the corresponding quantity after infinite time, expressed as a function of time (or $\alpha t/R^2$), for various values of the ratio k.

4.3 Non-steady Conduction with Finite Coefficient of Surface Heat Transfer

The sphere, initially at the uniform temperature T_{in} is immersed in a surrounding atmosphere of infinite volume at constant temperature T_{ext}. The coefficient of surface heat transfer by convection is finite. The equation of heat conduction within the sphere is thus

$$\frac{\partial T}{\partial t} = \alpha\left[\frac{\partial^2 T}{\partial r^2} + \frac{2}{r}\frac{\partial T}{\partial r}\right] \tag{4.2}$$

with the initial and boundary conditions

$$t = 0, \qquad 0 \leqslant r < R, \qquad T = T_{in} \tag{4.26}$$

$$t > 0, \qquad r = R, \qquad -\lambda\frac{\partial T}{\partial r} = h(T_{R,t} - T_{ext}) \tag{4.27}$$

where $T_{R,t}$ is the temperature at the surface of the sphere at time t.
On putting

$$U = (T - T_{ext})r \tag{4.28}$$

as in Eq. (4.3), the equation of heat conduction becomes

$$\frac{\partial U}{\partial t} = \alpha\frac{\partial^2 U}{\partial r^2}, \qquad 0 < r < R \tag{4.4}$$

with the conditions

$$t = 0, \qquad U = (T_{in} - T_{ext})r \tag{4.29}$$

$$\frac{\partial U}{\partial r} + \left(\bar{h} - \frac{1}{R}\right)U = 0, \qquad r = R \tag{4.30}$$

$$U = 0, \qquad r = 0 \tag{4.31}$$

The spherical problem is thus reduced to the problem of linear heat flow in a sheet, one side being kept at the external temperature zero, the other side being in contact with the external atmosphere at zero, with a finite coefficient of heat transfer.

The solution for the profile of temperature within the sphere, is given by

$$\frac{T_{r,t} - T_{ext}}{T_{in} - T_{ext}} = \frac{2MR}{r} \sum_{n=1}^{\infty} \frac{\sin \beta_n \frac{r}{R}}{[\beta_n^2 + M^2 - M]\sin \beta_n} \exp\left(-\frac{\beta_n^2}{R^2}\alpha t\right) \tag{4.32}$$

where the β_ns are the roots of

$$\beta_n \cot \beta_n + M - 1 = 0 \tag{4.33}$$

and the dimensionless number M is

$$M = \frac{Rh}{\lambda} = R\bar{h} \tag{4.34}$$

The roots of Eq. (4.33) are given in Table 4.2.

For the centre, $(R/r)\sin(\beta_n r/R)$ can be replaced by β_n and the following equation is obtained:

$$\frac{T_{0,t} - T_{ext}}{T_{in} - T_{ext}} = 2M \sum_{n=1}^{\infty} \frac{\beta_n}{[\beta_n^2 + M^2 - M]\sin \beta_n} \exp\left(-\beta_n^2 \frac{\alpha t}{R^2}\right) \tag{4.32'}$$

This series is alternated; it converges for any time $t \geqslant 0$.

The profiles of temperature developed within the sphere are drawn at various times (or rather $\alpha t/R^2$) for two values of the dimensionless number M:

$M = 0.2$ in Fig. 4.8
$M = 1$ in Fig. 4.9

Table 4.2. Roots of $\beta \cot \beta + M - 1 = 0$

M	β_1	β_2	β_3	β_4	β_5	β_6	β_7	β_8	β_9	β_{10}
0.00	0.0000	4.4934	7.7253	10.904	14.066	17.221	20.371	23.519	26.666	29.812
0.01	0.1730	4.4956	7.7265	10.905	14.067	17.221	20.372	23.520	26.666	29.812
0.10	0.5423	4.5157	7.7382	10.913	14.073	17.227	20.376	23.524	26.670	29.815
0.20	0.7593	4.5379	7.7511	10.922	14.080	17.232	20.381	23.528	26.674	29.818
0.50	1.1656	4.6042	7.7899	10.950	14.102	17.250	20.396	23.541	26.685	29.828
1.00	1.5708	4.7124	7.8540	10.996	14.137	17.279	20.420	23.562	26.704	29.845
2.00	2.0288	4.9132	7.9787	11.086	14.207	17.336	20.469	23.604	26.741	29.879
5.00	2.5704	5.3540	8.3029	11.335	14.408	17.503	20.612	23.729	26.851	29.978
10.0	2.8363	5.7172	8.6587	11.653	14.687	17.748	20.828	23.922	27.025	30.135
100	3.1102	6.2204	9.3308	12.441	15.552	18.663	21.775	24.886	27.999	31.111
1000	3.1385	6.2769	9.4154	12.554	15.692	18.831	21.969	25.108	28.246	31.385
∞	3.1416	6.2832	9.4248	12.566	15.708	18.850	21.991	25.133	28.274	31.416

Of course, steep gradients of temperature are obtained next to the surface of the sphere, for the higher value of M, and of the coefficient of surface heat transfer.

The amount of heat which has left (or has entered) the sphere at time t can be

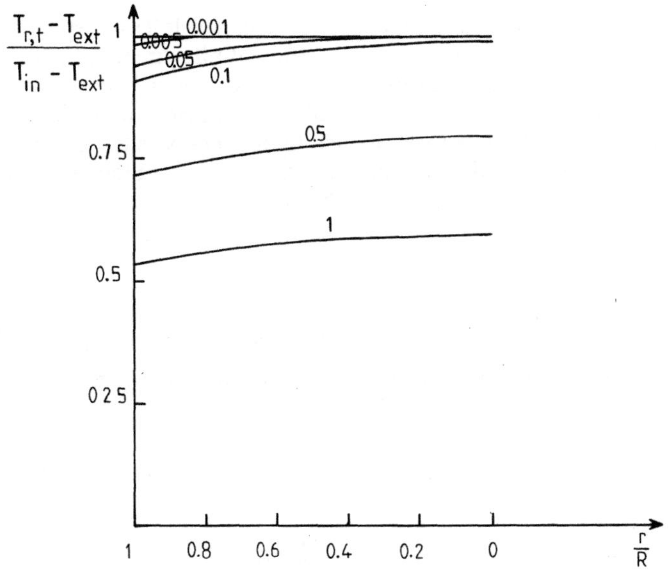

Fig. 4.8. Heat transferred with a finite coefficient of surface heat transfer. Temperature distribution within the sphere at difference times (or $\alpha t/R^2$), with the dimensionless number $M = 0.2$.

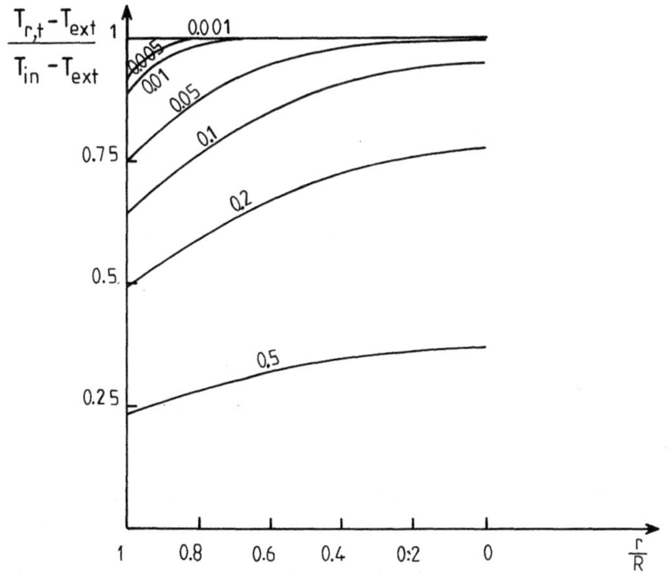

Fig. 4.9. Heat transferred with a finite coefficient of surface heat transfer. Temperature distribution within the sphere at different times (or $\alpha t/R^2$), with $M = 1$.

deduced from the profile of temperature developed within the sphere. It can be expressed in terms of time t as a fraction of Q_∞, the amount of heat transferred after infinite time, when equilibrium is reached:

$$\frac{Q_\infty - Q_t}{Q_\infty} = \sum_n \frac{6M^2}{\beta_n^2[\beta_n^2 + M^2 - M]} \exp\left(-\frac{\beta_n^2}{R^2}\alpha t\right) \tag{4.35}$$

The total amount of heat Q_t which has entered (or left) the sphere at time t, as a fraction of the corresponding quantity after infinite time, is expressed in terms of time (or of $\alpha t/R^2$), for various values of the dimensionless number M (Fig. 4.10). The effect of the number M, and of the coefficient of surface heat transfer, on the process is clearly illustrated in this figure. For a value of M of 10 000, the kinetics of heat transferred is about the same as that obtained in the case of a constant temperature on the surface of the sphere.

4.4 Steady Radial Conduction within a Hollow Sphere with Constant Temperature at Each Surface

In the case of steady temperature, the equation of radial heat conduction is reduced to

$$\frac{d}{dr}\left(r^2 \frac{dT}{dr}\right) = 0 \tag{4.36}$$

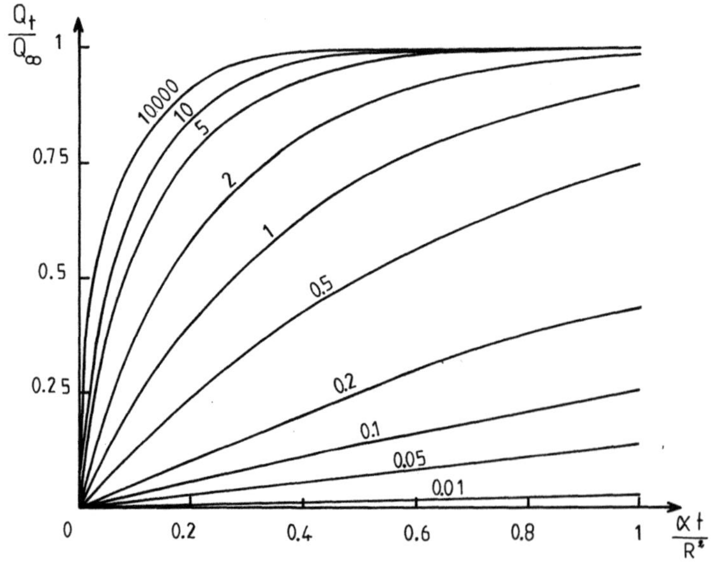

Fig. 4.10. Effect of the coefficient of surface heat transfer on the kinetics of heat transfer. Heat transferred at time t as a fraction of the corresponding quantity after infinite time, expressed in terms of time (or $\alpha t/R^2$), for various values of M.

The general solution of this equation is

$$T = \frac{A}{r} + B \tag{4.37}$$

where the constants A and B can be determined by considering the boundary conditions (Fig. 4.11):

$$r = R_1, \qquad T = T_1, \qquad r = R_2, \qquad T = T_2 \tag{4.38}$$

The temperature at radial position r is thus given by

$$T_r = \frac{R_1 T_1 (R_2 - r) + R_2 T_2 (r - R_1)}{r(R_2 - R_1)} \tag{4.39}$$

Remark. The steady state conduction is the limit of a non-steady conduction process after infinite time. If the initial temperature is uniform, equal to T_{in}, the temperature during the unsteady state varies with time according to the following expression:

$$T_{r,t} = \frac{R_1 T_1}{r} + \frac{(R_2 T_2 - R_1 T_1)(r - R_1)}{r(R_2 - R_1)} + \frac{2}{r\pi} \sum \frac{R_2(T_2 - T_{in}) \cos n\pi - R_1(T_1 - T_{in})}{n} \times \sin\left(\frac{n\pi(r - R_1)}{R_2 - R_1}\right) \exp\left(\frac{-n^2\pi^2}{(R_2 - R_1)^2}\, \alpha t\right) \tag{4.40}$$

By drawing the amount of heat passing through the outer surface of radius R_2, a curve is obtained showing the two steps: non-steady state, followed by the steady state. A result of interest appears with the intercept of the straight line on the time axis, which is given by (Fig. 4.12):

$$\frac{\alpha t}{(R_2 - R_1)^2} = \frac{1}{6} \tag{4.41}$$

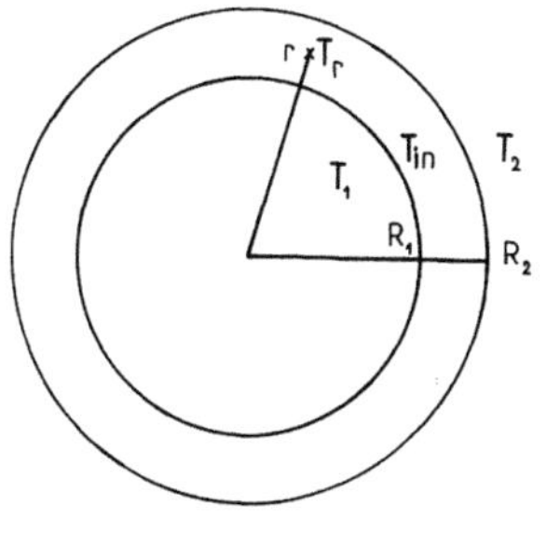

Fig. 4.11. Steady radial heat conduction in the hollow sphere of radii R_1 and R_2.

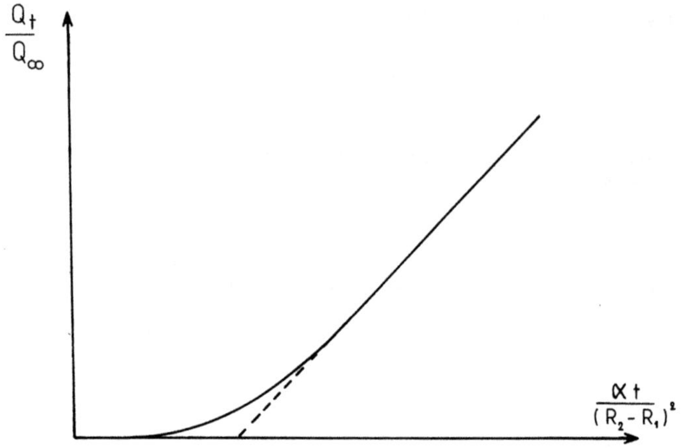

Fig. 4.12 Approach to steady state in a hollow sphere of radii R_1 and R_2.

Reference

1. Carslaw HS, Jaeger JC. Conduction of heat in solids, Clarendon Press, Oxford, 2nd edn, 1978, p 240

Chapter 5

Heat Conduction in a Cylinder

5.1 Introduction

The equation of heat conduction in cylindrical coordinates (r, θ, z) is expressed as follows:

$$\rho C \frac{\partial T}{\partial t} = \frac{1}{r}\frac{\partial}{\partial r}\left(r\lambda \frac{\partial T}{\partial r}\right) + \frac{1}{r^2}\frac{\partial}{\partial \theta}\left(\lambda \frac{\partial T}{\partial \theta}\right) + \frac{\partial}{\partial z}\left(\lambda \frac{\partial T}{\partial z}\right) \qquad (5.1)$$

The case where temperature depends only on r, z and t is considered. There is thus a symmetry of revolution about the Oz axis. The first equation becomes

$$\rho C \frac{\partial T}{\partial t} = \frac{1}{r}\frac{\partial}{\partial r}\left(r\lambda \frac{\partial T}{\partial r}\right) + \frac{\partial}{\partial z}\left(\lambda \frac{\partial T}{\partial z}\right) \qquad (5.2)$$

where r is the radial abscissa.

When heat conductivity λ is constant, the above equation can be rewritten in the form

$$\frac{\partial T}{\partial t} = \frac{\alpha}{r}\frac{\partial}{\partial r}\left(r \frac{\partial T}{\partial r}\right) + \alpha \frac{\partial^2 T}{\partial z^2} \qquad (5.2\ )$$

where α, defined by the ratio

$$\alpha = \frac{\lambda}{\rho C} \cdot \qquad (5.3)$$

is the heat diffusivity (expressed in $\mathrm{cm^2\,s^{-1}}$).

Two cases are of interest in the non-steady state for the solid cylinder:

1. When the length of the cylinder is infinite (or rather is very much longer than the radius).
2. When the length of the cylinder is finite (or is not very much longer than the radius).

In the steady state, the case where the hollow cylinder is of infinite length is of great interest.

5.1.1 Case of a Circular Cylinder of Infinite Length

The Oz axis is the axis of the cylinder. Heat conduction is radial everywhere in this simple case. The temperature in the cylinder is then a function of the radial abscissa r and time t only, and the equation of heat conduction becomes

$$\rho C \frac{\partial T}{\partial t} = \frac{1}{r}\frac{\partial}{\partial r}\left(r\lambda \frac{\partial T}{\partial r}\right) \tag{5.4}$$

With constant heat conductivity, this equation becomes

$$\frac{\partial T}{\partial t} = \frac{\alpha}{r}\frac{\partial}{\partial r}\left(r \frac{\partial T}{\partial r}\right) \tag{5.4'}$$

or

$$\frac{\partial T}{\partial t} = \alpha \left[\frac{\partial^2 T}{\partial r^2} + \frac{1}{r}\frac{\partial T}{\partial r}\right] \tag{5.4''}$$

In all cases, heat diffusivity is assumed to be constant, and no reaction takes place within the solid.

5.2 Non-Steady Conduction in a Solid Cylinder of Infinite Length

Various problems on the cylinder of infinite length with different boundary conditions are considered in succession. Following the method of the separation of variables, the expression

$$T = U \exp(-\alpha\gamma^2 t) \tag{5.5}$$

is a solution of the equation of radial heat conduction

$$\frac{\partial T}{\partial t} = \alpha \left[\frac{\partial^2 T}{\partial r^2} + \frac{1}{r}\frac{\partial T}{\partial r}\right] \tag{5.4''}$$

with constant heat diffusivity α, when U is a function of radius r only, satisfying the following equation:

$$\frac{d^2U}{dr^2} + \frac{1}{r}\frac{dU}{dr} + \gamma^2 U = 0 \tag{5.6}$$

which is Bessel's equation of order zero. Solutions of Bessel's equation are obtained in terms of the Bessel functions determined in such a way that the initial and boundary conditions are satisfied.

First the case where the surrounding medium is a fluid of infinite volume is studied. In addition, the fluid is strongly stirred, so that its temperature is uniform and constant, and constantly equal to the temperature at the surface of the sample. Thus the coefficient of surface heat convection is very high and can be considered as infinite.

In the general problem, with the initial temperature in the cylinder given by $f(r)$, and the surface $r = R$ kept at zero temperature, the initial and boundary conditions are

$$t = 0, \qquad 0 \leqslant r < R, \qquad T = f(r) \tag{5.7}$$

$$t > 0, \qquad r = R, \qquad T = 0 \tag{5.8}$$

The solution of the second kind being infinite for $r = 0$, the integral of Bessel's equation is thus expressed in terms of the Bessel function of the first kind of order zero J_0:

$$T = AJ_0(\gamma r)\exp(-\alpha\gamma^2 t) \tag{5.9}$$

The boundary condition is satisfied by the following series:

$$T = \sum_{n=1}^{\infty} A_n J_0(\gamma_n r)\exp(-\alpha\gamma_n^2 t) \tag{5.9'}$$

provided that the γ_ns are roots of

$$J_0(R\gamma) = 0 \tag{5.10}$$

This Bessel function of the first kind of order zero is finite at $r = 0$.

The initial condition is

$$f(r) = \sum_{m=1}^{\infty} A_m J_0(r\gamma_m) \tag{5.11}$$

If the function $f(r)$ can be expanded in a series of Bessel functions of order zero, coefficients A_n are obtained by multiplying both sides of Eq. (5.11) by $rJ_0(\gamma_n r)$ and integrating from 0 to R.

When γ_n and γ_m are different roots of Bessel's equation (5.10), the integral is equal to zero:

$$\int_0^R rJ_0(\gamma_n r)J_0(\gamma_m r)\,dr = 0 \tag{5.12}$$

When $m = n$, the integral becomes

$$\int_0^R rJ_0^2(\gamma_n r)\,dr = \frac{R^2}{2}J_1^2(\gamma_n R) \tag{5.13}$$

where J_1 is the Bessel function of the first order. The coefficient A_n is thus given by the relation

$$A_n \frac{R^2}{2} J_1^2(R\gamma_n) = \int_0^R rf(r)J_0(r\gamma_n)\,dr \tag{5.14}$$

and the final solution given by the series in Eq. (5.9′) becomes

$$T_{r,t} = \frac{2}{R^2}\sum_{n=1}^{\infty}\frac{J_0(r\gamma_n)}{J_1^2(R\gamma_n)}\exp(-\gamma_n^2\alpha t)\int_0^R rf(r)J_0(r\gamma_n)\,dr \tag{5.15}$$

where the γ_ns are the positive roots of Eq. (5.10).

5.2.1 Constant Initial Temperature in the Cylinder and Constant Surface Temperature

The simple case where the temperature is initially uniform throughout the cylinder is of great interest. For the cylinder of radius R, the initial and boundary conditions are

$$t = 0, \qquad 0 \leqslant r < R, \qquad T = T_{\text{in}} \tag{5.16}$$

$$t \geqslant 0, \qquad r = R, \qquad T = T_{\text{ext}} \tag{5.17}$$

For $f(r) = 1$, the value of the integral is

$$\int_0^R rJ_0(r\gamma_n)\,dr = \frac{RJ_1(R\gamma_n)}{\gamma_n} \tag{5.18}$$

In fact, upon putting $q_n = R\gamma_n$, Eq. (5.18) is rewritten in the following form:

$$\int_0^R rJ_0\left(\frac{r}{R}\,q_n\right) dr = \frac{R^2}{q_n}\,J_1(q_n) \tag{5.18'}$$

The solution for the profile of temperature within the cylinder deduced from Eq. (5.15) becomes

$$\frac{T_{r,t} - T_{\text{ext}}}{T_{\text{in}} - T_{\text{ext}}} = 2\sum_{n=1}^{\infty} \frac{J_0\left(\frac{r}{R}\,q_n\right)}{q_n J_1(q_n)} \exp\left(-q_n^2\,\frac{\alpha t}{R^2}\right) \tag{5.19}$$

where the numbers q_n are the positive roots of the Bessel function of the first kind of order zero:

$$J_0(q_n) = 0 \tag{5.10'}$$

Positive roots of the Bessel function of the first kind of order zero (Eq. (5.10')) are given in Table 5.1 for $M \to \infty$. In the same way, roots of the Bessel function of the first order are given in Table 5.2 for $\beta \to 0$.

The quantity of heat which has entered or left the cylinder in time t can be calculated in the following general way. The amount of heat remaining in the cylinder, per unit length, is at time t

$$Q'_t = 2\pi\rho C \int_0^R (T_{r,t} - T_{\text{ext}})\,r\,dr \tag{5.20}$$

By replacing the temperature $T_{r,t}$ by its value in Eq. (5.19), the amount of heat Q'_t becomes:

$$Q'_t = 4\pi(T_{\text{in}} - T_{\text{ext}})\rho C \sum_{n=1}^{\infty} \frac{1}{q_n J_1(q_n)} \exp\left(-q_n^2\,\frac{\alpha t}{R^2}\right) \int_0^R rJ_0\left(\frac{r}{R}\,q_n\right) dr \tag{5.21}$$

When $t \to \infty$, $Q'_t \to 0$.

Upon putting Q_∞, the amount of heat which has left the cylinder after infinite time per unit length, as Q_∞ is also the amount of heat initially located in the cylinder, there is:

$$Q_\infty = Q'_0 = \pi R^2 \rho C (T_{\text{in}} - T_{\text{ext}}) \tag{5.22}$$

The quantity of heat which has left the cylinder in time t is Q_t, with the relation

$$Q_\infty - Q_t = Q'_t \tag{5.23}$$

The following is obtained:

$$\frac{Q_\infty - Q_t}{Q_\infty} = 4\sum \frac{1}{q_n^2} \exp\left(-q_n^2\,\frac{\alpha t}{R^2}\right) \tag{5.24}$$

The quantity of heat which has left (or entered) the cylinder at time t, Q_t, as a fraction of the corresponding quantity after infinite time, is thus expressed in

Table 5.1. Roots of $\beta q J_0(q) = 2J_1(q) = 0$

β	q_1	q_2	q_3	q_4	q_5	q_6
∞	2.4048	5.5201	8.6537	11.7915	14.9309	18.0711
9.0000	2.4922	5.5599	8.6793	11.8103	14.9458	18.0833
4.0000	2.5888	5.6083	8.7109	11.8337	14.9643	18.0986
2.3333	2.6962	5.6682	8.7508	11.8634	14.9879	18.1183
1.5000	2.8159	5.7438	8.8028	11.9026	15.0192	18.1443
1.0000	2.9496	5.8411	8.8727	11.9561	15.0623	18.1803
0.6667	3.0989	5.9692	8.9709	12.0334	15.1255	18.2334
0.4286	3.2645	6.1407	9.1156	12.1529	15.2255	18.3188
0.2500	3.4455	6.3710	9.3397	12.3543	15.4031	18.4754
0.1111	3.6374	6.6694	9.6907	12.7210	15.7646	18.8215
0	3.8317	7.0156	10.1735	13.3237	16.4706	19.6159

Remark. The top and the bottom rows contain the roots of J_0 and J_1 respectively.

Table 5.2. Roots of $\beta J_1(\beta) = M J_0(\beta)$

M	β_1	β_2	β_3	β_4	β_5	β_6
0	0	3.8317	7.0156	10.1735	13.3237	16.4706
0.01	0.1412	3.8343	7.0170	10.1745	13.3244	16.4712
0.10	0.4417	3.8577	7.0298	10.1833	13.3312	16.4767
0.20	0.6170	3.8835	7.0440	10.1931	13.3387	16.4828
0.50	0.9408	3.9594	7.0864	10.2225	13.3611	16.5010
1.00	1.2558	4.0795	7.1558	10.2710	13.3984	16.5312
2.00	1.5994	4.2910	7.2884	10.3658	13.4719	16.5910
5.00	1.9898	4.7131	7.6177	10.6223	13.6786	16.7630
10.0	2.1795	5.0332	7.9569	10.9363	13.9580	17.0099
100	2.3809	5.4652	8.5678	11.6747	14.7834	17.8931
∞	2.4048	5.5201	8.6537	11.7915	14.9309	18.0711

Remark. The top and the bottom rows contain the roots of J_1 and J_0 respectively.

terms of the dimensionless number $\alpha t/R^2$, which is proportional to time (Fig. 5.1).

The numbers q_n are also the positive roots of Eq. (5.32) (see Sect. 5.2.2) for an infinite value of the ratio β of the thermal capacities of the surrounding fluid and of the cylinder.

Remark 1. Use of Error Functions [1]. The profiles of temperature developed within the cylinder can also be determined by using the error function and the following equation:

$$\frac{T_{r,t} - T_{\text{in}}}{T_{\text{ext}} - T_{\text{in}}} = \sqrt{\frac{R}{r}}\,\text{erfc}\left(\frac{R-r}{2\sqrt{\alpha t}}\right) + \frac{(R-r)\sqrt{R\alpha t}}{4R\sqrt{r^3}}\,\text{ierfc}\left(\frac{R-r}{2\sqrt{\alpha t}}\right)$$
$$+ \frac{(9R^2 - 7r^2 - 2Rr)\alpha t}{32\sqrt{R^3 r^5}}\, i^2\text{erfc}\left(\frac{R-r}{2\sqrt{\alpha t}}\right) + \ldots \qquad (5.25)$$

The amount of heat which has entered or left the cylinder in time t is expressed as a fraction of the corresponding amount of heat after infinite time:

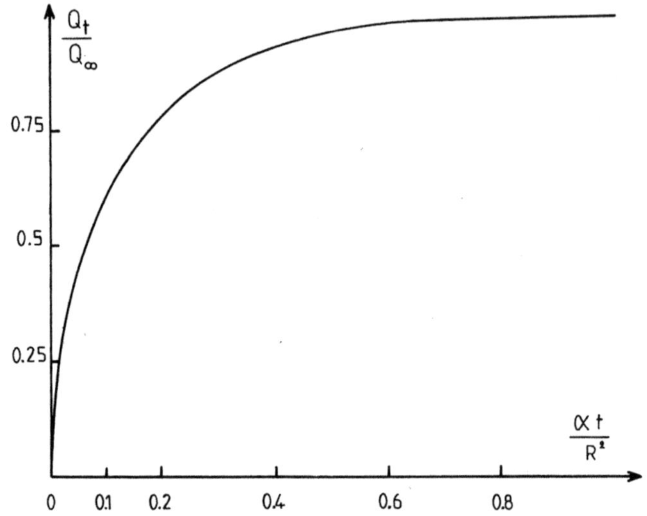

Fig. 5.1. Cylinder of infinite length with radial transfer, uniform initial temperature and constant temperature on the surface with surrounding atmosphere of infinite volume. Heat transferred at time t, Q_t, as a fraction of Q_∞, expressed in terms of $\alpha t/R^2$.

$$\frac{Q_t}{Q_\infty} = \frac{4}{\sqrt{\pi}}\sqrt{\frac{\alpha t}{R^2}} - \frac{\alpha t}{R^2} - \frac{1}{3\sqrt{\pi}}\left(\frac{\alpha t}{R^2}\right)^{1.5} + \ldots \tag{5.26}$$

Remark 2. Expression for Short Times. For short times, when the ratio Q_t/Q_∞ is small, for instance smaller than 0.3, the Eq. (5.26) can be reduced to

$$\frac{Q_t}{Q_\infty} = \frac{4}{R}\sqrt{\frac{\alpha t}{\pi}} \tag{5.27}$$

showing a linear relationship between Q_t and the square root of time. When Q_t/Q_∞ does not exceed 0.3, the relative error given by Eq. (5.27) is less than 0.07.

Remark 3. Very Long Times. For long times, for instance when $Q_t/Q_\infty > 0.7$, the first term of the series in Eq. (5.24) becomes preponderent, and a simple relationship is thus obtained:

$$\ln\left(\frac{Q_\infty - Q_t}{Q_\infty}\right) = -q_1^2\frac{\alpha t}{R^2} + 2\ln\frac{2}{q_1} \tag{5.28}$$

5.2.2 Conduction in a Strongly Stirred Fluid of Limited Volume (Very High Coefficient of Surface Convection)

The cylinder of infinite length occupies the space $r < R$, and the cross-section of the surrounding fluid is finite, equal to S (if this surrounding fluid is located in a cylinder having the same axis as the cylinder of radius R, the radius of this cylinder can be considered).

There is no transfer of heat through the external surface of the surrounding fluid. The fluid is strongly stirred, so that the temperature in the fluid is always uniform and always equal to that of the surface of the cylinder. Thus the coefficient of surface convection is very high and can be considered as infinite. The temperature in the fluid is initially T_{ie}, while the temperature in the cylinder is initially uniform and equal to T_{is}.

The initial conditions are thus as follows:

$$t = 0, \qquad 0 \leq x < R, \qquad T = T_{is} \qquad \text{(solid cylinder)}$$

$$x > R, \qquad T = T_{ie} \qquad \text{(surroundings)} \tag{5.29}$$

while the boundary conditions express the fact that the rate at which heat enters the fluid by convection is equal to the rate at which heat leaves the cylinder by conduction, per unit length of cylinder.

$$t > 0, \qquad \rho' C' S \frac{\partial T}{\partial t} = -\lambda 2\pi R \frac{\partial T}{\partial r}, \qquad r = R \tag{5.30}$$

The problem of heat conduction through the solid cylinder is about the same as the corresponding problem for the sheet (Sect. 3.2.3).

By analogy with matter transport controlled by diffusion [1], the total amount of heat Q_t which has entered (or left) the solid cylinder up to time t is expressed as a fraction of the corresponding amount Q_∞ after infinite time by the equation

$$\frac{Q_\infty - Q_t}{Q_\infty} = \sum_{n=1}^{\infty} \frac{4\beta(1+\beta)}{4 + 4\beta + \beta^2 q_n^2} \exp\left(-q_n^2 \frac{\alpha t}{R^2}\right) \tag{5.31}$$

where the q_ns are the positive, non-zero roots of

$$\beta q_n J_0(q_n) + 2J_1(q_n) = 0 \tag{5.32}$$

and β is the ratio of the thermal capacities of the surrounding fluid and of the solid cylinder:

$$\beta = \frac{S\rho' C'}{\pi R^2 \rho C} = \frac{T_\infty - T_{is}}{T_{ie} - T_\infty} \tag{5.33}$$

The roots of Eq. (5.32) are given in Table 5.1 for various values of the ratio β.

The amount of heat transferred at time t through the unit length of the cylinder, as a fraction of the corresponding amount after infinite time, is drawn as a function of time, or rather of the dimensionless number $\alpha t/R^2$, for various values of the ratio β (Fig. 5.2). Of course, for an infinite value of the ratio β, the curve obtained is the same as that obtained in Fig. 5.1.

The profile of temperature through the solid cylinder is given by

$$\frac{T_{r,t} - T_{is}}{T_\infty - T_{is}} = 1 + \sum_{n=1}^{\infty} \frac{4(\beta+1) J_0\left(q_n \frac{r}{R}\right)}{[4 + 4\beta + \beta^2 q_n^2] J_0(q_n)} \exp\left(-q_n^2 \frac{\alpha t}{R^2}\right) \tag{5.34}$$

5.2.3 Convection in a Slightly Stirred Fluid of Infinite Volume (Finite Coefficient of Surface Convection)

The cylinder of infinite length is initially at a uniform temperature T_{is}, and the surface condition describes the fact that the rate at which heat enters the cylinder

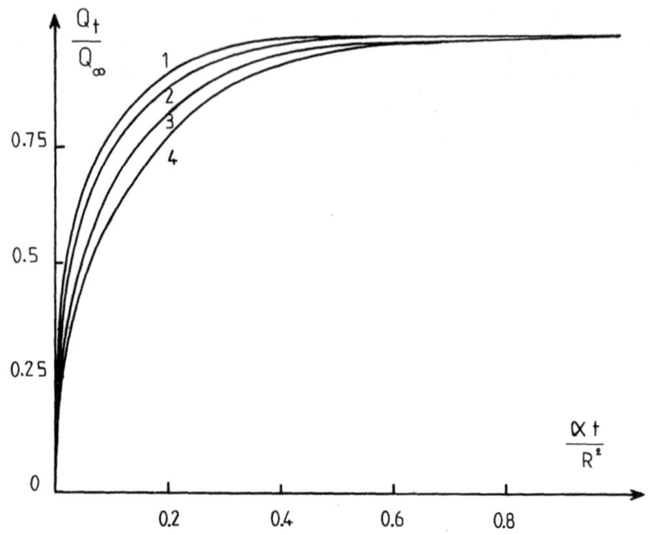

Fig. 5.2. Cylinder of infinite length immersed in a surrounding well-stirred fluid of finite cross-section. Heat transferred at time t, Q_t, as a fraction of Q_∞, expressed in terms of $\alpha t/R^2$, for various values of the ratio β.

by convection is constantly equal to the rate at which heat is brought to the surface by conduction within the cylinder:

$$-\lambda \frac{\partial T}{\partial t}(R, t) = h(T_{R,t} - T_{\text{ext}}) \qquad \text{(surface)} \tag{5.35}$$

where T_{ext} is the constant temperature of the surrounding atmosphere, far from the sample (T_{ext} may be considered constant as the volume of the surrounding fluid is very large) and $T_{R,t}$ is the temperature at the surface of the cylinder.

If the coefficient of surface heat transfer h is constant, the solution for the temperature within the cylinder, $T_{r,t}$, is

$$\frac{T_{r,t} - T_{\text{is}}}{T_{\text{ext}} - T_{\text{is}}} = 1 - \sum_{n=1}^{\infty} \frac{2MJ_0\left(\beta_n \dfrac{r}{R}\right)}{(\beta_n^2 + M^2)J_0(\beta_n)} \exp\left(-\beta_n^2 \frac{\alpha t}{R^2}\right) \tag{5.36}$$

where the β_ns are the roots of

$$\beta J_1(\beta) = MJ_0(\beta) \tag{5.37}$$

and the dimensionless number M is

$$M = R\bar{h} = R\frac{h}{\lambda} \tag{5.38}$$

Roots of Eq. (5.37) are given in Table 5.2 for various values of the dimensionless number M.

The total amount of heat entering or leaving the cylinder up to time t, Q_t, is expressed as a fraction of the corresponding quantity after infinite time Q_∞, per unit length of the cylinder, by the equation

$$\frac{Q_\infty - Q_t}{Q_\infty} = \sum_{n=1}^{\infty} \frac{4M^2}{\beta_n^2(\beta_n^2 + M^2)} \exp\left(-\beta_n^2 \frac{\alpha t}{R^2}\right) \tag{5.39}$$

The effect of the value of the coefficient of surface heat transfer on the kinetics of heat transferred through the surface of the cylinder (per unit length) is illustrated in Fig. 5.3. Heat transferred through the surface of the cylinder per unit length, as a fraction of the corresponding quantity after infinite time, is plotted for various times, or rather various values of $\alpha t/R^2$, and for various values of the dimensionless number M.

Of course, for the infinite value of M, the kinetics obtained is the same as in Fig. 5.1 with a constant temperature on the surface of the cylinder.

5.3 Non-Steady Conduction in a Hollow Cylinder of Infinite Length

Hollow cylinders are of great interest, especially from a practical point of view when their length can be considered as infinite. The most important application for hollow cylinders or tubings is in steady-state heat conduction when the internal and external surfaces of the solid are kept at constant temperatures. Nevertheless, some information is given on the non-steady heat conduction in two cases:

1. When the internal and external surfaces of the solid are kept at the same constant temperature
2. When the internal and external surfaces are maintained at constant temperatures T_i and T_e, these temperatures being different

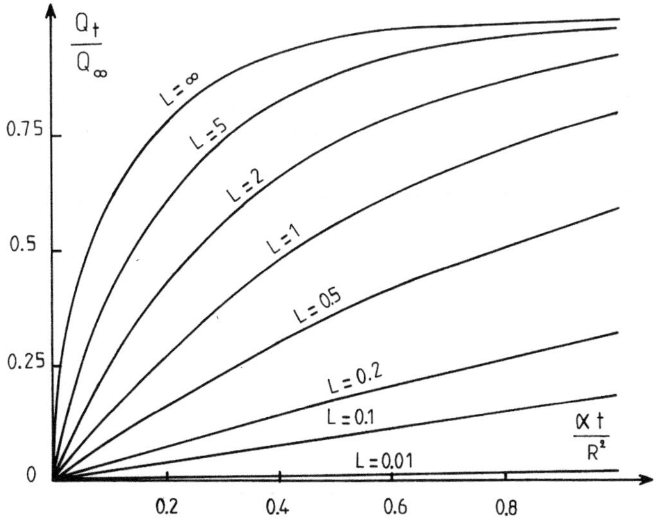

Fig. 5.3. Cylinder of infinite length immersed in a fluid of infinite volume, with a finite coefficient of surface heat convection. Heat transferred at time t, Q_t, as a fraction of Q_∞, expressed as a function of $\alpha t/R^2$, for various values of the dimensionless number $M = Rh/\lambda$.

5.3.1 Surfaces Maintained at the Same Constant Temperature

The hollow cylinder located between the radii R_i and R_e, with a constant initial temperature T_{in}, has its surfaces $r = R_i$ and $r = R_e$ maintained at a constant temperature T_{ext}. As the length is infinite, heat conduction is radial. The coefficient of surface heat convection is infinite, as the surfaces of the solid are kept at constant temperature.

The solution for the profile of temperature is given by [2]:

$$\frac{T_{r,t} - T_{ext}}{T_{in} - T_{ext}} = \pi \sum_{n=1}^{\infty} \frac{J_0(R_i\gamma_n)U_0(r\gamma_n)}{J_0(R_i\gamma_n) + J_0(R_e\gamma_n)} \exp(-\gamma_n^2\alpha t) \tag{5.40}$$

where

$$U_0(r\gamma_n) = J_0(r\gamma_n)Y_0(R_e\gamma_n) - J_0(R_e\gamma_n)Y_0(r\gamma_n) \tag{5.41}$$

and the γ_ns are the positive roots of

$$U_0(R_i\gamma_n) = 0 \tag{5.42}$$

α is the thermal diffusivity of the solid, and Y_0 is the Bessel function of second kind, of order zero. Roots of Eq. (5.42) are given in Table 5.3 for various values of the ratio of the radius R_e/R_i.

The amount of heat entering or leaving the hollow cylinder up to time t, Q_t, is expressed as a fraction of the corresponding value after infinite time, Q_∞, by

$$\frac{Q_\infty - Q_t}{Q_\infty} = \frac{4}{R_e^2 - R_i^2} \sum \frac{J_0(R_i\gamma_n) - J_0(R_e\gamma_n)}{\gamma_n^2[J_0(R_i\gamma_n) + J_0(R_e\gamma_n)]} \exp(-\gamma_n^2\alpha t) \tag{5.43}$$

5.3.2 Surface Maintained at Different Constant Temperatures

The hollow cylinder $R_i < r < R_e$, is initially at uniform temperature T_{in}, and the surface R_i is maintained at T_i, and R_e at T_e, the temperatures T_i and T_e being constant. The temperature in the hollow cylinder, $T_{r,t}$, approaches the steady-state distribution according to the expression

$$\begin{aligned} T_{r,t} = {} & \frac{T_i \ln(R_e/r) + T_e \ln(r/R_i)}{\ln(R_e/R_i)} \\ & + \pi T_{in} \sum_{n=1}^{\infty} \frac{J_0(R_i\gamma_n)U_0(r\gamma_n)}{J_0(R_i\gamma_n) + J_0(R_e\gamma_n)} \exp(-\gamma_n^2\alpha t) \\ & - \pi \sum_{n=1}^{\infty} \frac{J_0(R_i\gamma_n)U_0(r\gamma_n)[T_eJ_0(R_i\gamma_n) - T_iJ_0(R_e\gamma_n)]}{J_0^2(R_i\gamma_n) - J_0^2(R_e\gamma_n)} \exp(-\gamma_n^2\alpha t) \end{aligned} \tag{5.44}$$

where the γ_ns are the positive roots of Eq. (5.42).

The amount of heat leaving the hollow cylinder, per unit length, is obtained by expressing the rate of heat passing through the surface and integrating with respect to time t:

Table 5.3. Roots of $J_0(R_i\gamma)Y_0(R_e\gamma) = J_0(R_e\gamma)Y_0(R_i\gamma)$

$\dfrac{R_e}{R_i}$	$R_i\gamma_1$	$R_i\gamma_2$	$R_i\gamma_3$	$R_i\gamma_4$	$R_i\gamma_5$
1.2	15.7014	31.4126	47.1217	62.8302	78.5385
1.5	6.2702	12.5598	18.8451	25.1294	31.4133
2.0	3.1230	6.2734	9.4182	12.5614	15.7040
2.5	2.0732	4.1773	6.2754	8.3717	10.4672
3.0	1.5485	3.1291	4.7038	6.2767	7.8487
3.5	1.2339	2.5002	3.7608	5.0196	6.2776
4.0	1.0244	2.0809	3.1322	4.1816	5.2301

$$\pm 2\pi R\lambda \int_0^t \frac{\partial T}{\partial r}(R, \theta)\, d\theta, \qquad \begin{array}{l}\text{sign} - \text{for } R = R_e \\ \text{sign} + \text{for } R = R_i\end{array} \tag{5.45}$$

5.4 Steady Radial Conduction in a Hollow Cylinder of Infinite Length

The hollow cylinder is located between the inner and outer radii R_i and R_0, and the thermal parameters are constant. In this case, the temperature profile within the hollow cylinder is expressed by Eq. (5.4′) with $\partial T/\partial t = 0$:

$$\frac{d}{dr}\left(r\frac{dT}{dr}\right) = 0 \qquad R_i < r < R_0 \tag{5.46}$$

The general solution to this equation is

$$T = A + B \ln r \tag{5.47}$$

where A and B are constants to be evaluated from the boundary conditions at $r = R_i$ and $r = R_0$.

5.4.1 Given Constant Temperatures at Each Surface

The surface $r = R_i$ is kept at a constant temperature T_i, and $r = R_0$ at T_0. The profile of temperature is easily obtained within the hollow cylinder:

$$T_r = \frac{T_i \ln\left(\dfrac{R_0}{r}\right) + T_0 \ln\left(\dfrac{r}{R_i}\right)}{\ln\left(\dfrac{R_0}{R_i}\right)} \tag{5.48}$$

The constant rate of flow of heat per unit length of the hollow cylinder is

$$-2\pi r\lambda \frac{dT}{dr} = \frac{2\pi\lambda(T_i - T_0)}{\ln\left(\dfrac{R_0}{R_i}\right)} \tag{5.49}$$

Eq. (5.48) can also be obtained from Eq. (5.44), when the time is long enough for the exponential terms to vanish.

5.4.2 Infinite Coefficient of Convection at Internal Surface and Finite Coefficient at External Surface

The hollow cylinder with the internal surface $r = R_i$ kept at a constant temperature T_i and with a finite coefficient of surface heat convection at the external surface is an interesting case. The boundary condition at the external surface $r = R_0$ is

$$-\frac{dT}{dr} = \frac{h}{\lambda}(T_0 - T_{ext}) = \bar{h}(T_0 - T_{ext}) \tag{5.50}$$

where T_{ext} is the temperature of the fluid surrounding the external surface. This equation can be rewritten as follows:

$$\frac{T_i - T_0}{R_0 \ln\left(\dfrac{R_0}{R_i}\right)} = \bar{h}(T_0 - T_{ext}) \tag{5.50'}$$

The value of the temperature T_0 is thus obtained:

$$T_0 = \frac{T_i + T_{ext}\bar{h}R_0 \ln\left(\dfrac{R_0}{R_i}\right)}{1 + \bar{h}R_0 \ln\left(\dfrac{R_0}{R_i}\right)} \tag{5.50''}$$

and the temperature T_r at the radial abscissa r can thus be calculated using Eq. (5.48), giving

$$T_r = \frac{T_i\left[1 + \bar{h}R_0 \ln\left(\dfrac{R_0}{r}\right)\right] + T_{ext}\bar{h}R_0 \ln\left(\dfrac{r}{R_i}\right)}{1 + \bar{h}R_0 \ln\left(\dfrac{R_0}{R_i}\right)} \tag{5.51}$$

The outward rate of flow of heat per unit length of the cylinder is

$$2\pi\lambda(T_i - T_{ext})\frac{\bar{h}R_0}{1 + \bar{h}R_0 \ln\left(\dfrac{R_0}{R_i}\right)} \tag{5.52}$$

5.5 Non-Steady Conduction in a Cylinder of Finite Length with a Constant Temperature at the Surface (Infinite Coefficient of Surface Convection)

The solid cylinder of radius R and length $2Z$, initially at uniform temperature T_{in}, is immersed in a strongly stirred fluid of infinite volume kept at T_{ext}. The coefficient of surface heat convection is very high, so that the temperature at the surface is T_{ext}. The equation of heat conduction within the solid cylinder, with the constant thermal diffusivity α, is

$$\frac{\partial T}{\partial t} - \frac{\alpha}{r}\frac{\partial}{\partial r}\left(r\frac{\partial T}{\partial r}\right) + \alpha\frac{\partial^2 T}{\partial z^2} = 0 \tag{5.2'}$$

with the initial and boundary conditions

$$t = 0, -Z < z < +Z \text{ and } 0 \leqslant r < R, \quad T = T_{\text{in}} \text{ (solid cylinder)} \tag{5.53}$$

$$t > 0, z = \pm Z \text{ or } r = R, \quad T = T_{\text{ext}} \text{ (surface)} \tag{5.54}$$

The solution to the problem is obtained by the separation of spatial variables. On putting

$$T(r, z, t) = f(r, t)g(z, t) \tag{5.55}$$

the following relation is obtained:

$$\frac{\partial T}{\partial t} - \frac{\alpha}{r}\frac{\partial}{\partial r}\left(r\frac{\partial T}{\partial r}\right) - \alpha\frac{\partial^2 T}{\partial z^2} = g(z, t)\left[\frac{\partial f}{\partial t} - \frac{\alpha}{r}\frac{\partial}{\partial r}\left(r\frac{\partial f}{\partial r}\right)\right] + f(r, t)\left[\frac{\partial g}{\partial t} - \alpha\frac{\partial^2 g}{\partial z^2}\right] \tag{5.56}$$

It is sufficient that f and g verify respectively

$$\frac{\partial f}{\partial t} - \frac{\alpha}{r}\frac{\partial}{\partial r}\left(r\frac{\partial f}{\partial r}\right) = 0 \tag{5.4'}$$

$$f(R, t) = 0 \quad \text{for } t > 0, \qquad f(r, 0) = 1 \quad \text{for } 0 \leqslant r < R \tag{5.57}$$

$$\frac{\partial g}{\partial t} - \alpha\frac{\partial^2 g}{\partial z^2} = 0 \tag{3.1}$$

$$g(\pm Z, t) = 0 \quad \text{for } t > 0, \qquad g(z, t) = 1 \quad \text{for } -Z < z < Z \tag{5.58}$$

Thus f and g are given by Sects 5.2.1 and 3.2.1 respectively:

$$f(r, t) = 2\sum_{n=1}^{\infty}\frac{J_0\left(\frac{r}{R}q_n\right)}{q_n J_1(q_n)}\exp\left(-q_n^2\frac{\alpha t}{R^2}\right) \tag{5.19}$$

where the q_ns are the positive roots of the Bessel function J_0.

$$g(z, t) = \frac{4}{\pi}\sum_{p=0}^{\infty}\frac{(-1)^p}{2p+1}\cos\frac{(2p+1)\pi z}{2Z}\exp\left(-(2p+1)^2\pi^2\frac{\alpha t}{4Z^2}\right) \tag{3.26}$$

The solution to the present problem is

$$\frac{T_{r,z,t} - T_{\text{ext}}}{T_{\text{in}} - T_{\text{ext}}} = 2\sum_{n=1}^{\infty}\frac{J_0\left(\frac{r}{R}q_n\right)}{q_n J_1(q_n)}\exp\left(-q_n^2\frac{\alpha t}{R^2}\right) \times \frac{4}{\pi}\sum_{p=0}^{\infty}\frac{(-1)^p}{2p+1}\cos\frac{(2p+1)\pi z}{2Z}\exp\left(-(2p+1)^2\pi^2\frac{\alpha t}{4Z^2}\right) \tag{5.59}$$

The profile of temperature developed within the solid cylinder is thus given by the previous equation.

The amount of heat which has left the cylinder after time t, Q'_t, is given by

$$Q'_t = \rho C \int_0^R 2\pi r\,dr \int_{-Z}^{+Z} (T_{r,z,t} - T_{\text{ext}})\,dz \tag{5.60}$$

This amount of heat Q'_t is thus equal to [3]

$$Q'_t = \frac{64(T_{\text{in}} - T_{\text{ext}})\rho C Z R^2}{\pi^2} \sum_{m=1}^{\infty} \frac{1}{q_m^2} \exp\left[-\frac{q_m^2}{R^2}\alpha t\right] \times \sum_{n=0}^{\infty} \frac{1}{(2n+1)^2} \exp\left[-\frac{(2n+1)^2\pi^2}{4Z^2}\alpha t\right] \tag{5.61}$$

The amount of heat which can leave (or enter) the cylinder up to infinite time is

$$Q_\infty = 2\pi R^2 Z (T_{\text{in}} - T_{\text{ext}})\rho C \tag{5.62}$$

The amount of heat which has left (or entered) the cylinder after time t, Q_t, is thus expressed as a fraction of the corresponding amount after infinite time, by the following relationship:

$$\frac{Q_\infty - Q_t}{Q_\infty} = \frac{32}{\pi^2} \sum_{m=1}^{\infty} \frac{1}{q_m^2} \exp\left[-\frac{q_m^2}{R^2}\alpha t\right] \sum_{n=0}^{\infty} \frac{1}{(2n+1)^2} \times \exp\left[-\frac{(2n+1)^2\pi^2}{4Z^2}\alpha t\right] \tag{5.63}$$

The value of the heat which has entered (or left) the cylinder of finite length at time t, as a fraction of the corresponding quantity after infinite time, is expressed

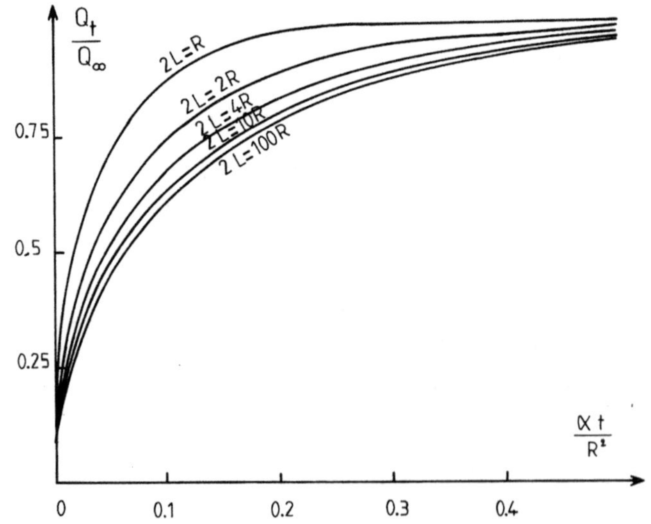

Fig. 5.4. Cylinder of finite length with a uniform initial temperature and a constant temperature on the surface. Heat transferred at time t, Q_t, as a fraction of Q_∞, expressed as a function of $\alpha t/R^2$, for various values of the length (the total length $2Z$ is expressed as a fraction of the radius R).

in terms of time, or rather of the dimensionless number $\alpha t/R^2$ (Fig. 5.4). Various values of the length are used, ranging from $2Z = R$ for a flat cylinder to $2Z = 50R$ for a long cylinder.

Of course, the kinetics of the heat transferred is faster for the flat cylinder, when the surface per unit volume is higher. For a value of the total length around $2Z = 100R$, the kinetics of the heat transferred is about the same as that obtained with a cylinder of infinite length (Fig. 5.1).

References

1. Wilson AH. Phil Mag 1984; 39: 48
2. Carslaw HS, Jaeger JC. Conduction of heat in solids, 2nd edn, Clarendon Press, Oxford, 1978, p 207
3. Adda Y, Philibert J. La diffusion dans les solides, Tome I, Presses Univ. de France, Paris, 1966, p 166

Chapter 6

Heat Conduction in a Rectangular Parallelepiped

6.1 Introduction

Conduction of heat in solids of simple geometrical shape such as the rectangular parallelepiped is of great importance as such shapes are very common in practice. Of course, in this case three space variables are involved.

The profile of temperature developed within the solid can be determined in different ways. The most important problems and also the simplest are when the solutions can be expressed as a product of solutions of one-variable problems.

Two problems are considered in a non-steady state:

1. When the solid, initially at a uniform temperature, is immersed in a well-stirred fluid of large volume kept at a constant temperature, so that the coefficient of surface heat convection is so high that the temperature on the surfaces of the solid are maintained at the constant temperature of the surrounding fluid
2. When the solid, initially at a uniform temperature, is immersed in a slightly stirred fluid of large volume kept at a constant temperature, with a finite coefficient of surface heat convection

In all cases, the thermal parameters are constant, and there is no change in the dimensions of the solid.

6.2 Non-Steady Conduction in a Rectangular Parallelepiped with a Uniform Initial Temperature

6.2.1 Infinite Coefficient of Surface Convection

The rectangular parallelepiped initially at a uniform temperature T_{in} is immersed in a well-stirred fluid of a volume so large that its temperature T_{ext} remains constant. The coefficient of surface heat convection is so high that the temperature on the surface of the solid is maintained equal to that of the surrounding atmosphere.

The initial and boundary conditions are

$$t = 0, \quad -X < x < X, \quad -Y < y < Y, \quad -Z < z < Z, \quad T = T_{\text{in}} \tag{6.1}$$

$$t > 0, \quad x = \pm X \quad \text{or} \quad y = \pm Y \quad \text{or} \quad z = \pm Z, \quad T = T_{\text{ext}} \tag{6.2}$$

for the rectangular parallelepiped of dimensions $2X$, $2Y$ and $2Z$.

The equation of heat conduction in the solid is

$$\frac{\partial T}{\partial t} = \alpha \left[\frac{\partial^2 T}{\partial x^2} + \frac{\partial^2 T}{\partial y^2} + \frac{\partial^2 T}{\partial z^2} \right] \tag{6.3}$$

where α is the constant of thermal diffusivity.

The solution of heat conduction has already been obtained in the case of one-dimensional heat transfer in terms of the Fourier series

$$\frac{T_{x,t} - T_{\text{ext}}}{T_{\text{in}} - T_{\text{ext}}} = f_X(x, t) \tag{6.4}$$

with

$$f_X(x, t) = \frac{4}{\pi} \sum_{n=0}^{\infty} \frac{(-1)^n}{2n+1} \cos \frac{(2n+1)\pi x}{2X} \exp\left(-\frac{(2n+1)^2}{4X^2} \alpha t \right) \tag{6.5}$$

where the sheet, through which heat is transferred, occupies the space $-X < x < X$. This function f_X verifies Eq. (3.1) and the conditions

$$f_X(\pm X, t) = 0 \quad \text{for } t > 0$$

$$f_X(x, 0) = 1 \quad \text{for } -X < x < X \tag{6.6}$$

In the case of a rectangular parallelepiped, the profile of temperature developed in the solid is obtained by taking the product of solutions of one-dimensional transfer:

$$\frac{T_{x,y,z,t} - T_{\text{ext}}}{T_{\text{in}} - T_{\text{ext}}} = f_X(x, t) f_Y(y, t) f_Z(z, t) \tag{6.7}$$

If Q_t is the amount of heat which has entered or left the parallelepiped at time t, and Q_∞ the corresponding quantity after infinite time, the following solution is obtained by taking the product of solutions of one-dimensional transfer (as shown for a sheet):

$$\frac{Q_\infty - Q_t}{Q_\infty} = \frac{512}{\pi^6} \sum_{m=0}^{\infty} \frac{1}{(2m+1)^2} \exp\left(-\frac{(2m+1)^2 \pi^2}{4X^2} \alpha t \right) \sum_{n=0}^{\infty} \frac{1}{(2n+1)^2} \times \exp\left(-\frac{(2n+1)^2 \pi^2}{4Y^2} \alpha t \right) \sum_{p=0}^{\infty} \frac{1}{(2p+1)^2} \exp\left(-\frac{(2p+1)^2 \pi^2}{4Z^2} \alpha t \right) \tag{6.8}$$

When the solid is a cube of sides $2X$ the equation becomes

$$\frac{Q_\infty - Q_t}{Q_\infty} = \frac{512}{\pi^6} \left(\sum_{p=0}^{\infty} \frac{1}{(2p+1)^2} \exp\left(-\frac{(2p+1)^2 \pi^2}{4X^2} \alpha t \right) \right)^3 \tag{6.9}$$

Parallelepipeds with $X = Y$, and various values of Z are considered. The amount of heat which has entered (or left) each parallelepiped at time t, as a fraction of the corresponding quantity after infinite time, is expressed as a function of time, or rather of $\alpha t/X^2$, for various values of Z. The effect of the relative value of Z is clearly shown in Fig. 6.1. Of course, the cube is obtained

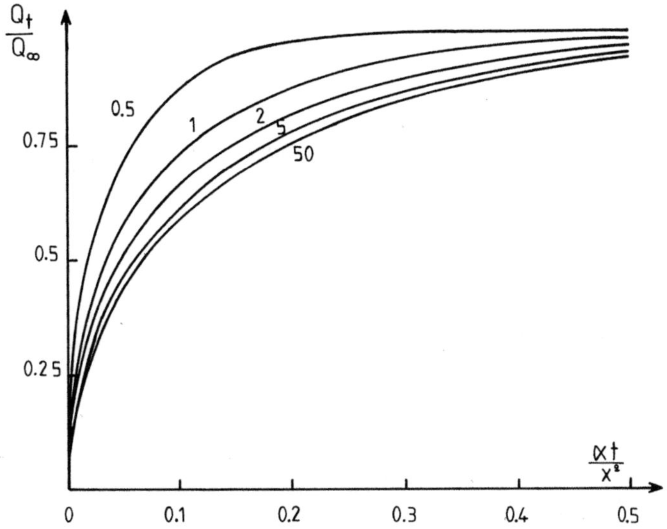

Fig. 6.1. Kinetics of heat transferred through the surface of parallelepipeds with $X = Y$, for various values of Z/X. The dimensionless number $\alpha t/X^2$ is taken instead of time.

for $Z = X$. For Z higher than $50X$, the kinetics are about the same. The following statement can be established: the higher the surface per unit volume of the parallelepiped, the faster the kinetics of heat transferred.

6.2.2 Finite Coefficient of Surface Convection

The rectangular parallelepiped initially at uniform temperature T_{in} is immersed in a slightly stirred fluid of large volume kept at constant temperature T_{ext}. The coefficient of surface heat convection is finite.

The initial conditions are thus for the rectangular parallelepiped of dimensions $2X$, $2Y$ and $2Z$:

$$t = 0, \quad -X < x < X, \quad -Y < y < Y, \quad -Z < z < Z, \quad T = T_{\text{in}} \tag{6.10}$$

The boundary conditions express the fact that on each side the rate of heat transmitted by convection is constantly equal to the rate at which heat is brought to the surface by internal conduction and also proportional to the difference between the temperature at the surface and the temperature in the fluid far from the sample:

$$t > 0, \quad -\lambda \frac{\partial T}{\partial x}(X, y, z, t) = h(T_{X,y,z,t} - T_{\text{ext}}) \tag{6.11}$$

The solution for this three-dimensional heat transfer is the product of the solutions obtained with one-dimensional heat transfer in a sheet under the same conditions (Sect. 3.3 with thickness $2R$).

The profile of temperature developed within the rectangular parallelepiped is thus expressed by the relationship

$$\frac{T_{x,y,z,t} - T_{\text{ext}}}{T_{\text{in}} - T_{\text{ext}}} = g(X, \bar{h}, x, t)g(Y, \bar{h}, y, t)g(Z, \bar{h}, z, t) \tag{6.12}$$

where each function is given by

$$g(X, \bar{h}, x, t) = \sum_{m=1}^{\infty} \frac{2M \cos \omega'_m \dfrac{x}{X}}{(\omega'^2_m + M^2 + M)\cos \omega'_m} \exp\left(\frac{\omega'^2_m \alpha t}{-X^2}\right) \tag{6.13}$$

where ω'_m are the positive roots of

$$\omega' \tan \omega' = M \tag{6.14}$$

with the dimensionless number M

$$M = X\bar{h} = \frac{Xh}{\lambda} \tag{6.15}$$

The total amount of heat which has left (or entered) the rectangular parallelepiped after time t, Q_t, is expressed as a fraction of the corresponding quantity after infinite time, Q_∞, by

$$\begin{aligned}\frac{Q_\infty - Q_t}{Q_\infty} = &\left(\sum_{m=1}^{\infty} \frac{2M^2}{\omega'^2_m[\omega'^2_m + M^2 + M]} \exp - \frac{\omega'^2_m}{X^2}\alpha t\right)\\ &\times\left(\sum_{n=1}^{\infty} \frac{2N^2}{\omega''^2_n[\omega''^2_n + N^2 + N]} \exp - \frac{\omega''^2_n}{Y^2}\alpha t\right)\\ &\times\left(\sum_{p=1}^{\infty} \frac{2P^2}{\omega'''^2_p[\omega'''^2_p + P^2 + P]} \exp - \frac{\omega'''^2_p}{Z^2}\alpha t\right)\end{aligned} \tag{6.16}$$

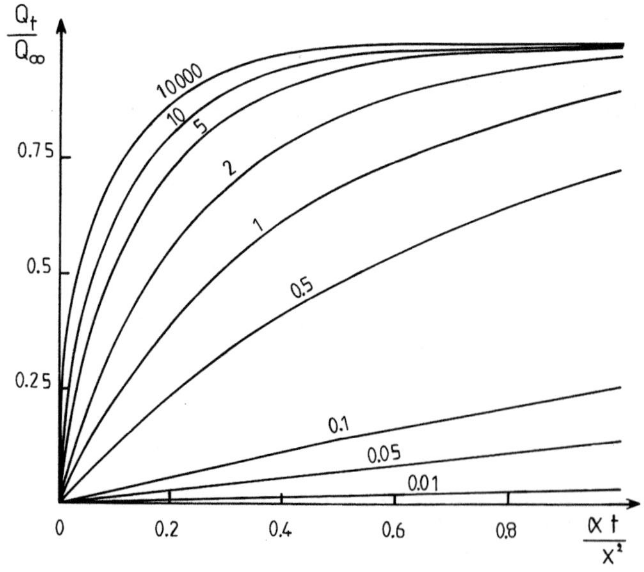

Fig. 6.2. Kinetics of heat transferred through the surface of a cube of sides $2X$, for various values of the dimensionless number $M = hX/\lambda$. The dimensionless number $\alpha t/X^2$ is taken instead of time.

where ω''_n and ω'''_p are the positive roots of

$$\omega'' \tan \omega'' = N, \qquad \omega''' \tan \omega''' = P \tag{6.14'}$$

with the other dimensionless numbers N and P:

$$N = Y\bar{h} = Y\frac{h}{\lambda}, \qquad P = Z\bar{h} = Z\frac{h}{\lambda} \tag{6.15'}$$

The effect of the value of the coefficient of surface heat convection, for a given solid (dimension, thermal diffusivity) on the kinetics of heat transferred through the surface of a cube of sides $2X$ is illustrated in Fig. 6.2. The amount of heat which has entered (or left) the cube up to time t, Q_t, as a fraction of the corresponding quantity after infinite time, is expressed in terms of time, or rather $\alpha t/Z^2$, for various values of the dimensionless number M. Of course, a value of M around 10000 is so high that it can be considered as infinite, as the kinetics curve for this value in Fig. 6.2 is about the same as that obtained in Fig. 6.1 for the cube with a constant temperature on the surface.

Chapter 7

Numerical Analysis for a Plane Sheet. One-Dimensional Heat Transfer and Cure Reaction

7.1 Introduction

The actual operating conditions in curing thermoset polymers do not always correspond to those for which the analytical solutions have been found in the previous chapters. The following facts very often appear:

1. The physical properties (thermal conductivity λ, density ρ, heat capacity C) of the solid may vary during the process of cure. In fact, they may vary not only with the temperature in the solid but also with the state of cure of the thermoset.
2. The initial distribution in the solid may not be uniform; moreover, the function of temperature $f(x)$ may not be treated mathematically.
3. The conditions at the surface may be very complex, as for instance in the case encountered with free convection.
4. The external temperature, for instance that of the mould or that of the surrounding atmosphere, may vary with time and position in a complex way.
5. Internal heat is usually generated by cure reactions taking place within the solid, the rate of heat evolved varying with temperature and following an Arrhenius law.

Such complex cases can be handled by using numerical methods with finite differences, which give approximative solutions. In applying a numerical method with finite differences, the continuous process of heat transfer and reaction is replaced by a stepwise process. In the case of a sheet so thin that effectively heat enters or leaves the plane faces and a negligible part of heat through the edges, the increment method is applied to one-dimensional transient heat transport through the thickness of the plane sheet. Let L be this thickness.

In numerical analysis, finite increments of space Δx and time Δt are considered. The Ox axis perpendicular to the sheet is chosen, the two surfaces of the sheet having as abscissae $x = 0$ and $x = L$. The sheet is divided into X slices of equal thickness Δx, defined by the relationship:

$$L = X\Delta x \tag{7.1}$$

Thus the time at which an event occurs, as well as the position, are expressed in terms of these increments and of integers i and n respectively:

$$t = i\Delta t \qquad \text{(time)} \tag{7.2}$$

$$x = n\Delta x \qquad \text{(position)}, \qquad 0 \leqslant n \leqslant X \tag{7.3}$$

At abscissa $x = n\Delta x$ and time $t = i\Delta t$, the temperature is denoted $T_{n,i}$.

The progress of heat within the sheet, and the following progress of the state of cure are clearly shown in the two-dimensional space–time diagram (Fig. 7.1). At the beginning of the process, at time zero, the initial conditions for the temperature distribution within the sheet appear on the first horizontal line. For each position characterised by the integer n, the temperature is

$$T_{n,0} = f(n\Delta x) \qquad \text{initial condition at position } n \text{ (with abscissa } x = n\Delta x) \tag{7.4}$$

At time t, defined by integer i, the temperatures at each position in the solid are

$$T_{n-1,i} \qquad T_{n,i} \qquad T_{n+1,i} \qquad \text{at time } t = i\Delta t \tag{7.5}$$

After time Δt, the time becomes $t + \Delta t$ or $(i + 1)\Delta t$, and the temperatures appear as follows:

$$T_{n-1,i+1},\ T_{n,i+1},\ T_{n+1,i+1} \text{ at time } t + \Delta t \tag{7.6}$$

It is generally preferred to write the temperature at time t and the new temperature at $t + \Delta t$ in the following way:

$$\left.\begin{array}{ll} T_n & \text{instead of } T_{n,i} \\ \mathrm{TN}_n & \text{instead of } T_{n,i+1} \end{array}\right\} \tag{7.7}$$

Of course, the precision obtained for the approximate solution is better when the number of slices is increased.

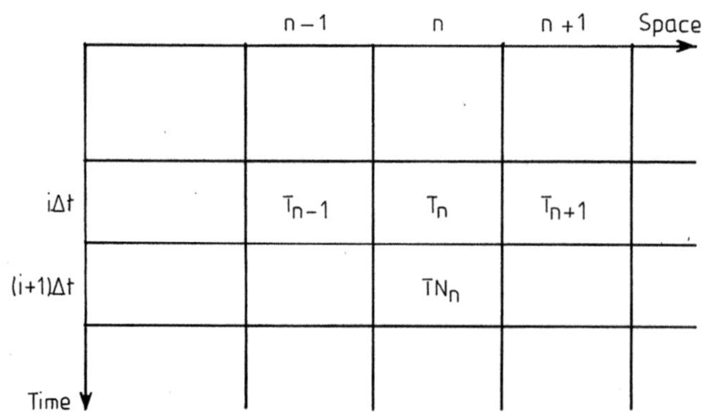

Fig. 7.1. Space–time diagram for numerical analysis.

7.2 No Reaction, Constant Thermal Parameters

Two positions are of interest:

Inside the solid, with heat conduction

On the surface of the sheet, generally with heat convection between the surface and the surrounding

7.2.1 Within the Sheet

As shown in Sect. 7.1, the solid sheet of thickness L is divided into X slices of thickness Δx, and each plane separating these slices is characterised by an abscissa along the x axis and by an integer n (Fig. 7.2). The heat balance is evaluated within the slice at position n (with $1 \leqslant n \leqslant X - 1$), located between the planes $n - 0.5$ and $n + 0.5$, during the interval of time $[t, t + \Delta t]$.

On the one hand, the heat variation within this slice during the time Δt is due to transient heat conduction:

$$A[\phi_{n-0.5,i} - \phi_{n+0.5,i}]\Delta t = \Delta Q \tag{7.8}$$

where A is the area of the parallel planes through which heat is conducted and $\phi_{n-0.5,i}$ is the heat flux by conduction at the position $n - 0.5$ and at time $i\Delta t$.

On the other hand, this heat variation within the slice is responsible for a change in temperature:

$$\Delta Q = A\Delta x\rho C[T_{n,i+1} - T_{n,i}] = A\Delta x\rho C[\mathrm{TN}_n - T_n] \tag{7.9}$$

where $T_{n,i}$ is approximately equal to the mean temperature within the slice n at time $i\Delta t$, and $T_{n,i+1}$ is the mean temperature in the same slice after an interval of time Δt, and thus at time $(i + 1)\Delta t$.

The heat flux by conduction is

$$\phi = -\lambda \frac{\partial T}{\partial x} \tag{7.10}$$

where $\partial T/\partial x$ is the gradient of temperature. This heat flux is evaluated by using the finite differences. The following main assumption is made by approximating the gradient of temperature at plane $n - 0.5$ by the chord slope between the

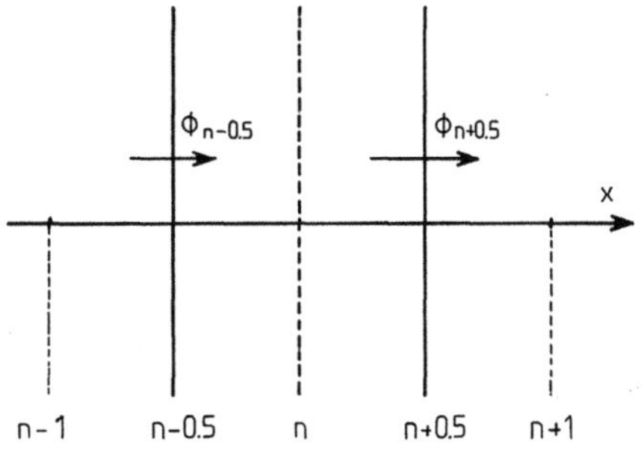

Fig. 7.2. Heat balance within the solid sheet, within the slice located between the planes $n - 0.5$ and $n + 0.5$.

planes $n - 1$ and n:

$$-\left(\frac{\partial T}{\partial x}\right)_{n-0.5,i} = \frac{T_{n-1,i} - T_{n,i}}{\Delta x} \quad \text{for } 1 \leqslant n \leqslant X \tag{7.11}$$

The resulting heat balance within the slice located between the planes $n - 0.5$ and $n + 0.5$ during the interval of time $[t, t + \Delta t]$, is

$$A\left[\lambda \frac{T_{n-1} - T_n}{\Delta x} - \lambda \frac{T_n - T_{n+1}}{\Delta x}\right] \Delta t = A \Delta x \rho C [\mathrm{TN}_n - T_n] \tag{7.12}$$

Upon replacing the thermal diffusivity $\lambda/\rho C$ by α, the new temperature after an interval of time Δt within the slice n becomes

$$\mathrm{TN}_n = \frac{1}{M}[T_{n-1} + (M - 2)T_n + T_{n+1}] \quad \text{for } 1 \leqslant n \leqslant X - 1 \tag{7.13}$$

where the dimensionless number M is

$$M = \frac{(\Delta x)^2}{\Delta t} \frac{1}{\alpha} \tag{7.14}$$

Remark. If λ remains constant, but if ρ and C are variable, then Eq. (7.13) remains valid, with variable values for M.

7.2.2 On the Surfaces of the Sheet

Two different cases must be considered:

When the temperature on the sheet surface is given
When the coefficient of surface heat convection is finite

They are considered successively.

Given Temperatures on the Sheet Surfaces

In the simplest case, the temperature on the sheet surfaces is constant and equal to the temperature of the surroundings at the start of the process.

$$T_0 = \text{constant}, \qquad T_X = \text{constant} \tag{7.15}$$

These constant temperatures T_0 and T_X may be the same or different. A programmed temperature can also be imposed at the surfaces of the sheet:

$$T_0 = g_1(t), \qquad T_X = g_2(t) \tag{7.16}$$

Heat Convection at the Surface

The surface of the sheet (for instance that at position zero) is heated (or cooled) by a fluid warmer (or colder) at instantaneous temperature T_{ext}. This temperature of the fluid taken at a place sufficiently far from the sheet is assumed to be uniform.

The slice in the solid next to the surface of thickness $\Delta x/2$ is considered (Fig. 7.3). The heat balance is evaluated in this half slice during the interval of time $[t, t + \Delta t]$ by considering heat convection through the surface and heat conduction through the plane at position 0.5.

$$A(\phi_0 - \phi_{0.5})\Delta t = A \frac{\Delta x}{2} \rho C(TN_{0.25} - T_{0.25}) \tag{7.17}$$

$TN_{0.25}$ being the new temperature at position 0.25.

At time t, the heat flux ϕ_0 which enters the sheet through the surface $x = 0$, by convection, is given by

$$\phi_0 = h(T_{ext} - T_0) \tag{7.18}$$

where h is the coefficient of surface heat transfer by convection, and T_{ext} can be either constant or variable, as a function of time.

The heat flux by conduction $\phi_{0.5}$, at time t, is as shown above:

$$\phi_{0.5} = -\lambda \frac{\partial T}{\partial x}(0.5, t) \tag{7.10}$$

The gradient of temperature at plane 0.5 is well approximated (at the second order) by the chord slope between the positions 0 and 1, so that the heat flux at position 0.5 becomes:

$$\phi_{0.5} = \lambda \frac{T_0 - T_1}{\Delta x} \tag{7.19}$$

The heat flux by convection at the surface is also given by Eq. (7.18). The heat balance within this half slice can thus be written as follows:

$$\left[h(T_{ext} - T_0) - \lambda \frac{T_0 - T_1}{\Delta x}\right] \Delta t = \frac{\Delta x}{2} \rho C[TN_{0.25} - T_{0.25}] \tag{7.20}$$

By considering the dimensionless numbers M and N:

$$M = \frac{(\Delta x)^2}{\Delta t} \frac{1}{\alpha} \tag{7.14}$$

$$N = \frac{h \Delta x}{\lambda} \tag{7.21}$$

the above equation, after multiplying both sides by $\Delta x/\lambda\Delta t$, becomes

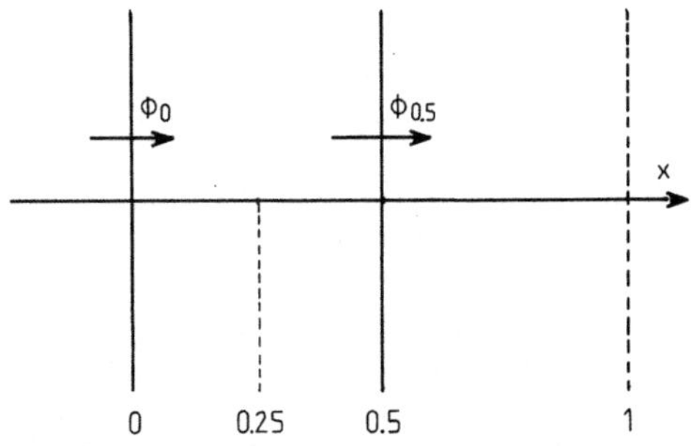

Fig. 7.3. Heat balance at the solid surface, with heat conduction within the solid and heat convection to the surface. Adjacent slice of thickness $\frac{1}{2}\Delta x$.

$$\frac{M}{2}[TN_{0.25} - T_{0.25}] = N(T_{ext} - T_0) - (T_0 - T_1) \qquad (7.22)$$

Two kinds of assumptions can be made for evaluating the temperature at plane 0.25:

1. By making the simple assumption

$$TN_{0.25} - T_{0.25} = TN_0 - T_0 \qquad (7.23)$$

the new temperature at the surface after an interval of time Δt becomes

$$TN_0 = \frac{1}{M}[2NT_{ext} + (M - 2 - 2N)T_0 + 2T_1] \qquad (7.24)$$

2. By making a linear interpolation, a more accurate relation is obtained:

$$TN_{0.25} - T_{0.25} = \tfrac{3}{4}(TN_0 - T_0) + \tfrac{1}{4}(TN_1 - T_1) \qquad (7.25)$$

and the new temperature TN_0 at the surface is thus

$$TN_0 = T_0 - \frac{1}{3}(TN_1 - T_1) + \frac{8}{3M}[NT_{ext} - (N + 1)T_0 + T_1] \quad (7.26)$$

In this last case, the new temperature (at time $t + \Delta t$), TN_0, is expressed in terms of the previous temperatures at time t as well as of the new temperature at plane 1, TN_1. The new temperature TN_1 must be calculated beforehand by using Eq. (7.13) available for positions located within the sheet. The new temperature TN_0 can thus be obtained with the help of Eq. (7.26).

Change of Step of Calculation

For each increment of time, the initial temperatures are the final temperatures of the preceding increment. In PASCAL language, for instance, this is written as

For i := 0 to X do $T(i)$:= $TN(i)$

7.2.3 Amount of Heat Transferred

The amount of heat which has entered the sheet up to time t, or after infinite time, per unit area, can be deduced from the mean temperature of the sheet at time t, TM_t:

$$Q_t = \rho CL(TM_t - TM_0) \qquad (7.27)$$

$$Q_\infty = \rho CL(T_{ext} - TM_0) \qquad (7.28)$$

Of course after infinite time, if the external temperature tends to a limit, equilibrium is established and the temperature in the sheet becomes uniform and equal to the external temperature T_{ext}.

The mean temperature TM_t, can be evaluated by using the profile of temperature developed within the sheet.

For $1 \leqslant n \leqslant X - 1$, T_n is approximatively the mean temperature in the slice $[n - 0.5, n + 0.5]$, while the mean temperature in the half slice $[0, 0.5]$, is approximately equal to $T_{0.25}$. Hence

$$TM_t = \frac{1}{X}\left[\frac{1}{2}T_{0.25} + \sum_{n=1}^{X-1} T_n + \frac{1}{2}T_{X-0.25}\right] \tag{7.29}$$

By expressing $T_{0.25}$ and $T_{X-0.25}$ by a linear interpolation, the following is obtained:

$$TM_t = \frac{1}{X}\left[\frac{3}{8}(T_0 + T_X) + \frac{9}{8}(T_1 + T_{X-1}) + \sum_{n=2}^{X-2} T_n\right] \tag{7.30}$$

Symmetry

If the initial and boundary conditions exhibit a symmetry with respect to the midplane of the sheet, then the duration of calculation can be reduced, by taking an even number of slices $2X$ and by putting at each step

$$T_{X+1} = T_{X-1} \tag{7.31}$$

Stability of Calculations

If Δt is too large, the successive values of the calculated functions may oscillate with an increasing amplitude. In order to avoid this troublesome phenomenon, the following conditions must be fulfilled. In the equation giving the new temperature TN_n, for each value of n, the whole coefficient of the temperature T_n must be positive. Thus Eqs (7.13), (7.24) and (7.26) impose respectively

$$M - 2 \geqslant 0 \quad \text{or} \quad \Delta t \leqslant \frac{\Delta x^2}{2\alpha} \tag{7.32}$$

$$M - 2N - 2 \geqslant 0 \quad \text{or} \quad \Delta t \leqslant \frac{\Delta x^2}{2\alpha(N+1)} \tag{7.33}$$

$$1 - \frac{8}{3M}(N+1) \geqslant 0 \quad \text{or} \quad \Delta t \leqslant \frac{3\Delta x^2}{8\alpha(N+1)} \tag{7.34}$$

7.3 No Reaction, Temperature-Dependent Thermal Parameters

Thermal parameters such as the thermal conductivity λ, as well as the other parameters such as heat capacity C and density ρ, may vary with temperature.

7.3.1 Within the Sheet

The solid sheet of thickness L is divided into X slices of thickness Δx, as shown in Fig. 7.2. The thickness of the sheet L and of the slices Δx may vary with temperature. In order to simplify the problem, a frame of reference fixed with respect to the original sheet is taken. Then the thickness of the sheet, measured in these units, is constant and equal to that of the original sheet. Calculation is made in the same way as for constant thermal parameters but by considering the temperature dependency of these parameters.

The heat balance evaluated in the slice at position n, during the interval of time $[t, t + \Delta t]$, gives

$$A[\phi_{n-0.5} - \phi_{n+0.5}]\Delta t = A\Delta x(\rho C)_n(\mathrm{TN}_n - T_n) \tag{7.35}$$

$$\phi_{n-0.5} = -\lambda_{n-0.5}\left(\frac{\partial T}{\partial x}\right)_{n-0.5} \tag{7.36}$$

Upon defining the function G,

$$G(n) = \lambda_{n-0.5}(T_{n-1} - T_n) \tag{7.37}$$

with the temperature-dependent thermal conductivity λ, it becomes

$$\phi_{n-0.5} = \frac{G(n)}{\Delta x} \quad \text{and} \quad \phi_{n+0.5} = \frac{G(n+1)}{\Delta x} \quad \text{for } n = 1 \text{ to } X - 1 \tag{7.38}$$

The new temperature at position n after an interval of time Δt is thus expressed in terms of the previous temperature obtained at the same position n and adjacent positions:

$$\mathrm{TN}_n = T_n + \frac{\Delta t}{\Delta x^2}\frac{1}{\rho_n C_n}[G(n) - G(n+1)] \tag{7.39}$$

with the function G defined in Eq. (7.37).

7.3.2 On the Surfaces of the Sheet

Given Temperature on the Sheet Surfaces

In the simplest case, the temperature on each surface of the sheet is always equal to the constant temperature of the surrounding atmosphere. The temperature on the surfaces is thus

$$T_0 = T_X = T_{\mathrm{ext}} = \text{constant} \tag{7.15}$$

A programmed temperature can also be imposed at the surfaces of the sheet:

$$T_0 = g_1(t), \qquad T_X = g_2(t) \tag{7.16}$$

Heat Convection at the Surfaces

The heat balance during the increment of time Δt on the adjacent slice is evaluated by considering heat convection between the surface and the surrounding and heat conduction within the solid.

The slice of thickness $\Delta x/2$ next to the surface is considered (Fig. 7.3). The heat balance during the increment of time Δt is written as follows:

$$A(\phi_0 - \phi_{0.5})\Delta t = A\frac{\Delta x}{2}(\rho C)_{0.25}[\mathrm{TN}_{0.25} - T_{0.25}] \tag{7.40}$$

The heat flux at positions 0 and 0.5 is

$$\phi_0 = h(T_{ext} - T_0) \qquad \text{(convection)} \tag{7.41}$$

$$\phi_{0.5} = -\lambda_{0.5} \left(\frac{\partial T}{\partial x}\right)_{0.5} \qquad \text{(conduction)} \tag{7.42}$$

where the thermal conductivity $\lambda_{0.5}$ is a function of temperature, and also a function of time.

With the simple assumption for temperatures at position 0.25, and at each time t

$$TN_{0.25} - T_{0.25} = TN_0 - T_0 \tag{7.23}$$

the new temperature, after an interval of time Δt, at the surface becomes

$$TN_0 = T_0 + \frac{\Delta t}{\Delta x} \frac{2}{(\rho C)_{0.25}} [\phi_0 - \phi_{0.5}] \tag{7.43}$$

The more complex but more accurate assumption for these temperatures at plane 0.25,

$$TN_{0.25} - T_{0.25} = \tfrac{3}{4}(TN_0 - T_0) + \tfrac{1}{4}(TN_1 - T_1) \tag{7.25}$$

yields to the following expression for the new temperature at the surface, TN_0

$$TN_0 = T_0 - \frac{1}{3}(TN_1 - T_1) + \frac{4\Delta t}{3\Delta x(\rho C)_{0.25}} [\phi_0 - \phi_{0.5}] \tag{7.44}$$

As shown previously in the case of constant thermal parameters, the new temperature TN_0 at the surface is expressed in terms not only of the previous temperatures appearing in the flux ϕ_0 and ϕ_1, but also of the new temperature at plane 1, TN_1. This new temperature TN_1 must thus be calculated beforehand by using Eq. (7.39) which needs only the previous temperature $T_{0,i}$ at the surface.

Heat-Tight Plane or Symmetry Plane

If the surface, for instance at $x = L$, does not allow heat to get through it, the temperature gradient at this surface is zero. To express this condition, the abscissa numbers are prolonged up to $X + 1$:

$$T_{X+1} = T_{X-1} \tag{7.31}$$

Likewise, if the initial and boundary conditions exhibit a symmetry with respect to the midplane of the sheet, then the duration of calculation can be reduced, by taking an even number of slices $2X$ and by putting at each step

$$T_{X+1} = T_{X-1} \tag{7.31}$$

7.3.3 Calculation of Thermal Parameters

Values of temperature are obtained for each slice, when the abscissa characterising the position is an integer. It is thus easy to determine the value of all thermal parameters at these positions, even though they are temperature-dependent, providing that the temperature dependency of these parameters is known.

When these thermal parameters must be used in other places where the value of x is not an integer, the value of these thermal parameters has to be determined beforehand. Two ways of working can be used, depending on the degree of accuracy desirable.

(i) Thermal Parameters Vary Linearly with Temperature

By assuming that each thermal parameter varies about linearly with temperature, the values of thermal parameters are easily obtained for each position, by linear interpolation, giving the following relations:

$$\lambda_{n-0.5} = \tfrac{1}{2}(\lambda_{n-1} + \lambda_n) \qquad \text{for } 1 \leqslant n \leqslant X \tag{7.45}$$

$$(\rho C)_{0.5} = \tfrac{1}{2}[(\rho C)_0 + (\rho C)_1] \tag{7.46}$$

$$(\rho C)_{0.25} = \tfrac{3}{4}(\rho C)_0 + \tfrac{1}{4}(\rho C)_1 \tag{7.46'}$$

(ii) Thermal Parameters Do Not Vary Linearly with Temperature

In this case, temperature must be calculated first, at each position. The easiest way is followed by assuming that temperature varies linearly with space within a narrow space of the order of the half slice $\Delta x/2$ or less. Thus the temperature at position $n - 0.5$ is obtained by

$$T_{n-0.5} = \tfrac{1}{2}(T_{n-1} + T_n) \qquad \text{for } 1 \leqslant n \leqslant X \tag{7.47}$$

The value of the thermal parameters can then be calculated at this position and time, when the temperature dependency of these parameters is known, and expressed by way of relationships.

7.3.4 Amount of Heat Transferred

The amount of heat transferred in the sheet up to time t or after infinite time is evaluated in the same way as for constant thermal parameters, by taking into account the temperature-dependence of the thermal parameters ρC.

First Method of Calculation

The amount of heat which enters the slice of area A, located between the abscissae $x - \frac{1}{2}\Delta x$ and $x + \frac{1}{2}\Delta x$, during the interval of time $[t, t + \Delta t]$ is given by

$$\Delta Q = A \Delta x \rho(x, t) C(x, t) \frac{\partial T}{\partial t}(x, t) \Delta t \tag{7.48}$$

$$\Delta Q_n = A \rho_n C_n (\mathrm{TN}_n - T_n) \Delta x \tag{7.49}$$

The amount of heat which enters the total sheet in the same interval of time is given by

$$Q_{t+\Delta t} - Q_t = \frac{3}{8}(\Delta Q_0 + \Delta Q_X) + \frac{9}{8}(\Delta Q_1 + \Delta Q_{X-1}) + \sum_{n=2}^{X-2} \Delta Q_n \quad (7.50)$$

$$Q_{t+\Delta t} - Q_t = A\Delta t \int_0^L \rho(x, t)C(x, t) \frac{\partial T}{\partial t}(x, t)\, dx \quad (7.51)$$

The amount of heat which has entered the sheet up to time t is

$$Q_t = A \int_0^L \int_0^t \rho(x, \theta)C(x, \theta) \frac{\partial T}{\partial t}(x, \theta)\, dx\, d\theta \quad (7.52)$$

Second Method of Calculation

The heat flux which enters the sheet, at time t, through the two surfaces $x = 0$ and $x = L$ is

$$\lambda(L, t) \frac{\partial T}{\partial x}(L, t) - \lambda(0, t) \frac{\partial T}{\partial x}(0, t) = \lambda_X \frac{3T_X - 4T_{X-1} + T_{X-2}}{2\Delta x} + \lambda_0 \frac{3T_0 - 4T_1 + T_2}{2\Delta x} \quad (7.53)$$

where the flux at the surfaces is calculated by parabolic approximation. Thus the amount of heat which has entered the sheet during the interval of time $[0, t]$ is

$$Q_t = A \int_0^t \left[\lambda(L, \theta) \frac{\partial T}{\partial x}(L, \theta) - \lambda(0, \theta) \frac{\partial T}{\partial x}(0, \theta)\right] d\theta, \qquad 0 \leqslant \theta \leqslant t \quad (7.54)$$

If the amount of heat which has entered the sheet, Q_t, is negative, then the amount of heat which has left the sheet is positive.

Unlike the case where the parameters λ, ρ and C are constant, knowledge of the initial and final conditions in the sheet is not sufficient to deduce the amount of heat which has entered the sheet. A step by step calculation for temperature and these parameters is needed, for each increment of time. The heat which has entered the sheet during each time increment is thus obtained by integration with respect to space (Eq. (7.50)). The total heat which has entered the sheet up to time t is finally obtained by adding these amounts of heat.

7.4 Heat Evolved from Reaction

Very often the kinetics of heat evolved from the overall cure reaction is expressed in terms of absolute temperature T by a simple Arrhenius relation:

$$\frac{1}{Q_\infty} \frac{dQ_t}{dt} = k_0\left(1 - \frac{Q_t}{Q_\infty}\right)^n \exp - \frac{E}{RT} = k_T\left(1 - \frac{Q_t}{Q_\infty}\right)^n \quad (7.55)$$

where

Q_t is the heat evolved from reaction up to time t

Q_∞ is the heat evolved from reaction after infinite time
k_0 is a pre-exponential factor which depends only on the type of reaction
n is the apparent order of the overall reaction
E is the activation energy of the overall reaction

The state of cure (SOC) at time t is expressed by the ratio of heat evolved up to time t, Q_t, by the enthalpy of cure Q_∞.

$$\text{SOC} = 100 \frac{Q_t}{Q_\infty} \tag{7.56}$$

The kinetics of cure are generally determined by calorimetry. In calorimetry techniques, enthalpy changes accompanying chemical events in the sample are monitored as a function of temperature or time. The result is the profile of the rate of enthalpy change either as a function of temperature as the sample is heated at a known linear rate in differential scanning calorimetry (DSC), or as a function of time, the calorimeter being held at constant temperature in differential calorimetry (DC). The assumption in calorimetry technique calculation is that the partial enthalpy evolved up to time t as a fraction of the total enthalpy evolved, represents the conversion up to time t. This assumption is the basis for the kinetic analysis of reaction exotherm. The enthalpy associated with the overall exotherm is determined by measuring the area under the exotherm after the selection of a baseline. The simplest equation for the kinetics of cure is thus expressed by

$$\frac{dY}{dt} = k_0(1 - Y)^n \exp\left(-\frac{E}{RT}\right) = k_T(1 - Y)^n \tag{7.57}$$

where Y represents the conversion at time t.

7.4.1 Apparent Order of Reaction Different from One

Under isothermal conditions, the equation representing the kinetics of cure can be integrated with respect to time. The conversion up to time t is thus given by

$$Y = \frac{Q_t}{Q_\infty} = 1 - [1 + (n - 1)k_T t]^{1/1-n} \tag{7.58}$$

Generally the temperature T at a given place depends on time t. Eq. (7.57) can be rewritten

$$(1 - Y)^{-n}\, dY = k_{T_t}\, dt \tag{7.59}$$

where T_t is the temperature in the considered position at time t. By integrating this equation and putting

$$S_t = \int_0^t k_{T_\theta}\, d\theta, \qquad 0 \leqslant \theta \leqslant t \tag{7.60}$$

the following is obtained, if $n \neq 1$:

$$\frac{-(1 - Y)^{1-n} + 1}{1 - n} = S_t \tag{7.61}$$

$$Y = 1 - [1 + (n - 1)S_t]^{1/1-n} \tag{7.62}$$

The integral S_t can be calculated approximatively step by step, for each interval of time $[t, t + \Delta t]$, as follows:

$$S_{t+\Delta t} = S_t + k_{T_t}\Delta t \tag{7.63}$$

$$S_{t+\Delta t} = S_t + k_0 \Delta t \exp\left(\frac{-E}{RT_t}\right) \tag{7.64}$$

Of course, at time zero, there is

$$S_0 = 0 \tag{7.65}$$

The state of cure is thus given by

$$\mathrm{SOC}_t = 100 Y_t \tag{7.66}$$

and heat evolved during the interval of time $[t, t + \Delta t]$ is

$$\Delta Q = Q_\infty (Y_{t+\Delta t} - Y_t) \tag{7.67}$$

7.4.2 Apparent Order of Reaction Equal to One

The apparent order of the overall reaction is one when the rate of cure is proportional to the heat of reaction remaining to be evolved.

$$\frac{dQ_t}{Q_\infty dt} = k_0\left(1 - \frac{Q_t}{Q_\infty}\right)\exp\left(-\frac{E}{RT}\right) = k_T\left(1 - \frac{Q_t}{Q_\infty}\right) \tag{7.68}$$

or by considering the conversion Y up to time t:

$$\frac{dY}{dt} = k_0(1 - Y)\exp\left(-\frac{E}{RT}\right) \tag{7.68'}$$

By integrating the Eq. (7.59) and using Eq. (7.60), the conversion Y is obtained:

$$-\ln(1 - Y) = S_t \tag{7.69}$$

$$Y = 1 - \exp(-S_t) \tag{7.70}$$

with the same function S:

$$S_t = \int_0^t k_0 \exp\left(-\frac{E}{RT_\theta}\right) d\theta \qquad 0 \leqslant \theta \leqslant t \tag{7.71}$$

and the relation enabling one to obtain $S_{t+\Delta t}$ after an interval of time Δt:

$$S_{t+\Delta t} = S_t + k_0 \Delta t \exp\left(-\frac{E}{RT_t}\right) \tag{7.72}$$

with

$$S_0 = 0 \tag{7.73}$$

The value of the conversion Y can thus be calculated at any time and any position, and heat evolved during the increment of time Δt, at time t is given by

$$\Delta Q = Q_\infty (Y_{t+\Delta t} - Y_t) \tag{7.74}$$

7.5 Heat Transfer and Reaction Heat

The equation for the heat transfer rate by conduction within the sheet is as follows:

$$\rho C \frac{\partial T}{\partial t} = \frac{\partial}{\partial x}\left(\lambda \frac{\partial T}{\partial x}\right) + \rho \frac{\partial Q}{\partial t} \tag{7.75}$$

where $\partial Q/\partial t$ expresses the rate of heat evolved from the overall cure reaction (per gram of material, as it is usual in calorimetry). The equations giving the temperature at various places and times in the resin are thus rewritten, by considering not only heat transferred by conduction through the resin but also heat evolved from the overall reaction.

7.5.1 Constant Thermal Parameters

Within the Solid

The heat balance during the increment of time Δt is calculated within the slice of thickness Δx located between the planes at the positions $n - 0.5$ and $n + 0.5$ (Fig. 7.2), by considering not only heat transferred by conduction through these planes but also heat evolved from the cure reaction.

Eq. (7.12) is rewritten as follows:

$$A\left[\lambda \frac{T_{n-1} - T_n}{\Delta x} - \lambda \frac{T_n - T_{n+1}}{\Delta x}\right] \Delta t + A\Delta x \rho \Delta Q_n = A\Delta x \rho C[\mathrm{TN}_n - T_n] \tag{7.76}$$

where ΔQ_n is the amount of heat evolved from cure reaction per mass unit at position n during the period of time between t and $t + \Delta t$, and ρ is the density of the resin ($\mathrm{g\,cm^{-3}}$). As ΔQ is expressed in calories per gram of resin, the term $A\Delta x \rho \Delta Q$ represents the amount of heat evolved from the cure reaction in the slice of volume $A\Delta x$. By dividing both terms of Eq. (7.76) by $A\Delta x \rho C$, the following is obtained:

$$\mathrm{TN}_n = \frac{1}{M}[T_{n-1} + (M - 2)T_n + T_{n+1}] + \frac{\Delta Q_n}{C} \tag{7.77}$$

where the dimensionless number M is given by

$$M = \frac{(\Delta x)^2}{\Delta t} \frac{\rho C}{\lambda} \tag{7.14}$$

and the term $\Delta Q_n/C$ represents the contribution to the increase of temperature due to the cure reaction during the increment of time Δt, between the times t and $t + \Delta t$, within the slice of thickness Δx located at the position n.

Heat Convection at the Surface

The heat balance during the increment of time Δt within the slice of thickness $\frac{1}{2}\Delta x$ located between the planes 0 and 0.5 must be rewritten by considering not only the heat conduction through the plane 0.5 and heat convection through the suface 0, but also the heat evolved from the cure reaction within this slice (Fig. 7.3).

$$A\left[h(T_{\text{ext}} - T_0) - \lambda \frac{T_0 - T_1}{\Delta x}\right]\Delta t + A\frac{\Delta x}{2}\rho\Delta Q_{0.25}$$

$$= A\frac{\Delta x}{2}\rho C[\text{TN}_{0.25} - T_{0.25}] \tag{7.78}$$

Dividing by $A\frac{1}{2}\Delta x\rho C$, this equation becomes

$$\text{TN}_{0.25} - T_{0.25} = \frac{2N}{M}(T_{\text{ext}} - T_0) - \frac{2}{M}(T_0 - T_1) + \frac{\Delta Q_{0.25}}{C} \tag{7.79}$$

where M and N are the dimensionless numbers

$$M = \frac{(\Delta x)^2}{\Delta t}\frac{\rho C}{\lambda} \tag{7.14}$$

$$N = \frac{h\Delta x}{\lambda} \tag{7.21}$$

Of course the temperatures at the abscissa 0.25 can be evaluated by making one of the two assumptions already made (Sect. 7.2.2):

1. With the simple assumption

$$\text{TN}_{0.25} - T_{0.25} = \text{TN}_0 - T_0 \tag{7.23}$$

the new temperature at the surface after an interval of time Δt becomes

$$\text{TN}_0 = \frac{1}{M}[2NT_{\text{ext}} + (M - 2 - 2N)T_0 + 2T_1] + \frac{\Delta Q_{0.25}}{C} \tag{7.80}$$

2. With the more accurate assumption

$$\text{TN}_{0.25} - T_{0.25} = \tfrac{3}{4}(\text{TN}_0 - T_0) + \tfrac{1}{4}(\text{TN}_1 - T_1) \tag{7.25}$$

the new temperature at the surface after an interval of time Δt becomes

$$\text{TN}_0 = -\frac{1}{3}(\text{TN}_1 - T_1) + \frac{8}{3M}\left[NT_{\text{ext}} + \left(\frac{3M}{8} - 1 - N\right)T_0 + T_1\right]$$

$$+ \frac{4}{3}\frac{\Delta Q_{0.25}}{C} \tag{7.81}$$

$\Delta Q_{0.25}$ can be replaced by ΔQ_0 or it can be expressed by linear interpolation as follows:

$$\Delta Q_{0.25} = \tfrac{3}{4}\Delta Q_0 + \tfrac{1}{4}\Delta Q_1 \tag{7.82}$$

Heat Conduction at the Surface of a Resin and Mould (Fig. 7.4)

The mould is also divided in slices of equal thickness Δx_m. Within the mould the calculations are the same as within the resin, with other values of parameters. It is assumed that there is no production of heat in the mould.

As shown in Fig. 7.4, the resin and the mould are in good contact. Half slices of resin and mould are considered on each side of the composite–mould interface. The heat balance during the increment of time Δt is evaluated within these two half slices, by considering heat conduction in the mould and the resin and heat generated by the cure reaction in the resin. As the resin and mould are in perfect contact, no additional heat resistance is considered at the interface, and the temperature is the same on each side of the interface ($x = L, n = X$); resin for $x < L$; mould for $x > L$.

$$A(\phi_{X-0.5} - \phi_{X+0.5})\Delta t + A\,\frac{\Delta x_r}{2}\,\rho_r \Delta Q_{X-0.25}$$

$$= A\left[\frac{\Delta x_r}{2}(\rho C)_r + \frac{\Delta x_m}{2}(\rho C)_m\right][\mathrm{TN}_X - T_X] \tag{7.83}$$

where ϕ is the heat flux by conduction. This equation is rewritten as follows:

$$\left(\lambda_r \frac{T_{X-1} - T_X}{\Delta x_r} - \lambda_m \frac{T_X - T_{X+1}}{\Delta x_m}\right)\Delta t + \frac{\Delta x_r}{2}\,\rho_r \Delta Q_{X-0.25} = F(\mathrm{TN}_X - T_X) \tag{7.84}$$

The new temperature at the interface is thus obtained:

$$\mathrm{TN}_X = T_X + B(T_{X-1} - T_X) + A(T_{X+1} - T_X) + \frac{\Delta x_r \rho_r}{2F}\left(\tfrac{3}{4}\Delta Q_X + \tfrac{1}{4}\Delta Q_{X-1}\right) \tag{7.85}$$

with the coefficients F, A, B given by

$$F = \frac{\Delta x_m}{2}(\rho C)_m + \frac{\Delta x_r}{2}(\rho C)_r \tag{7.86}$$

$$A = \frac{\lambda_m}{\Delta x_m}\,\frac{\Delta t}{F} \tag{7.86'}$$

$$B = \frac{\lambda_r}{\Delta x_r}\,\frac{\Delta t}{F} \tag{7.86''}$$

where the subscripts m and r characterise the mould and resin, respectively.

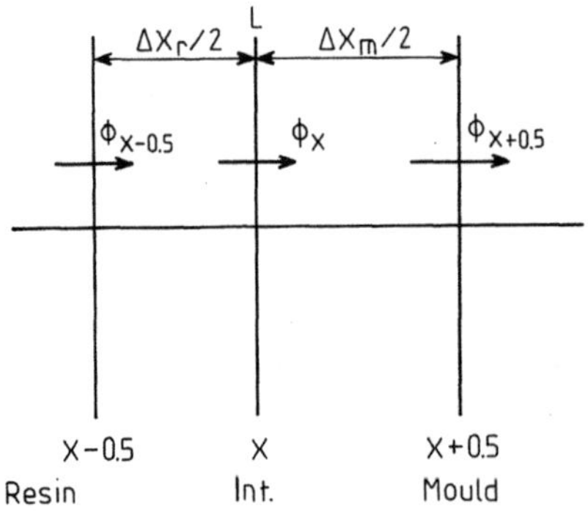

Fig. 7.4. Interface between resin and mould. Half slice of resin and mould.

7.5.2 Temperature-Dependent Thermal Parameters

Within the Resin

The heat balance in the slice at position n, is evaluated during the increment of time Δt by considering not only the heat conduction but also heat generated by the cure reaction. Thus Eq. (7.76) becomes

$$A[\phi_{n-0.5} - \phi_{n+0.5}]\Delta t + A\Delta x \rho_n \Delta Q_n = A\Delta x(\rho C)_n[\mathrm{TN}_n - T_n] \quad (7.87)$$

where $\phi_{n-0.5}$ is the heat flux by conduction through the plane $n - 0.5$ at time t, TN_n and T_n are the new temperature after an interval of time Δt and the previous temperature, at position n, and $(\rho C)_n$ is the value at position n and temperature T_n. The heat flux by conduction is

$$\phi_{n-0.5} = -\lambda_{n-0.5}\left(\frac{\partial T}{\partial x}\right)_{n-0.5} = \lambda_{n-0.5}\frac{T_{n-1} - T_n}{\Delta x} = \frac{G(n)}{\Delta x} \quad (7.36)$$

After division by $A\Delta x(\rho C)_n$ in Eq. (7.83), the new temperature TN_n is thus obtained:

$$\mathrm{TN}_n = T_n + \frac{\Delta t}{(\Delta x)^2}\frac{1}{(\rho C)_n}[G(n) - G(n+1)] + \frac{\Delta Q_n}{C_n} \quad (7.88)$$

where C_n is the heat capacity at position n and at temperature T_n and ΔQ_n is the heat generated by the cure reaction, per unit mass, during the interval of time $[t, t + \Delta t]$.

Heat Convection at the Surface (Fig. 7.3)

The heat balance during the increment of time Δt is evaluated within the slice of thickness $\frac{1}{2}\Delta x$ next to the surface by considering heat conduction on one surface and heat convection on the external surface, as well as heat evolved from the cure reaction. Eq. (7.78) is still valid, the values of parameters being dependent on the position ($\lambda_{0.5}$, $\rho_{0.25}$ and $C_{0.25}$):

$$A(\phi_0 - \phi_{0.5})\Delta t + A\frac{\Delta x}{2}\rho_{0.25}\Delta Q_{0.25} = A\frac{\Delta x}{2}(\rho C)_{0.25}[\mathrm{TN}_{0.25} - T_{0.25}] \quad (7.89)$$

After division by $A\frac{1}{2}\Delta x(\rho C)_{0.25}$, the new temperature $\mathrm{TN}_{0.25}$ at position 0.25 can thus be obtained.

By making the simple assumption

$$\mathrm{TN}_{0.25} - T_{0.25} = \mathrm{TN}_0 - T_0 \quad (7.23)$$

the new temperature on the surface TN_0 after an interval of time can thus be expressed in terms of the previous temperatures by

$$\mathrm{TN}_0 = T_0 + \frac{\Delta t}{\Delta x}\frac{2}{(\rho C)_{0.25}}[\phi_0 - \phi_{0.5}] + \frac{\Delta Q_{0.25}}{C_{0.25}} \quad (7.90)$$

As temperature is only obtained at positions defined by an integer $(0, 1, 2, \ldots, X)$, $\Delta Q_{0.25}$ must be evaluated by the following:

$$\Delta Q_{0.25} = \tfrac{3}{4}\Delta Q_0 + \tfrac{1}{4}\Delta Q_1 \quad (7.91)$$

and likewise for the thermal parameters ρC and C at position 0.25.

Heat Conduction at the Surface (Fig. 7.4)

The resin and mould are in good contact at the interface. The heat balance during the increment of time Δt is calculated within the two half slices on both sides of the interface, by considering heat conduction in the mould and the resin, as well as heat generated by the cure reaction, in the same way as for constant thermal parameters.

The new temperature at the interface ($x = L, n = X$) is thus obtained as follows:

$$\mathrm{TN}_X = T_X + \frac{\Delta t}{F}[\phi_{X-0.5} - \phi_{X+0.5}] + \frac{\Delta x_r \rho_r}{2F}\Delta Q_{X-0.25} \tag{7.92}$$

with the variable coefficient F defined by

$$F = \frac{\Delta x_m}{2}(\rho_m C_m)_{X+0.5} + \frac{\Delta x_r}{2}(\rho_r C_r)_{X-0.25} \tag{7.86}$$

where $\Delta Q_{X-0.25}$ can be evaluated by interpolation:

$$\Delta Q_{X-0.25} = \tfrac{3}{4}\Delta Q_X + \tfrac{1}{4}\Delta Q_{X-1} \tag{7.93}$$

Chapter 8

Numerical Analysis for a Cylinder

8.1 Introduction

Perhaps only a few kinds of cylindrical parts are usually prepared in industry by using thermosets. Generally, solid cylinders or hollow cylinders and tubes are made of elastomers. However, this kind of shape is of great interest for all materials for at least two reasons:

1. In the calorimetry technique, the small sample size used necessitates that all ingredients in the sample are well dispersed. Dispersion problems do arise on occasion and are generally observed by lack of reproducibility in calorimetric curves. The question which always arises is how representative is a small sample taken from a large batch of the homogeneity of the batch. This is the reason why samples up to 150–200 mg are used, the resin being located in a hollow cylinder.
2. Cylindrical samples and moulds, either of long length or of finite length can be studied easily, for instance by recording the temperature on the longitudinal axis of the cylinder.

The following cases are considered successively:

1. Solid cylinder of infinite length with radial heat transfer
2. Solid cylinder of finite length with radial and longitudinal heat transfer
3. Hollow cylinder of infinite length with radial heat transfer
4. Hollow cylinder of finite length with radial and longitudinal heat transfer

In each case, the thermal parameters are either constant or temperature-dependent.

On the surface of the material, three cases are studied:

When the temperature is constant or is given as a function of time and position

When there is a finite coefficient of surface heat transfer, the surface being in contact with a fluid

When the solid is in contact with another solid

In order to reduce the number of studied cases, the heat evolved from cure reaction is introduced in the equations. Of course, these equations can be used either when the reaction takes place or when there is no reaction. In the second case, the equations obtained with a reaction can be used, by putting the constant k_0 (see Sect. 7.4) equal to zero, and thus eliminating the heat evolved from reaction.

8.2 Solid Cylinder of Infinite Length with Radial Heat Transfer, with or Without Internal Reaction, with Constant Thermal Parameters

If the initial and boundary conditions depend only on the radial abscissa, heat transfer is radial only. Likewise for a circular cylinder of finite length limited by two planes perpendicular to axis, if these plane surfaces are heat-tight. The circular cross-section of the cylinder is thus considered (Fig. 8.1). It is divided into N circular membranes of constant thickness Δr. Let R be the radius of the cylinder.

Each circle limiting two of these membranes has a radius defined by an integer n such that

$$r = n\Delta r, \qquad 0 \leqslant n \leqslant N, \qquad R = N\Delta r \tag{8.1}$$

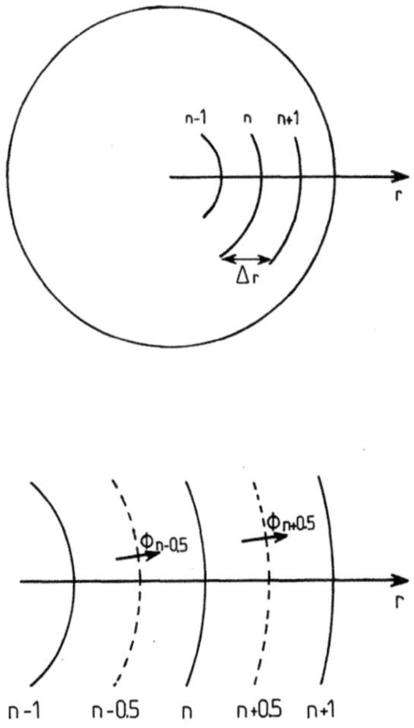

Fig. 8.1. Radial heat transfer within the solid cylinder of infinite length.

8.2.1 Within the Cylinder

The heat balance is evaluated during the interval of time $[t, t + \Delta t]$ within the circular membrane of thickness Δr and height 1, located between the radial abscissae $r - \frac{1}{2}\Delta r$ and $r + \frac{1}{2}\Delta r$:

$$[(A\phi)_{n-0.5} - (A\phi)_{n+0.5}]\Delta t + B_n\rho\left(\frac{\partial Q}{\partial t}\right)_n \Delta t = B_n\rho C(\mathrm{TN}_n - T_n) \quad (8.2)$$

where

$\phi_{n-0.5}$ is the heat flux at position $(n - 0.5)\Delta r$ and time $t = i\Delta t$
$A_w = 2\pi w\Delta r$ is the area of the cylindrical surface of radius $w\Delta r$ and unit height
$B_n = (2\pi n\Delta r)\Delta r$ is the area of the cross-section of the membrane and also the volume of this membrane of unit height

By assuming that the gradient of temperature at position $n - 0.5$ is equal to the chord slope between positions $n - 1$ and n, the heat flux becomes

$$\phi_{n-0.5} = -\lambda\left(\frac{\partial T}{\partial r}\right)_{n-0.5} = \lambda\frac{T_{n-1} - T_n}{\Delta r} \quad (8.3)$$

The equation expressing the heat balance thus becomes

$$\left[2\pi(n - 0.5)\Delta r\lambda\frac{T_{n-1} - T_n}{\Delta r} - 2\pi(n + 0.5)\Delta r\lambda\frac{T_n - T_{n+1}}{\Delta r}\right]\Delta t$$
$$= (2\pi n\Delta r)\Delta r\rho C\left(\mathrm{TN}_n - T_n - \frac{\Delta t}{C}\frac{\partial Q_n}{\partial t}\right) \quad (8.4)$$

On defining the dimensionless number M:

$$M = \frac{(\Delta r)^2}{\Delta t}\frac{\rho C}{\lambda} = \frac{(\Delta r)^2}{\alpha\Delta t} \quad (8.5)$$

Eq. (8.4) becomes

$$(n - 0.5)(T_{n-1} - T_n) - (n + 0.5)(T_n - T_{n+1}) = nM\left(\mathrm{TN}_n - T_n - \frac{\Delta Q_n}{C}\right) \quad (8.6)$$

The new temperature at radial abscissa $n\Delta r$, after an interval of time Δt, is thus expressed in terms of the previous temperatures obtained at the same position and adjacent positions:

$$\mathrm{TN}_n = T_n + \frac{1}{M}\left[\left(\frac{n - 0.5}{n}\right)(T_{n-1} - T_n) - \left(\frac{n + 0.5}{n}\right)(T_n - T_{n+1})\right] + \frac{\Delta Q_n}{C} \quad (8.7)$$

Another expression for this new temperature is also obtained:

$$\mathrm{TN}_n = \frac{1}{M}[T_{n-1} + (M - 2)T_n + T_{n+1}] + \frac{1}{2Mn}(T_{n+1} - T_{n-1}) + \frac{\Delta Q_n}{C} \quad (8.8)$$

where ΔQ_n represents the heat evolved from the cure reaction per unit mass at position n, during the interval of time $[t, t + \Delta t]$. This quantity may be calculated as shown in Sect. 7.4. On the right-hand side of this equation, the first term is the same as that obtained in the plane sheet, and the second term expresses the effect of the concavity of the isothermal surface.

Centre of the Cylinder

Eqs (8.7) and (8.8) cannot be used for the centre of the cylinder. The small cylinder of radius $\frac{1}{2}\Delta r$, and height H, is considered (Fig. 8.2). The heat balance during the increment of time Δt is calculated within this small cylinder:

$$-H(A\phi)_{0.5}\Delta t + \pi\left(\frac{\Delta r}{2}\right)^2 H\rho\Delta Q_0 = \pi\left(\frac{\Delta r}{2}\right)^2 H\rho C(\mathrm{TN}_0 - T_0) \tag{8.9}$$

Since the heat flux at radial abscissa $\frac{1}{2}\Delta r$ and time $i\Delta t$ is approximated by

$$\phi_{0.5} = -\lambda\left(\frac{\partial T}{\partial r}\right)_{0.5} = \lambda\,\frac{T_0 - T_1}{\Delta r} \tag{8.10}$$

the new temperature after an interval of time Δt at the middle of the cylinder becomes

$$\mathrm{TN}_0 = T_0 + \frac{4}{M}(T_1 - T_0) + \frac{\Delta Q_0}{C} \tag{8.11}$$

with the dimensionless number M defined in Eq. (8.5).

Of course, when no cure reaction takes place in the solid, the last term in Eqs (8.8) and (8.11) must be deleted.

8.2.2 On the Surface

Three different cases are considered:

1. When the temperature on the surface is constant or is given as a function of time
2. When the coefficient of surface heat convection is finite
3. When there is a coefficient of heat transfer between the solid cylinder (resin) and another solid (mould)

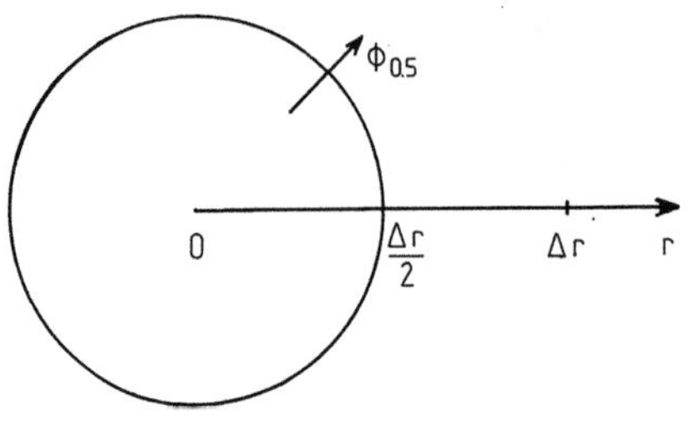

Fig. 8.2. Heat transfer at the centre of the solid cylinder of infinite length.

Given Temperature

This case is very simple:

$$T_s = T(R, t) = g(t) \tag{8.12}$$

T_s being the temperature on the surface of the cylinder. The following conditions may be observed for this case: the coefficient of heat transfer between the cylinder and the surrounding medium, fluid or metal, is very high and the temperature of this medium is uniform and either constant or given as a function of time. For instance the surrounding fluid is strongly stirred or the surrounding metal has a very high thermal conductivity.

Heat Convection

The boundary condition is

$$\phi_N = h(T_N - T_{\text{ext}}) \tag{8.13}$$

The cylindrical membrane of thickness $\frac{1}{2}\Delta r$, located next to the surface of the cylinder, is considered (Fig. 8.3). The heat balance is evaluated during the increment of time Δt within this membrane by considering heat conduction through one surface and heat convection through the external surface:

$$H[(A\phi)_{N-0.5} - (A\phi)_N]\Delta t + 2\pi\left(R - \frac{\Delta r}{4}\right)\frac{\Delta r}{2} H\rho\Delta Q_{N-0.25}$$

$$= 2\,\pi\left(R - \frac{\Delta r}{4}\right)\frac{\Delta r}{2} H\rho C(\text{TN}_{N-0.25} - T_{N-0.25}) \tag{8.14}$$

H being the height of the truncated cylinder. As the heat flux is given by

$$\phi_{N-0.5} = -\,\lambda\left(\frac{\partial T}{\partial r}\right)_{N-0.5} = \lambda\,\frac{T_{N-1} - T_N}{\Delta r} \tag{8.15}$$

the heat balance becomes after simplification by $2\pi H$

$$(N - 0.5)\Delta r\lambda\,\frac{(T_{N-1} - T_N)}{\Delta r} - N\Delta r h(T_N - T_{\text{ext}})$$

$$= (N - 0.25)\,\frac{\Delta r}{\Delta t}\,\frac{\Delta r}{2}\,\rho\, C\left(\text{TN}_{N-0.25} - T_{N-0.25} - \frac{\Delta Q_{N-0.25}}{C}\right) \tag{8.16}$$

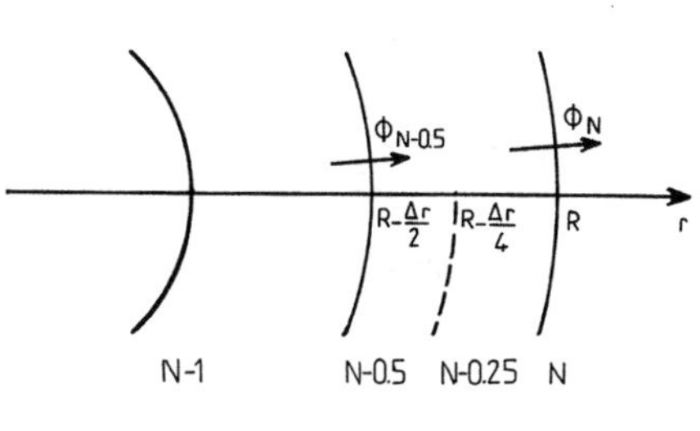

Fig. 8.3. Heat transfer by convection at the surface of the solid cylinder of infinite length.

By making the simple assumption

$$\mathrm{TN}_{N-0.25} - T_{N-0.25} = \mathrm{TN}_N - T_N \tag{8.17}$$

the new temperature after an interval of time Δt on the surface of the solid cylinder is given by

$$\mathrm{TN}_N = T_N + \frac{2}{M}\frac{N-0.5}{N-0.25}(T_{N-1} - T_N) + \frac{2P}{M}(T_{\mathrm{ext}} - T_N) + \frac{\Delta Q_{N-0.25}}{C} \tag{8.18}$$

where the dimensionless numbers M and P are

$$M = \frac{(\Delta r)^2}{\Delta t}\frac{\rho C}{\lambda} = \frac{(\Delta r)^2}{\alpha \Delta t} \tag{8.5}$$

$$P = \frac{h\Delta r}{\lambda}\frac{N}{N-0.25} \tag{8.19}$$

By making the more accurate assumption

$$\mathrm{TN}_{N-0.25} - T_{N-0.25} = \tfrac{3}{4}(\mathrm{TN}_N - T_N) + \tfrac{1}{4}(\mathrm{TN}_{N-1} - T_{N-1}) \tag{8.20}$$

the new temperature on the surface becomes

$$\mathrm{TN}_N = T_N - \frac{1}{3}(\mathrm{TN}_{N-1} - T_{N-1}) + \frac{8}{3M}\frac{N-0.5}{N-0.25}(T_{N-1} - T_N) + \frac{8P}{3M}(T_{\mathrm{ext}} - T_N) + \frac{4}{3}\frac{\Delta Q_{N-0.25}}{C} \tag{8.21}$$

In this case, the new temperature on the surface TN_N is expressed in terms of the new temperature obtained at position $N-1$, TN_{N-1}. This new temperature TN_{N-1} must be calculated beforehand, by using Eq. (8.8).

8.2.3 Stability of Calculations

In the equation giving the new temperature TN_n, for each value of n ($0 \leqslant n \leqslant N$), the whole coefficient of the temperature T_n must be positive. Thus Eqs (8.8), (8.11) and (8.18) impose respectively

$$M - 2 \geqslant 0 \qquad 1 - \frac{4}{M} \geqslant 0 \qquad 1 - \frac{2}{M}\frac{N-0.5}{N-0.25} - \frac{2P}{M} \geqslant 0$$

which give

$$\Delta t \leqslant \frac{(\Delta r)^2}{4\alpha} \qquad \Delta t \leqslant \frac{(N-0.25)(\Delta r)^2}{(2N-1)\alpha + 2N\dfrac{h\Delta r}{\rho C}} \tag{8.22}$$

Remark. If the parameters ρ, C and h vary with the temperature or the state of cure, and λ is constant, the previous equations (8.8), (8.11) and (8.18) are still valid, with variable values for M, C and P.

8.3 Solid Cylinder of Infinite Length with Radial Heat Transfer, with or Without Internal Reaction, with Temperature-Dependent Thermal Parameters

8.3.1 Within the Cylinder

Eq (8.2) is still valid and it is now written

$$((A\phi)_{n-0.5} - (A\phi)_{n+0.5})\Delta t + B_n \rho_n \Delta Q_n = B_n(\rho C)_n(\text{TN}_n - T_n) \quad (8.23)$$

The heat flux by conduction is given by

$$\phi_{n-0.5} = + \lambda_{n-0.5} \frac{T_{n-1} - T_n}{\Delta r} \quad (8.24)$$

where $\lambda_{n-0.5}$ is the temperature-dependent thermal conductivity at position $(n - 0.5)\Delta r$ and time $t = i\Delta t$. The new temperature after an interval of time Δt is thus expressed in terms of the previous temperatures by

$$\text{TN}_n = T_n + \frac{1}{n} \frac{\Delta t}{(\Delta r)^2} \frac{1}{(\rho C)_n} [\Gamma_{n+1} - \Gamma_n] + \frac{\Delta Q_n}{C_n} \quad (8.25)$$

where $(\rho C)_n$ are the thermal parameters at position n and time $t = i\Delta t$ and the function Γ is defined by

$$\Gamma_\text{n} = (n - 0.5)\lambda_{n-0.5}(T_n - T_{n-1}) \qquad \text{for } 1 \leqslant n \leqslant N \quad (8.26)$$

Centre of the Cylinder

In the same way as for constant thermal parameters, the small cylinder of radius $\frac{1}{2}\Delta r$ and height H is considered (Fig. 8.2). The heat balance during the increment of time Δt is thus written

$$-H(A\phi)_{0.5}\Delta t + \pi \left(\frac{\Delta r}{2}\right)^2 H\rho_0 \Delta Q_0 = \pi \left(\frac{\Delta r}{2}\right)^2 H(\rho C)_0(\text{TN}_0 - T_0) \quad (8.27)$$

with temperature-dependent parameters. The new temperature at the middle of the solid cylinder after an interval of time Δt is then given by the relation

$$\text{TN}_0 = T_0 + 8 \frac{\Delta t}{(\Delta r)^2} \frac{\Gamma_1}{(\rho C)_0} + \frac{\Delta Q_0}{C_0} \quad (8.28)$$

where the function Γ_1 is given by Eq. (8.26):

$$\Gamma_1 = \tfrac{1}{2}\lambda_{0.5}(T_1 - T_0) \quad (8.26')$$

The thermal parameters must be determined at positions which are not defined by an integer. As temperatures are only given at positions associated with an integer, the temperature at position $n - 0.5$ must be evaluated. The best way is to take the mean value between the temperatures at the adjacent places T_{n-1} and T_n, and then to use relationships expressing the temperature dependency of the thermal parameters.

8.3.2 On the Surface

The following three cases are also of interest:

1. When temperature on the surface is given
2. When the coefficient of surface heat convection is finite
3. When the resin is in contact with a mould

Given Temperature

In this case, there is during the whole process

$$T_s = T(R, t) = g(t) \tag{8.12}$$

As stated previously, the coefficient of heat transfer between the solid cylinder and the surroundings is very high, and the temperature of the surroundings is uniform and either constant or given as a function of time.

Heat Convection

The same method as in Sect. 8.2.2 gives the same equation:

$$H[(A\phi)_{N-0.5} - (A\phi)_N]\Delta t + 2\pi\left(R - \frac{\Delta r}{4}\right)\frac{\Delta r}{2} H\rho_{N-0.25}\Delta Q_{N-0.25}$$

$$= 2\pi\left(R - \frac{\Delta r}{4}\right)\frac{\Delta r}{2} H(\rho C)_{N-0.25}(\mathrm{TN}_{N-0.25} - T_{N-0.25}) \tag{8.14'}$$

The heat flux is given by

$$\phi_{N-0.5} = \lambda_{N-0.5}\frac{T_{N-1} - T_N}{\Delta r} \tag{8.29}$$

$$\phi_N = h(T_N - T_{\text{ext}}) \tag{8.13}$$

With the function Γ defined by Eq. (8.26), the new temperature at position $N - 0.25$ is thus given by

$$\mathrm{TN}_{N-0.25} =$$

$$T_{N-0.25} + \frac{2\Delta t}{(N - 0.25)(\Delta r)^2}\frac{1}{(\rho C)_{N-0.25}}[-\Gamma_N - N\Delta r h(T_N - T_{\text{ext}})]$$

$$+ \left(\frac{\Delta Q}{C}\right)_{N-0.25} \tag{8.30}$$

The new temperature on the surface is thus easily obtained, either by making the approximation

$$\mathrm{TN}_{N-0.25} - T_{N-0.25} = \mathrm{TN}_N - T_N \tag{8.17}$$

or by making the other more accurate approximation

$$\mathrm{TN}_{N-0.25} - T_{N-0.25} = \tfrac{3}{4}(\mathrm{TN}_N - T_N) + \tfrac{1}{4}(\mathrm{TN}_{N-1} - T_{N-1}) \tag{8.20}$$

For instance, with the latter assumption, there is

$$\mathrm{TN}_N = T_N - \tfrac{1}{3}[\mathrm{TN}_{N-1} - T_{N-1}]$$
$$+ \frac{8}{3(N-0.25)} \frac{\Delta t}{(\Delta r)^2} \frac{1}{(\rho C)_{N-0.25}} [N \Delta r h(T_{\mathrm{ext}} - T_N) - \Gamma_N]$$
$$+ \frac{4}{3}\left(\frac{\Delta Q}{C}\right)_{N-0.25} \tag{8.31}$$

8.4 Solid Cylinder of Finite Length with or Without Reaction, with Constant Thermal Parameters

In the case of a solid cylinder of finite length, heat transfer is radial and longitudinal (Fig. 8.4), as heat transfer along the longitudinal axis cannot be disregarded as for the cylinder of infinite length.

The radius of the cylinder is R and its height is $2H$. The radius is divided into NR intervals of equal length Δr, and the height into 2NH intervals of equal length Δz. Various positions in the cylinder are considered successively by using the notations

$$r = n\Delta r \qquad 0 \leqslant n \leqslant \mathrm{NR} \qquad R = \mathrm{NR}\Delta r \qquad \text{(radial)}$$
$$z = h\Delta z \qquad 0 \leqslant h \leqslant 2\mathrm{NH} \qquad 2H = 2\mathrm{NH}\Delta z \qquad \text{(longitudinal)} \tag{8.32}$$

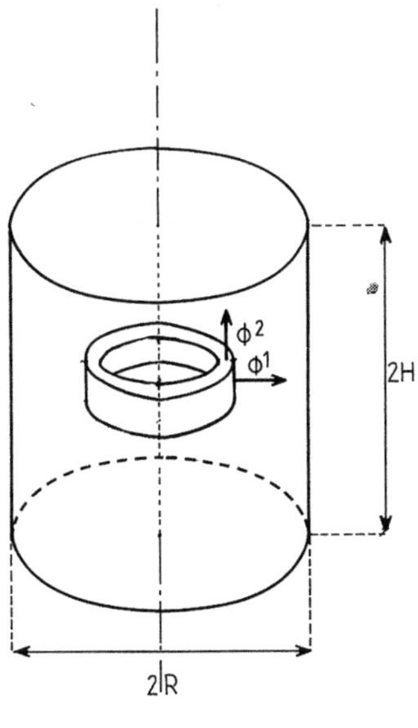

Fig. 8.4. Cylinder of finite length, of radius R and length $2H$.

8.4.1 Within the Cylinder

The annulus of thickness Δr, height Δz, radius r and longitudinal position h is considered (Fig. 8.5). The heat balance during the increment of time Δt is obtained by considering the radial and longitudinal heat transfer by conduction and the heat generated by the reaction:

$$2\pi r\Delta r\Delta z\rho C(\mathrm{TN}_{n,h} - T_{n,h}) =$$

$$\Delta z\left[2\pi\left(r - \frac{\Delta r}{2}\right)\phi^1_{n-0.25,h} - 2\pi\left(r + \frac{\Delta r}{2}\right)\phi^1_{n+0.5h}\right]\Delta t$$

$$+ 2\pi r\Delta r[\phi^2_{n,h-0.5} - \phi^2_{n,h+0.5}]\Delta t + 2\pi r\Delta r\Delta z\rho\Delta Q_{n,h} \tag{8.33}$$

where $\phi^1_{u,v}$ and $\phi^2_{u,v}$ are the radial and longitudinal fluxes, respectively, at time $t = i\Delta t$ and position ($r = u\Delta r$, $z = v\Delta z$), and $\mathrm{TN}_{n,h}$ is the new temperature at position n, h, after an interval of time Δt. The values of ϕ^1 and ϕ^2 are given by

$$\phi^1 = -\lambda\frac{\partial T}{\partial r} \quad \text{at } n \pm 0.5,\ h \tag{8.34}$$

$$\phi^2 = -\lambda\frac{\partial T}{\partial z} \quad \text{at } n,\ h \pm 0.5 \tag{8.34'}$$

The gradients of temperature are approximated by the chord slope, as shown previously. The heat balance is thus rewritten as follows

$$\frac{r\Delta r\Delta z}{\Delta t}\rho C(\mathrm{TN}_{n,h} - T_{n,h}) = r\Delta r\Delta t\left[\lambda\frac{T_{n,h-1} - T_{n,h}}{\Delta z} - \lambda\frac{T_{n,h} - T_{n,h+1}}{\Delta z}\right]$$

$$+ \Delta z\Delta t\left[\left(r - \frac{\Delta r}{2}\right)\lambda\frac{T_{n-1,h} - T_{n,h}}{\Delta r} - \left(r + \frac{\Delta r}{2}\right)\lambda\frac{T_{n,h} - T_{n+1,h}}{\Delta r}\right]$$

$$+ r\Delta r\Delta z\rho\Delta Q_{n,h} \tag{8.35}$$

The new temperature after an interval of time Δt within the annulus, $\mathrm{TN}_{n,h}$ is thus expressed in terms of the previous temperatures in the adjacent places:

$$\mathrm{TN}_{n,h} = T_{n,h} + \frac{1}{\mathrm{MR}}[T_{n-1,h} - 2T_{n,h} + T_{n+1,h}] + \frac{\Delta r}{2r\mathrm{MR}}[T_{n+1,h} - T_{n-1,h}]$$

$$+ \frac{1}{\mathrm{MH}}[T_{n,h-1} - 2T_{n,h} + T_{n,h+1}] + \frac{\Delta Q_{n,h}}{C} \tag{8.36}$$

where the dimensionless numbers MR and MH are

$$\mathrm{MR} = \frac{(\Delta r)^2}{\Delta t}\frac{\rho C}{\lambda} \qquad \mathrm{MH} = \frac{(\Delta z)^2}{\Delta t}\frac{\rho C}{\lambda} \tag{8.37}$$

By recollecting that $r = n\Delta r$, the preceding equation becomes

$$\mathrm{TN}_{n,h} = T_{n,h} + \frac{1}{\mathrm{MR}}[T_{n-1,h} - 2T_{n,h} + T_{n+1,h}] + \frac{1}{2n\mathrm{MR}}[T_{n+1,h} - T_{n-1,h}]$$

$$+ \frac{1}{\mathrm{MH}}[T_{n,h-1} - 2T_{n,h} + T_{n,h+1}] \tag{8.38}$$

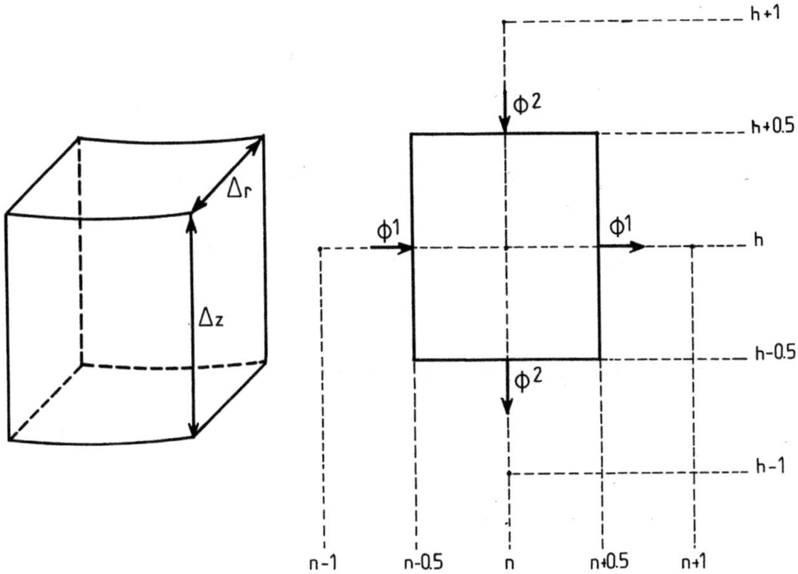

Fig. 8.5. Circular annulus of mean radius $r = n\Delta r$, of thickness Δr and height Δz. Radial and longitudinal heat transfer.

Longitudinal Axis of the Cylinder: $n = 0,\ 1 \leqslant h \leqslant 2\mathrm{NH} - 1$

Another calculation must be made for the temperature on the longitudinal axis of the cylinder, when $r = 0$. The small cylinder of radius $\frac{1}{2}\Delta r$ and height Δz is considered (Fig. 8.6). The heat balance is calculated during the increment of time Δt by considering the radial and longitudinal heat transfer:

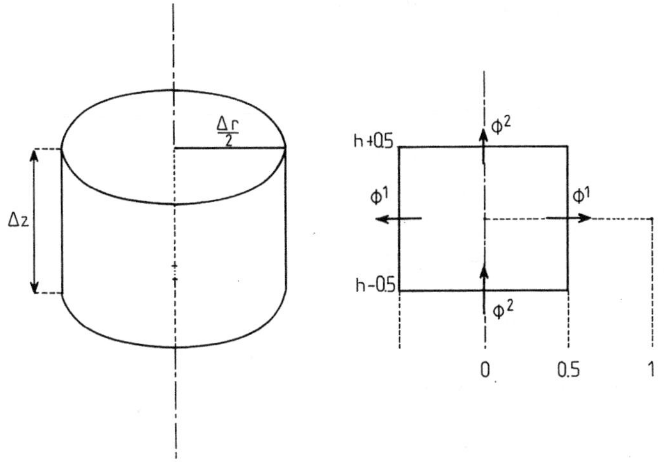

Fig. 8.6. Small cylinder of radius $\frac{1}{2}\Delta r$ and height Δz, at the centre of the cylinder.

$$\Delta z 2\pi \frac{\Delta r}{2} \lambda \frac{T_{1,h} - T_{0,h}}{\Delta r} + \pi \left(\frac{\Delta r}{2}\right)^2 \left[\lambda \frac{T_{0,h-1} - T_{0,h}}{\Delta z} - \lambda \frac{T_{0,h} - T_{0,h+1}}{\Delta z}\right]$$

$$+ \pi \left(\frac{\Delta r}{2}\right)^2 \Delta z \rho \frac{\Delta Q_{0,h}}{\Delta t} = \pi \left(\frac{\Delta r}{2}\right)^2 \Delta z \rho C \frac{\mathrm{TN}_{0,h} - T_{0,h}}{\Delta t} \tag{8.39}$$

The new temperature $\mathrm{TN}_{0,h}$ on the longitudinal axis is thus given by

$$\mathrm{TN}_{0,h} = T_{0,h} + \frac{4}{\mathrm{MR}}(T_{1,h} - T_{0,h}) + \frac{1}{\mathrm{MH}}(T_{0,h-1} - 2T_{0,h} + T_{0,h+1}) + \frac{\Delta Q_{0,h}}{C} \tag{8.40}$$

8.4.2 Given Temperature on the Surface

The temperature on the surface during the whole process is given as follows:

$$\left.\begin{array}{lll} T_s = g_1(R, z, t) & \text{on the cylindrical surface} & (0 \leqslant z \leqslant 2H) \\ T_s = g_2(r, 0, t) & \text{on the first plane surface} & (0 \leqslant r \leqslant R) \\ T_s = g_3(r, 2H, t) & \text{on the second plane surface} & (0 \leqslant r \leqslant R) \end{array}\right\} \tag{8.41}$$

These boundary conditions become for the numerical analysis

$$\left.\begin{array}{ll} T_{\mathrm{NR},h} = g_1(R, h\Delta z, t) & 0 \leqslant h \leqslant 2\mathrm{NH} \\ T_{n,0} = g_2(n\Delta r, 0, t) & 0 \leqslant n \leqslant \mathrm{NR} \\ T_{n,2\mathrm{NH}} = g_3(n\Delta r, 2\mathrm{NH}, t) & 0 \leqslant n \leqslant \mathrm{NR} \end{array}\right\} \tag{8.41'}$$

8.4.3 Finite Coefficient of Convection at the Surface

The boundary condition expresses the leaving flux at the surface:

$$\phi_s = k(T_s - T_{\mathrm{ext}}) \tag{8.42}$$

where T_s is the temperature at the surface.

The following positions are successively examined:

1. Circular surface
2. Plane surface
3. Plane surface at the centre, for $r = 0$
4. Corner

Cylindrical Surface, $n = \mathrm{NR}$, $1 \leqslant h \leqslant 2\mathrm{NH} - 1$

The annulus defined by $R - \frac{1}{2}\Delta r \leqslant r \leqslant R$ and $(h - 0.5)\Delta z \leqslant z \leqslant (h + 0.5)\Delta z$ is considered (Fig. 8.7). The heat balance during the increment of time Δt is evaluated within this annulus by taking into account the radial heat conduction through the cylindrical surface of radius $(\mathrm{NR} - 0.5)\Delta r$ and the radial heat convection at the cylindrical surface of radius $\mathrm{NR}\Delta r$, as well as the longitudinal heat conduction through the plane surfaces at position $(h - 0.5)\Delta z$ and $(h + 0.5)\Delta z$. It may be written as follows:

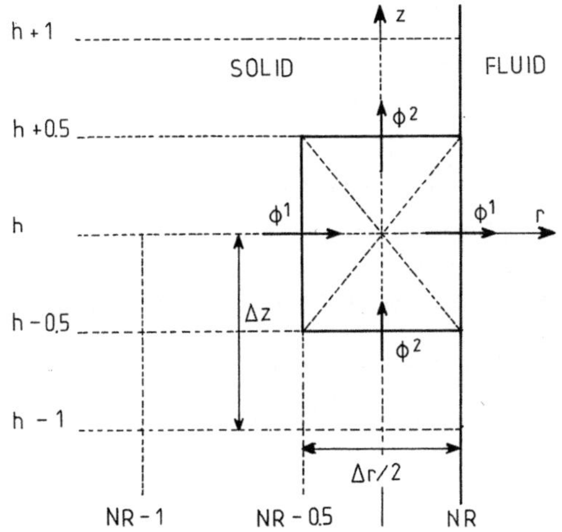

Fig. 8.7. Heat convection at the circular surface. Annulus of radius $(\mathrm{NR}-0.25)\Delta r$, of thickness $\frac{1}{2}\Delta r$ and height Δz.

$$2\pi(\mathrm{NR}-0.25)\Delta r\,\frac{\Delta r}{2}\,\Delta z\,\frac{\rho C}{\Delta t}\left(\mathrm{TN}_{\mathrm{NR}-0.25,h}-T_{\mathrm{NR}-0.25,h}-\frac{\Delta Q_{\mathrm{NR}-0.25,h}}{C}\right)$$
$$=(\underset{\mathrm{NR}-0.5,h}{\overset{}{A\phi^1}}-\underset{\mathrm{NR},h}{A\phi^1})+(\underset{\mathrm{NR}-0.25,h+0.5}{B\phi^2}-\underset{\mathrm{NR}-0.25,h-0.5}{B\phi^2}) \tag{8.43}$$

(at $\mathrm{NR}-0.5,h$; at NR,h; at $\mathrm{NR}-0.25,h+0.5$; at $\mathrm{NR}-0.25,h-0.5$)

By expressing all the surfaces A, B and heat fluxes ϕ^1, ϕ^2 by their value:

$$2\pi(\mathrm{NR}-0.25)\Delta r\,\frac{\Delta r}{2}\,\Delta z\,\frac{\rho C}{\Delta t}\left(\mathrm{TN}_{\mathrm{NR}-0.25,h}-T_{\mathrm{NR}-0.25,h}-\frac{\Delta Q_{\mathrm{NR}-0.25,h}}{C}\right)$$
$$=2\pi(\mathrm{NR}-0.5)\Delta r\Delta z\lambda\,\frac{T_{\mathrm{NR}-1,h}-T_{\mathrm{NR},h}}{\Delta r}-2\pi\mathrm{NR}\Delta r\Delta zk(T_{\mathrm{NR},h}-T_{\mathrm{ext}})$$
$$+\,2\pi(\mathrm{NR}-0.25)\Delta r\,\frac{\Delta r}{2}\,\lambda$$
$$\times\left(\frac{T_{\mathrm{NR}-0.25,h-1}-T_{\mathrm{NR}-0.25,h}}{\Delta z}-\frac{T_{\mathrm{NR}-0.25,h}-T_{\mathrm{NR}-0.25,h+1}}{\Delta z}\right) \tag{8.44}$$

The new temperature within this annulus after an interval of time Δt can thus be expressed in terms of the previous temperatures obtained at various places:

$$\mathrm{TN}_{\mathrm{NR}-0.25,h}=T_{\mathrm{NR}-0.25,h}+\frac{1}{\mathrm{MH}}\,[T_{\mathrm{NR}-0.25,h-1}-2T_{\mathrm{NR}-0.25,h}+T_{\mathrm{NR}-0.25,h+1}]$$
$$+\,\frac{2}{\mathrm{MR}}\,\frac{\mathrm{NR}-0.5}{\mathrm{NR}-0.25}\,(T_{\mathrm{NR}-1,h}-T_{\mathrm{NR},h})-PR(T_{\mathrm{NR},h}-T_{\mathrm{ext}})$$
$$+\,\frac{\Delta Q_{\mathrm{NR}-0.25,h}}{C} \tag{8.45}$$

with the dimensionless numbers MR, MH and PR:

$$\mathrm{MR} = \frac{(\Delta r)^2}{\Delta r}\frac{\rho C}{\lambda} \qquad \mathrm{MH} = \frac{(\Delta z)^2}{\Delta t}\frac{\rho C}{\lambda} \tag{8.37}$$

$$\mathrm{PR} = \frac{\mathrm{NR}}{\mathrm{NR} - 0.25}\frac{2k\Delta t}{\Delta r \rho C} \tag{8.19'}$$

k being the coefficient of surface heat convection at time $t = i\Delta t$.

As temperatures are known only at positions defined by an integer, the temperature in other places must be approximated by using one of the following relationships:

$$T_{\mathrm{NR}-0.25,h} = T_{\mathrm{NR},h} \tag{8.46}$$

$$T_{\mathrm{NR}-0.25,h} = \tfrac{3}{4}T_{\mathrm{NR},h} + \tfrac{1}{4}T_{\mathrm{NR}-1,h} \tag{8.46'}$$

Plane Surface at Position $r = n\Delta r$ $\quad 1 \leqslant n \leqslant \mathrm{NR} - 1$

The annulus of mean radius $n\Delta r$, thickness Δr and height $\frac{1}{2}\Delta h$, located next to the plane surface, $z = 0$ of the cylinder, is shown in Fig. 8.8.

The heat balance within this annulus during the increment of time Δt is evaluated by considering the longitudinal heat conduction and heat convection at the plane surface, as well as the radial heat conduction.

$$\begin{aligned}2\pi n(\Delta r)^2 \frac{\Delta z}{2}\frac{\rho C}{\Delta t}(\mathrm{TN}_{n,0.25} - T_{n,0.25}) &= 2\pi n(\Delta r)^2 \frac{\Delta z}{2}\rho\frac{\Delta Q_{n,0.25}}{\Delta t}\\ &+ 2\pi(n - 0.5)\Delta r\frac{\Delta z}{2}\lambda\frac{T_{n-1,0.25} - T_{n,\mathrm{NH}-0.25}}{\Delta r}\\ &- 2\pi(n + 0.5)\Delta r\frac{\Delta z}{2}\lambda\frac{T_{n,0.25} - T_{n+1,0.25}}{\Delta r}\\ &+ 2\pi n(\Delta r)^2\left[\lambda\frac{T_{n,1} - T_{n,0}}{\Delta z} - k(T_{n,0} - T_{\mathrm{ext}})\right]\end{aligned} \tag{8.47}$$

The new temperature within this annulus is thus given by

$$\begin{aligned}\mathrm{TN}_{n,0.25} &= T_{n,0.25} + \frac{\Delta Q_{n,0.25}}{C} + \frac{1}{\mathrm{MR}}[T_{n-1,0.25} - 2T_{n,0.25} + T_{n+1,0.25}]\\ &+ \frac{1}{2n\mathrm{MR}}(T_{n+1,0.25} - T_{n-1,0.25})\\ &+ \frac{2}{\mathrm{MH}}[T_{n,1} - T_{n,0} - \mathrm{PH}(T_{n,0} - T_{\mathrm{ext}})]\end{aligned} \tag{8.48}$$

with the new dimensionless number PH defined by

$$\mathrm{PH} = \frac{k\Delta z}{\lambda} \tag{8.49}$$

The temperature at position (n, 0.25) can be approximated by one of the following relations

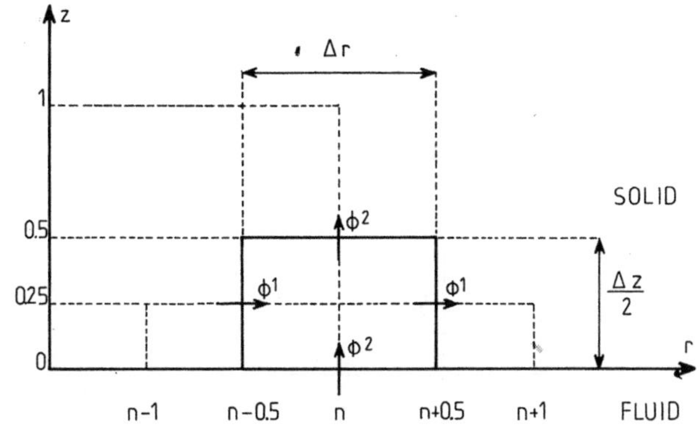

Fig. 8.8. Heat convection at the plane surface. Annulus of mean radius $n\Delta r$, of thickness Δr and height $\frac{1}{2}\Delta z$.

$$T_{n,0.25} = T_{n,0} \tag{8.50}$$

$$T_{n,0.25} = \tfrac{3}{4}T_{n,0} + \tfrac{1}{4}T_{n,1} \tag{8.51}$$

Plane Surface at the Centre n = 0, h = 0

The equation obtained for the plane surface at various values of the radius different from zero cannot be used for $r = 0$.

The cylinder of radius $\frac{1}{2}\Delta r$ and height $\frac{1}{2}\Delta h$, located next to the plane surface, $z = 0$, is considered. The heat balance within this cylinder during the time increment Δt is calculated by taking into account the longitudinal heat conduction and heat convection at the plane surface, the radial heat conduction and the heat generated by the reaction (Fig. 8.9):

$$\pi\left(\frac{\Delta r}{2}\right)^2 \frac{\Delta z}{2}\frac{\rho C}{\Delta t}(\mathrm{TN}_{0,0.25} - T_{0,0.25}) =$$

$$\pi\left(\frac{\Delta r}{2}\right)^2 \frac{\Delta z}{2}\rho\frac{\Delta Q_{0,0.25}}{\Delta t} + 2\pi\frac{\Delta r}{2}\frac{\Delta z}{2}\lambda\frac{T_{0,0.25} - T_{1,0.25}}{\Delta r}$$

$$+ \pi\left(\frac{\Delta r}{2}\right)^2\left[\lambda\frac{T_{0,1} - T_{0,0}}{\Delta h} - k(T_{0,0} - T_{\mathrm{ext}})\right] \tag{8.52}$$

After simplifying this equation, the new temperature within this small cylinder after an interval of time Δt is obtained as a function of the previous temperatures at the same and adjacent positions:

$$\mathrm{TN}_{0,025} = T_{0,0.25} + \frac{4}{\mathrm{MR}}(T_{1,0.25} - T_{0,0.25})$$

$$+ \frac{2}{\mathrm{MH}}[T_{0,1} - T_{0,0} - \mathrm{PH}(T_{0,0} - T_{\mathrm{ext}})]$$

$$+ \frac{\Delta Q_{0,0.25}}{C} \tag{8.53}$$

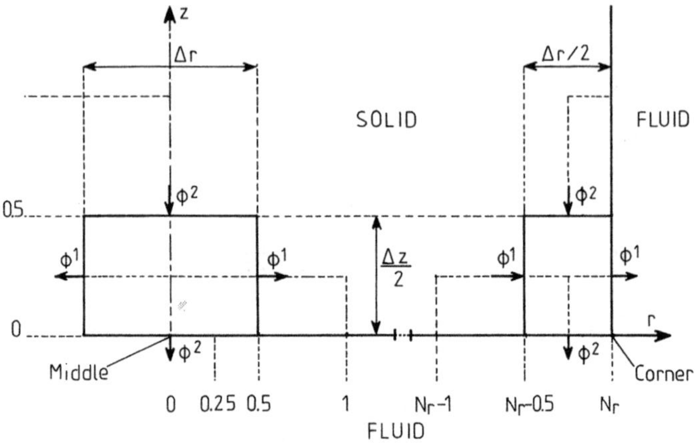

Fig. 8.9. Heat convection at the centre of the plane surface (left) and at the corner (right). (Left) Small cylinder of radius $\frac{1}{2}\Delta r$ and thickness $\frac{1}{2}\Delta z$, of centre $r = 0$, $z = \frac{1}{4}\Delta r$. (Right) Small annulus of thickness $\frac{1}{2}\Delta r$ and $\frac{1}{2}\Delta z$ at the corner.

Corner of the Cylinder, $n = \mathrm{NR}$, $h = 0$

The annulus of mean radius $(\mathrm{NR} - 0.25)\Delta r$, thickness $\frac{1}{2}\Delta r$ and height $\frac{1}{2}\Delta z$, at the corner of the cylinder of finite length, is considered (Fig. 8.9). The heat balance within this annulus during the increment of time Δt is evaluated, by considering radial and longitudinal heat by conduction as well as heat convection at the circular and plane surfaces next to the corner.

$$2\pi(\mathrm{NR} - 0.25)\Delta r \frac{\Delta r}{2} \frac{\Delta z}{2} \frac{\rho C}{\Delta t} (\mathrm{TN}_{\mathrm{NR}-0.25,0.25} - T_{\mathrm{NR}-0.25,0.25})$$

$$= 2\pi(\mathrm{NR} - 0.5)\Delta r \frac{\Delta z}{2} \lambda \frac{T_{\mathrm{NR}-1,0.25} - T_{\mathrm{NR},0.25}}{\Delta r}$$

$$- 2\pi \mathrm{NR}\Delta r \frac{\Delta z}{2} k(T_{\mathrm{NR},0.25} - T_{\mathrm{ext}})$$

$$+ 2\pi(\mathrm{NR} - 0.25)\Delta r \frac{\Delta r}{2} \left[\lambda \frac{T_{\mathrm{NR}-0.25,1} - T_{\mathrm{NR}-0.25,0}}{\Delta z} - k(T_{\mathrm{NR}-0.25,0} - T_{\mathrm{ext}})\right] \tag{8.54}$$

The new temperature within this annulus after time Δt is thus expressed in terms of the previous temperatures obtained at the same and adjacent places:

$$\mathrm{TN}_{\mathrm{NR}-0.25,0.25} = T_{\mathrm{NR}-0.25,0.25} + \frac{\mathrm{NR} - 0.5}{\mathrm{NR} - 0.25} \frac{2}{\mathrm{MR}} (T_{\mathrm{NR}-1,0.25} - T_{\mathrm{NR},0.25})$$

$$+ \frac{2}{\mathrm{MH}} [T_{\mathrm{NR}-0.25,1} - T_{\mathrm{NR}-0.25,0} - \mathrm{PH}(T_{\mathrm{NR}-0.25,0} - T_{\mathrm{ext}})]$$

$$- \mathrm{PR}(T_{\mathrm{NR},0.25} - T_{\mathrm{ext}}) + \frac{\Delta Q_{\mathrm{NR}-0.25,0.25}}{C} \tag{8.55}$$

where PR is the dimensionless number defined in this section by Eq. (8.19′).

The temperatures at position (NR − 0.25, 0.25) are not directly calculated, because only positions defined by integers are considered. These temperatures can be approximated by one of the following relations:

$$T_{\mathrm{NR}-0.25,0.25} = T_{\mathrm{NR},0} \tag{8.56}$$

$$T_{\mathrm{NR}-0.25,0.25} = \tfrac{3}{4}T_{\mathrm{NR},0} + \tfrac{1}{4}T_{\mathrm{NR}-1,1} \tag{8.57}$$

$$T_{\mathrm{NR}-0.25,0.25} = \tfrac{1}{2}T_{\mathrm{NR},0} + \tfrac{1}{4}T_{\mathrm{NR},1} + \tfrac{1}{4}T_{\mathrm{NR}-1,0} \tag{8.58}$$

as a function of the temperatures at the adjacent places.

Stability of Calculations

In the equation giving the new temperature $\mathrm{TN}_{n,h}$, for all values of (n, h), the whole coefficient of the corresponding temperature $T_{n,h}$ must be positive. Thus Eqs (8.36), (8.40), (8.45), (8.48), (8.53) and (8.55) impose respectively

$$\frac{2}{\mathrm{MR}} + \frac{2}{\mathrm{MH}} \leqslant 1 \tag{8.59a}$$

$$\frac{4}{\mathrm{MR}} + \frac{2}{\mathrm{MH}} \leqslant 1 \tag{8.59b}$$

$$\frac{2}{\mathrm{MH}} + \frac{2}{\mathrm{MR}}\,\frac{\mathrm{NR} - 0.5}{\mathrm{NR} - 0.25} + \mathrm{PR} \leqslant 1 \tag{8.59c}$$

$$\frac{2}{\mathrm{MR}} + \frac{2}{\mathrm{MH}} - \frac{2\mathrm{PH}}{\mathrm{MH}} \leqslant 1 \tag{8.59d}$$

$$\frac{4}{\mathrm{MR}} + \frac{2 + 2\mathrm{PH}}{\mathrm{MH}} \leqslant 1 \tag{8.59e}$$

$$\frac{2}{\mathrm{MR}}\,\frac{\mathrm{NR} - 0.5}{\mathrm{NR} - 0.25} + \frac{2 + 2\mathrm{PH}}{\mathrm{MH}} + \mathrm{PR} \leqslant 1 \tag{8.59f}$$

If the temperatures are given at the surface, Eq. (8.59b) alone is sufficient and it can be rewritten

$$\Delta t \leqslant \frac{1}{\dfrac{4\alpha}{\Delta r^2} + \dfrac{2\alpha}{\Delta z^2}} \qquad \text{with} \qquad \alpha = \frac{\lambda}{\rho C} \tag{8.60}$$

If there is a finite coefficient of transfer at the surface, k, Eqs (8.59e) and (8.59f) are sufficient and they can be rewritten

$$\Delta t \leqslant \frac{1}{\dfrac{4\alpha}{\Delta r^2} + \dfrac{2\alpha}{\Delta z^2}(1 + \mathrm{PH})} \tag{8.61}$$

$$\Delta t \leqslant \frac{1}{\dfrac{2\alpha}{\Delta r^2}\dfrac{\mathrm{NR}-0.5}{\mathrm{NR}-0.25}+\dfrac{2\alpha}{\Delta z^2}(1+\mathrm{PH})+\dfrac{2k}{\rho C \Delta r}\dfrac{\mathrm{NR}}{\mathrm{NR}-0.25}} \qquad (8.61')$$

Symmetry

If the initial and boundary conditions are symmetrical with respect to the midplane $z = H$, then the time of calculation can be reduced by putting at each step

$$T_{n,\mathrm{NH}+1} = T_{n,\mathrm{NH}-1} \qquad \text{for} \qquad 0 \leqslant n \leqslant \mathrm{NR}$$

and the calculations are made only for $0 \leqslant h \leqslant \mathrm{NH}$. The same procedure is available when there is a heat-tight plane perpendicular to the axis of equation $z = H$.

8.5 Solid Cylinder of Finite Length with or Without Reaction, with Temperature-Dependent Thermal Parameters

Thermal parameters may vary with temperature, and also with the value of the state of cure of the resin. The principle of the method is thus clear: at each position and time, the temperature is determined as well as the state of cure. The values of all thermal parameters must thus be calculated for each position and time and introduced in the equations expressing the new temperature and state of cure in terms of the previous temperatures. Of course, relationships describing the variation of each thermal parameters with temperature and state of cure must be known.

Calculations are made at various positions of the solid cylinder of finite length by rewriting the heat balance obtained with constant thermal parameters and using the same figures.

8.5.1 Within the Cylinder

The annulus of radius $r = n\Delta r$, thickness Δr and height Δh of Fig. 8.5 is considered. The heat balance within this annulus during the time increment Δt is written by taking into account the radial and longitudinal conduction heat transfers and the heat generated by the reaction

$$\begin{aligned} &2\pi r \Delta r \Delta z \rho C(\mathrm{TN}_{n,h} - T_{n,h}) \\ &= \Delta z\left[2\pi\left(r - \frac{\Delta r}{2}\right)\phi^1_{n-0.5,h} - 2\pi\left(r + \frac{\Delta r}{2}\right)\phi^1_{n+0.5,h}\right]\Delta t \\ &\quad + 2\pi r \Delta r[\phi^2_{n,h-0.5} - \phi^2_{n,h+0.5}]\Delta t + 2\pi r \Delta r \Delta z \rho \Delta Q_{n,h} \end{aligned} \qquad (8.33)$$

Upon defining the following functions,

$$G_{n,h} = \lambda_{n,h-0.5}(T_{n,h-1} - T_{n,h}) \tag{8.62}$$

$$\Gamma_{n,h} = (n - 0.5)\lambda_{n-0.5,h}(T_{n-1,h} - T_{n,h}) \tag{8.63}$$

the new temperature within this annulus after an interval of time Δt is thus expressed in terms of these functions and of the thermal parameter ρC:

$$\begin{aligned} \mathrm{TN}_{n,h} &= T_{n,h} + \frac{1}{\mathrm{n}} \frac{\Delta t}{(\Delta r)^2} \frac{1}{(\rho C)_{n,h}} [\Gamma_{n,h} - \Gamma_{n+1,h}] \\ &+ \frac{\Delta t}{(\Delta z)^2} \frac{1}{(\rho C)_{n,h}} [G_{n,h} - G_{n,h+1}] + \left(\frac{\Delta Q}{C}\right)_{n,h} \end{aligned} \tag{8.64}$$

Value of the Thermal Parameters

$(\rho C)_{n,h}$ is directly obtained, as it is found at a position (n, h) at which the temperature is given. However, the thermal conductivity λ appearing either in the function G or the function Γ must be also obtained at positions (either $n, h \pm 0.5$ or $n \pm 0.5,\ h$) for which the temperature is not given. Two ways of calculating λ exist:

1. By considering the mean value of λ, with

$$\lambda_{n,h-0.5} = \tfrac{1}{2}(\lambda_{n,h} + \lambda_{n,h-1}) \tag{8.65}$$

$$\lambda_{n-0.5,h} = \tfrac{1}{2}(\lambda_{n,h} + \lambda_{n-1,h}) \tag{8.65'}$$

   because the temperature being given at positions defined by integers, the thermal conductivity λ can be easily determined at these positions.

2. By calculating the temperature at each position using the following equations:

$$T_{n,h-0.5} = \tfrac{1}{2}(T_{n,h} + T_{n,h-1}) \tag{8.66}$$

$$T_{n-0.5,h} = \tfrac{1}{2}(T_{n,h} + T_{n-1,h}) \tag{8.66'}$$

   and then determining the value of the thermal conductivity which is associated with this temperature by using the law expressing the temperature dependency of the thermal conductivity. This method is more accurate than the first.

Axis of the Cylinder $n = 0,\ 1 \leqslant h \leqslant 2\mathrm{NH} - 1$

From the small cylinder of radius $\frac{1}{2}\Delta r$ and height Δz, located on the axis of the cylinder shown in Fig. 8.6, the heat balance is determined during the increment of time Δt.

Eq. (8.41) which expresses this heat balance enables one to obtain the new temperature $\mathrm{TN}_{0,h}$:

$$\begin{aligned} \mathrm{TN}_{0,\mathrm{h}} &= T_{0,\mathrm{h}} + \frac{\Delta t}{(\Delta z)^2} \frac{1}{(\rho C)_{0,\mathrm{h}}} [G_{0,\mathrm{h}} - G_{0,\mathrm{h}+1}] \\ &- \frac{\Delta t}{(\Delta r)^2} \frac{8}{(\rho C)_{0,\mathrm{h}}} \Gamma_{1,\mathrm{h}} + \left(\frac{\Delta Q}{C}\right)_{0,\mathrm{h}} \end{aligned} \tag{8.67}$$

with the functions G and Γ defined by Eqs (8.62) and (8.63).

The value of the thermal conductivity at position $(0.5, h)$ is determined by one of the two ways described above.

8.5.2 Given Temperature on the Surface

In this case, the boundary conditions are the same, whether the thermal parameters are constant or not. The solution can be seen in Sect. 8.4.2.

8.5.3 Finite Coefficient of Convection at the Surface

In the same way as for constant thermal parameters, the following positions are considered:

1. Cylindrical surface
2. Plane surface
3. Plane surface at the centre
4. Corner

The boundary condition expresses the flux leaving at the surface:

$$\phi_s = k(T_s - T_{ext}) \tag{8.42}$$

where T_s is the temperature at the surface.

Cylindrical Surface, $n = \mathrm{NR}$ $\quad 1 \leqslant h \leqslant 2\mathrm{NH} - 1$

The annulus defined by $R - \frac{1}{2}\Delta r \leqslant r \leqslant R$ $\quad (h - 0.5)\Delta z \leqslant z \leqslant (h + 0.5)\Delta z$ is considered (Fig. 8.7). From the heat balance evaluated during the time increment Δt within this annulus, expressed by Eq. (8.43), the new temperature after an interval of time Δt within this annulus is found:

$$\begin{aligned} TN_{\mathrm{NR}-0.25,h} = {} & T_{\mathrm{NR}-0.25,h} + \frac{\Delta t}{(\Delta z)^2} \frac{1}{(\rho C)_{\mathrm{NR}-0.25,h}} [G_{\mathrm{NR}-0.25,h} - G_{\mathrm{NR}-0.25,h+1}] \\ & + \frac{2\Delta t}{(\mathrm{NR} - 0.25)(\Delta r)^2 (\rho C)_{\mathrm{NR}-0.25,h}} \Gamma_{\mathrm{NR},h} \\ & - \frac{\mathrm{NR}}{\mathrm{NR} - 0.25} \frac{2\Delta t k}{\Delta r (\rho C)_{\mathrm{NR}-0.25,h}} (T_{\mathrm{NR},h} - T_{ext}) \\ & + \left(\frac{\Delta Q}{C}\right)_{\mathrm{NR}-0.25,h} \end{aligned} \tag{8.68}$$

with the dimensionless term $\Delta t k/\Delta r(\rho C)$ and the functions G and Γ defined by Eqs (8.62) and (8.63) in Sect. 8.5.1.

The values of the thermal parameters are approximated by following one of the two ways described in Sect. 8.5.1.

Plane Surface at Radial Abscissa $r = n\Delta r \qquad 1 \leq n \leq \mathrm{NR} - 1$

The annulus of mean radius $n\Delta r$, thickness Δr and height $\frac{1}{2}\Delta h$, located next to the plane surface, for instance $z = 0$, is considered (Fig. 8.8). The heat balance within this annulus during the time increment Δt is evaluated in Eq. (8.47), with constant thermal parameters.

The new temperature within this annulus after an interval of time Δt is given by

$$\mathrm{TN}_{n,0.25} = T_{n,0.25} + \frac{\Delta t}{n\Delta r^2(\rho C)_{n,0.25}}[\Gamma_{n,0.25} - \Gamma_{n+1,0.25}]$$
$$+ \frac{2\Delta t}{\Delta z^2}\frac{1}{(\rho C)_{n,0.25}}[k\Delta z(T_{\text{ext}} - T_{n,0}) - G_{n,1}] + \left(\frac{\Delta Q}{C}\right)_{n,0.25} \tag{8.69}$$

The temperatures at positions $(n, 0.25)$ and $(n - 0.5, 0.25)$ are approximated by

$$T_{n,0.25} = \tfrac{3}{4}T_{n,0} + \tfrac{1}{4}T_{n,1} \tag{8.70}$$
$$T_{n-0.5,0.25} = \tfrac{3}{8}(T_{n,0} + T_{n-1,0}) + \tfrac{1}{8}(T_{n,1} + T_{n-1,1}) \tag{8.71}$$

The thermal parameters can thus be evaluated by using the relations expressing their temperature dependency.

Plane Surface $h = 0$ *at Position* $r = 0$

For the constant thermal parameters, the cylinder of radius $\frac{1}{2}\Delta r$ and height $\frac{1}{2}\Delta h$, located next to the plane surface is considered (Fig. 8.9). From the heat balance evaluated during the time increment Δt within this small cylinder by considering the radial heat conduction and heat convection at the plane surface, the new temperature after an interval of time Δt within this cylinder is given by

$$\mathrm{TN}_{0,0.25} = T_{0,0.25} - \frac{\Delta t}{\Delta r^2}\frac{8}{(\rho C)_{0,0.25}}\Gamma_{1,0.25}$$
$$+ \frac{\Delta t}{\Delta z^2}\frac{2}{(\rho C)_{0,0.25}}[k\Delta z(T_{\text{ext}} - T_{0,0}) - G_{0,1}] + \left(\frac{\Delta Q}{C}\right)_{0,0.25} \tag{8.72}$$

The temperatures at positions (0, 0.25) and (0.5, 0.25) are approximated by Eqs (8.70) and (8.71) and all thermal parameters can be determined.

Corner of the Cylinder $n = \mathrm{NR},\ h = 0$

The annulus of mean radius $(\mathrm{NR} - 0.25)\Delta r$, of thickness $\frac{1}{2}\Delta r$ and height $\frac{1}{2}\Delta h$, at the corner of the cylinder is shown in Fig. 8.9. From the heat balance already evaluated in the case of constant thermal parameters, by considering the radial and longitudinal heat conduction as well as heat convection at the circular and plane surfaces, the new temperature within this annulus after an interval of time Δt can thus be found:

$$TN_{NR-0.25,0.25} = T_{NR-0.25,0.25} + \frac{2\Delta t}{(NR - 0.25)(\Delta r)^2} \frac{1}{(\rho C)_{NR-0.25,0.25}} \Gamma_{NR,0.25}$$

$$- \frac{2NR}{NR - 0.25} \frac{\Delta t}{\Delta r} \frac{k}{(\rho C)_{NR-0.25,0.25}} (T_{NR,0.25} - T_{ext})$$

$$+ \frac{\Delta t}{(\Delta z)^2} \frac{2}{(\rho C)_{NR-0.25,0.25}} [k\Delta z(T_{ext} - T_{NR-0.25,0}) - G_{NR-0.25,1}]$$

$$+ \left(\frac{\Delta Q}{C}\right)_{NR-0.25,0.25} \tag{8.73}$$

The temperatures at positions $(NR - 0.5, 0.25)$, $(NR - 0.25, 0.5)$ and $(NR - 0.25, 0.25)$ are approximated by

$$T_{NR-0.5,0.25} = \tfrac{3}{8}(T_{NR,0} + T_{NR-1,0}) + \tfrac{1}{8}(T_{NR,1} + T_{NR-1,1}) \tag{8.74}$$

$$T_{NR-0.25,0.5} = \tfrac{3}{8}(T_{NR,0} + T_{NR,1}) + \tfrac{1}{8}(T_{NR-1,0} + T_{NR-1,1}) \tag{8.75}$$

$$T_{NR-0.25,0.25} = \tfrac{1}{2}T_{NR,0} + \tfrac{1}{4}T_{NR,1} + \tfrac{1}{4}T_{NR-1,0} \tag{8.58}$$

and all thermal parameters can be calculated by using the relation expressing their temperature dependency.

8.6 Hollow Cylinder of Infinite Length with or Without Reaction, with Constant Thermal Parameters

For an infinitely long tube, if the initial and boundary conditions depend only on the radial abscissa, heat transfer is radial only; likewise for a tube of finite length limited by two planes perpendicular to the axis, if these plane surfaces are heat-tight.

Let R_1 and R_2 be the respective radii of the internal and external cylindrical surfaces of the tube. The tube is divided in N small tubes of equal thickness Δr, so that

$$R_2 - R_1 = N\Delta r \tag{8.76}$$

Each cylindrical surface separating two small tubes has for a radial abscissa

$$r = R_1 + n\Delta r \qquad 0 \leqslant n \leqslant N \qquad (n \text{ integer}) \tag{8.77}$$

The circular cross-section of the tube is shown in Fig. 8.10.

The problem of radial heat transfer is about the same as that studied in Sect. 8.2 for the solid cylinder of infinite length.

8.6.1 Within the Cylinder Wall: $1 \leqslant n \leqslant N - 1$

The new temperature after an interval of time Δt at radial position defined by the integer n is thus given by the following equation, analogous to Eq. (8.8):

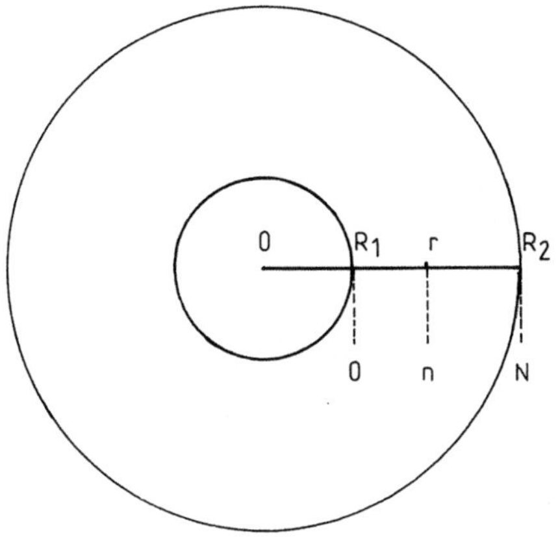

Fig. 8.10. Scheme of the circular cross-section of the tube of infinite length, with radial heat transfer.

$$\mathrm{TN}_n = \frac{1}{\mathrm{M}}[T_{n-1} + (M-2)T_n + T_{n+1}] + \frac{\Delta r}{2Mr}(T_{n+1} - T_{n-1}) + \frac{\Delta Q_n}{C} \tag{8.78}$$

with the dimensionless number M

$$M = \frac{(\Delta r)^2}{\Delta t}\frac{\rho C}{\lambda} \tag{8.5}$$

8.6.2 Given Temperature on Each Surface

The temperature on each surface is given as a function of time. This case is similar to the case considered in Sects 8.2.2 and 8.3.2 as well as in Sects 7.2.2 and 7.3.2 for the sheet.

$$T_0 = g_1(t) \qquad T_N = g_2(t) \tag{8.79}$$

8.6.3 Finite Coefficient of Convection at Each Surface

External Circular Surface

This problem has already been resolved with the solid cylinder, in the case of the external circular surface (Fig. 8.3). The new temperature on the external circular surface after an interval of time Δt is thus given either by the following equation, analogous to Eq. (8.18):

$$\mathrm{TN}_N = T_N + \frac{2}{\mathrm{M}}\frac{R_2 - \Delta r/2}{R_2 - \Delta r/4}(T_{N-1} - T_N) - \frac{2P_2}{M}(T_N - T_{\mathrm{ext2}}) + \frac{\Delta Q_{N-0.25}}{C} \tag{8.18'}$$

with the simple assumption

$$\mathrm{TN}_{N-0.25} - T_{N-0.25} = \mathrm{TN}_N - T_N \tag{8.17'}$$

or by

$$\mathrm{TN}_N = T_N - \frac{1}{3}(\mathrm{TN}_{N-1} - T_{N-1}) + \frac{8}{3M}\frac{R_2 - \Delta r/2}{R_2 - \Delta r/4}(T_{N-1} - T_N) - \frac{8P_2}{3M}(T_N - T_{\mathrm{ext2}}) + \frac{4}{3}\frac{\Delta Q_{N-0.25}}{C} \tag{8.21'}$$

with the more accurate assumption

$$\mathrm{TN}_{N-0.25} - T_{N-0.25} = \tfrac{3}{4}(\mathrm{TN}_N - T_N) + \tfrac{1}{4}(\mathrm{TN}_{N-1} - T_{N-1}) \tag{8.20'}$$

Here P_2 is expressed by

$$P_2 = \frac{h\Delta r}{\lambda}\frac{R_2}{R_2 - \Delta r/4} \tag{8.19'}$$

Internal Circular Surface

Calculation of the heat balance is made in the same way as in the case of the external circular surface (Fig. 8.11). The new temperature on the internal surface after an interval of time Δt is thus expressed by

$$\mathrm{TN}_0 = T_0 + \frac{2}{M}\frac{R_1 + \Delta r/2}{R_1 + \Delta r/4}(T_1 - T_0) - \frac{2P_1}{M}(T_0 - T_{\mathrm{ext1}}) + \frac{\Delta Q_{0.25}}{C} \tag{8.18''}$$

with the assumption

$$\mathrm{TN}_{0.25} - T_{0.25} = \mathrm{TN}_0 - T_0 \tag{8.17''}$$

or by

$$\mathrm{TN}_0 = T_0 - \frac{1}{3}(\mathrm{TN}_1 - T_1) + \frac{8}{3M}\frac{R_1 + \Delta r/2}{R_1 + \Delta r/4}(T_1 - T_0) - \frac{8P_1}{3M}(T_0 - T_{\mathrm{ext1}}) + \frac{4}{3}\frac{\Delta Q_{0.25}}{C} \tag{8.21''}$$

with the assumption

$$\mathrm{TN}_{0.25} - T_{0.25} = \tfrac{3}{4}(\mathrm{TN}_0 - T_0) + \tfrac{1}{4}(\mathrm{TN}_1 - T_1) \tag{8.20''}$$

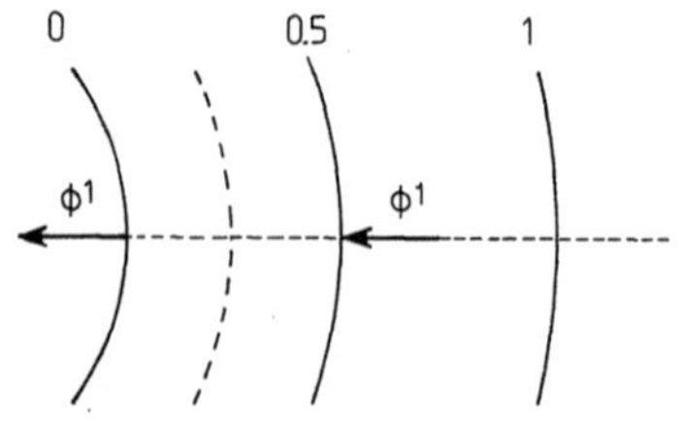

Fig. 8.11. Heat transfer by convection at the internal circular surface of the tube of infinite length.

Here P_1 is expressed by

$$P_1 = \frac{h\Delta r}{\lambda}\frac{R_1}{R_1 + \Delta r/4} \tag{8.19''}$$

8.7 Hollow Cylinder of Infinite Length with or without Reaction, with Temperature-Dependent Thermal Parameters

8.7.1 Within the Cylinder Wall: $1 \leqslant n \leqslant N-1$

This problem has already been resolved in the case of the solid cylinder of infinite length with radial heat transfer, with temperature-dependent thermal parameters (Fig. 8.1). The only difference is that the solid is located between two circular surfaces of radii R_1 and R_2 (Fig. 8.10). The new temperature within the solid after an interval of time Δt is thus given by

$$\mathrm{TN}_n = T_n + \frac{\Delta r}{r}\frac{\Delta t}{(\Delta r)^2}\frac{1}{(\rho C)_n}[\Gamma_{n+1} - \Gamma_n] + \frac{\Delta Q_n}{C_n} \tag{8.25'}$$

with $r = R_1 + n\Delta r$ and with the function Γ expressed by

$$\Gamma_n = \left(\frac{R_1}{\Delta r} + n - 0.5\right)\lambda_{n-0.5}(T_n - T_{n-1}) \qquad \text{for} \qquad 1 \leqslant n \leqslant N \tag{8.26'}$$

Temperature has to be determined at position $n - 0.5$, so that the value of thermal conductivity $\lambda_{n-0.5}$ can be evaluated. This temperature is easily obtained by taking the mean value of the temperatures at adjacent places, T_n and T_{n-1}. Relationships expressing the temperature dependency of the thermal conductivity can thus be used to obtain the right value of this parameter.

8.7.2 Given Temperature on Each Surface

During the whole process, the temperature is kept constant on each surface. These temperatures can be equal or different:

$$T_0 = \text{constant} \qquad T_N = \text{constant} \tag{8.80}$$

A programmed temperature can also be imposed at the two surfaces:

$$T_0 = g_1(t) \qquad T_N = g_2(t) \tag{8.81}$$

8.7.3 Finite Coefficient of Convection at the Surface

Two cases must be considered: one for the external circular surface, the other for the internal circular surface.

External Circular Surface

This case has already been studied in the chapter devoted to the solid cylinder (Fig. 8.3).

The new temperature after elapse of time Δt on the external circular surface, with $n = N$, is thus given either by the following expression, analogous to Eq. (8.30):

$$\mathrm{TN}_N = T_N + \frac{\Delta t}{\left(R_2 - \frac{\Delta r}{4}\right)\Delta r} \frac{2}{(\rho C)_{N-0.25}} [R_2 h(T_{\mathrm{ext2}} - T_N) - \Gamma_N] + \left(\frac{\Delta Q}{C}\right)_{N-0.25} \tag{8.82}$$

with the simple assumption

$$\mathrm{TN}_{N-0.25} - T_{N-0.25} = \mathrm{TN}_N - T_N \tag{8.17}$$

or by the expression

$$\mathrm{TN}_N = T_N - \frac{1}{3}(\mathrm{TN}_{N-1} - T_{N-1}) + \frac{4}{3}\left(\frac{\Delta Q}{C}\right)_{N-0.25} + \frac{8\Delta t}{3\left(R_2 - \frac{\Delta r}{4}\right)\Delta r} \frac{1}{(\rho C)_{N-0.25}} [R_2 h(T_{\mathrm{ext2}} - T_N) - \Gamma_N] \tag{8.83}$$

obtained with the more accurate assumption

$$\mathrm{TN}_{N-0.25} - T_{N-0.25} = \tfrac{3}{4}(\mathrm{TN}_N - T_N) + \tfrac{1}{4}(\mathrm{TN}_{N-1} - T_{N-1}) \tag{8.20}$$

The function Γ has been already defined in Sect. 8.7.1 by Eq. (8.26′).

Internal Circular Surface

Calculation of temperature at the internal circular surface is made in the same way as for the external circular surface. The annulus of thickness $\frac{1}{2}\Delta r$ and mean radius $R_1 + \frac{1}{4}\Delta r$, next to the internal circular surface is considered (Fig. 8.11). From the heat balance evaluated within this annulus during the time increment Δt, by taking into account heat conduction within the solid and heat convection at the surface, the new temperature after an interval of time Δt is thus expressed by

$$\mathrm{TN}_{0.25} = T_{0.25} + \frac{\Delta t}{\left(R_1 + \frac{\Delta r}{4}\right)\Delta r} \frac{2}{(\rho C)_{0.25}} [\Gamma_1 - R_1 h(T_0 - T_{\mathrm{ext1}})] + \left(\frac{\Delta Q}{C}\right)_{0.25} \tag{8.82′}$$

with the function Γ given by Eq. (8.26′) in Sect. 8.7.1.

With the simple assumption

$$\mathrm{TN}_{0.25} - T_{0.25} = \mathrm{TN}_0 - T_0 \tag{8.17″}$$

the new temperature at the surface becomes

$$\mathrm{TN}_0 = T_0 + \frac{\Delta t}{\left(R_1 + \frac{\Delta r}{4}\right)\Delta r} \frac{2}{(\rho C)_{0.25}} [\Gamma_1 - R_1 h(T_0 - T_{\mathrm{ext1}})] + \left(\frac{\Delta Q}{C}\right)_{0.25} \tag{8.84}$$

With the better assumption

$$\mathrm{TN}_{0.25} - T_{0.25} = \tfrac{3}{4}(\mathrm{TN}_0 - T_0) + \tfrac{1}{4}(\mathrm{TN}_1 - T_1) \tag{8.20''}$$

the new temperature at the internal surface is then

$$\mathrm{TN}_0 = T_0 - \tfrac{1}{3}(\mathrm{TN}_1 - T_1) + \frac{8\Delta t}{3\left(R_1 + \frac{\Delta r}{4}\right)\Delta r} \frac{1}{(\rho C)_{0.25}} [\Gamma_1 - R_1 h(T_0 - T_{\mathrm{ext1}})] + \frac{4}{3}\left(\frac{\Delta Q}{C}\right)_{0.25} \tag{8.85}$$

8.8 Hollow Cylinder of Finite Length with or Without Reaction, with Constant Thermal Parameters

The hollow cylinder of height $2H$, of internal and external radii R_i and R_e, is shown in Fig. 8.12. Various places are considered, within the solid and on the circular and plane surfaces. The results are given without demonstration, as the temperature has already been calculated in many places.

The radial thickness, $R_e - R_i$, and the mid-height, H, are divided respectively in NR and NH intervals of equal length, Δr and Δz, respectively. Various positions in the solid are thus defined by their radial abscissa r and longitudinal abscissa z, as follows:

$$r = R_i + n\Delta r, \quad 0 \leqslant n \leqslant \mathrm{NR}, \quad R_e - R_i = \mathrm{NR}\Delta r$$
$$z = h\Delta z, \quad 0 \leqslant h \leqslant 2\mathrm{NH}, \quad H = \mathrm{NH}\Delta z \tag{8.86}$$

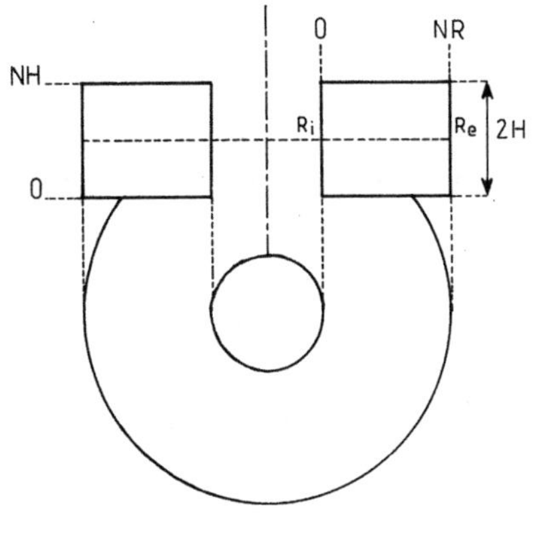

Fig. 8.12. Scheme of the cross-section of the tube of finite length.

8.8.1 Within the Cylinder Wall: $1 \leqslant n \leqslant \mathrm{NR}-1 \quad 1 \leqslant h \leqslant 2\mathrm{NH}-1$

The result has been given in the case of the solid cylinder of finite length (Fig. 8.5), in Sect. 8.4.1.

The new temperature after an interval of time Δt at position n, h is given by

$$\mathrm{TN}_{n,h} = T_{n,h} + \frac{1}{\mathrm{MR}}[T_{n-1,h} - 2T_{n,h} + T_{n+1,h}] + \frac{\Delta r}{2r\mathrm{MR}}[T_{n+1,h} - T_{n-1,h}]$$

$$+ \frac{1}{\mathrm{MH}}[T_{n,h-1} - 2T_{n,h} + T_{n,h+1}] + \frac{\Delta Q_{n,h}}{C} \tag{8.36}$$

with $r = R_\mathrm{i} + n\Delta r$ and with the dimensionless numbers MR and MH

$$\mathrm{MR} = \frac{(\Delta r)^2}{\Delta t}\frac{\rho C}{\lambda} \qquad \mathrm{MH} = \frac{(\Delta z)^2}{\Delta t}\frac{\rho C}{\lambda} \tag{8.37}$$

8.8.2 Given Temperature on Each Surface

The temperature on the surface during the whole process is given as follows:

$$\left.\begin{array}{lll} T_\mathrm{s} = g_1(R_\mathrm{i}, z, t) & \text{(internal cylindrical surface)} & 0 \leqslant z \leqslant 2H \\ T_\mathrm{s} = g_2(R_\mathrm{e}, z, t) & \text{(external cylindrical surface)} & 0 \leqslant z \leqslant 2H \\ T_\mathrm{s} = g_3(r, 0, t) & \text{(first plane surface)} & R_\mathrm{i} \leqslant r \leqslant R_\mathrm{e} \\ T_\mathrm{s} = g_4(r, 2H, t) & \text{(second plane surface)} & R_\mathrm{i} \leqslant r \leqslant R_\mathrm{e} \end{array}\right\} \tag{8.87}$$

8.8.3 Finite Coefficient of Convection at the Surface

The following positions are considered successively, with constant thermal parameters:

1. Circular external surface
2. Circular internal surface
3. Plane surface
4. Corner next to the external surface
5. Corner next to the internal surface

As calculations have already been made, especially for the solid cylinder of finite length, the new temperatures after an interval of time Δt are only given without demonstration in the following places.

External Circular Surface (Fig. 8.7), $n = \mathrm{NR} \quad 1 \leqslant h \leqslant 2\mathrm{NH}-1$

Eq. (8.45) is rewritten in the following more general form, available in the present case:

$$TN_{NR-0.25,h} = T_{NR-0.25,h} + \frac{1}{MH}[T_{NR-0.25,h-1} - 2T_{NR-0.25,h} + T_{NR-0.25,h+1}]$$

$$+ \frac{2}{MH}\frac{R_e - \frac{\Delta r}{2}}{R_e - \frac{\Delta r}{4}}(T_{NR-1,h} - T_{NR,h})$$

$$- \frac{R_e}{R_e - \frac{\Delta r}{4}}P(T_{NR,h} - T_{ext2}) + \frac{\Delta Q_{NR-0.25,h}}{C} \tag{8.88}$$

with the dimensionless numbers MR, MH and P:

$$P = \frac{2k\Delta t}{\Delta r \rho C} \tag{8.19''}$$

Temperatures at position (NR − 0.25, h) must be evaluated by one of the two ways

1. The simple expression

$$T_{NR-0.25,h} = T_{NR,h} \tag{8.46}$$

2. The more accurate relation

$$T_{NR-0.25,h} = \tfrac{3}{4}T_{NR,h} + \tfrac{1}{4}T_{NR-1,h} \tag{8.46'}$$

Circular Internal Surface (Fig. 8.11), $n = 0$ $\quad 1 \leqslant h \leqslant 2\text{NH} - 1$

In Eq. (8.88), R_e and $-\Delta r$ must be replaced by R_i and $+\Delta r$, respectively. Thus it follows that

$$TN_{0.25,h} = T_{0.25,h} + \frac{1}{MH}[T_{0.25,h-1} - 2T_{0.25,h} + T_{0.25,h+1}]$$

$$+ \frac{2}{MR}\frac{R_i + \frac{\Delta r}{2}}{R_i + \frac{\Delta r}{4}}(T_{1,h} - T_{0,h})$$

$$- \frac{R_i}{R_i + \frac{\Delta r}{4}}P(T_{0,h} - T_{ext1}) + \frac{\Delta Q_{0.25,h}}{C} \tag{8.89}$$

with the dimensionless numbers MR, MH and P, and the relationships

$$T_{0.25,h} = T_{0,h} \tag{8.46''}$$

or

$$T_{0.25,h} = \tfrac{3}{4}T_{0,h} + \tfrac{1}{4}T_{1,h} \tag{8.46'''}$$

Plane Surface, $h = 0$ $\quad 1 \leqslant n \leqslant \mathrm{NR} - 1$ (Fig. 8.8)

$$\mathrm{TN}_{n,0.25} = T_{n,0.25} + \frac{\Delta Q_{n,0.25}}{C} + \frac{1}{\mathrm{MR}}[T_{n-1,0.25} - 2T_{n,0.25} + T_{n+1,0.25}]$$

$$+ \frac{\Delta r}{2r\mathrm{MR}}(T_{n+1,0.25} - T_{n-1,0.25})$$

$$+ \frac{2}{\mathrm{MH}}[T_{n,1} - T_{n,0} - \mathrm{PH}(T_{n,0} - T_{\mathrm{ext}})] \tag{8.48'}$$

with $r = R_\mathrm{i} + n\Delta r$ and with the relationships to determine the temperature $T_{n,0.25}$

$$T_{n,0.25} = T_{n,0} \tag{8.50}$$

or

$$T_{n,0.25} = \tfrac{3}{4}T_{n,0} + \tfrac{1}{4}T_{n,1} \tag{8.51}$$

Corner Next to the External Circular Surface, $n = \mathrm{NR}$ $\quad h = 0$ (Fig. 8.9)

$$\mathrm{TN}_{\mathrm{NR}-0.25,0.25} = T_{\mathrm{NR}-0.25,0.25} + \frac{R_\mathrm{e} - \dfrac{\Delta r}{2}}{R_\mathrm{e} - \dfrac{\Delta r}{4}}\,\frac{2}{\mathrm{MR}}(T_{\mathrm{NR}-1,0.25} - T_{\mathrm{NR},0.25})$$

$$+ \frac{2}{\mathrm{MH}}[T_{\mathrm{NR}-0.25,1} - T_{\mathrm{NR}-0.25,0} - \mathrm{PH}(T_{\mathrm{NR}-0.25,0} - T_{\mathrm{ext}})]$$

$$- \mathrm{PR}_2(T_{\mathrm{NR},0.25} - T_{\mathrm{ext}}) + \frac{\Delta Q_{\mathrm{NR}-0.25,0.25}}{C} \tag{8.55'}$$

with the dimensionless numbers MR, MH, PH and PR_2. This last number is defined by

$$\mathrm{PR}_2 = \frac{R_\mathrm{e}}{R_\mathrm{e} - \dfrac{\Delta r}{4}}\,\frac{2k\Delta t}{\Delta r\rho C} \tag{8.19''}$$

The temperatures in position $(\mathrm{NR} - 0.25, 0.25)$ are approximated in the same way as for the solid cylinder by one of the relations (8.56), (8.57) or (8.58).

Corner Next to the Internal Circular Surface: $n = 0$, $h = 0$

$$\mathrm{TN}_{0.25,0.25} = T_{0.25,0.25} + \frac{R_\mathrm{i} + \dfrac{\Delta r}{2}}{R_\mathrm{i} + \dfrac{\Delta r}{4}}\,\frac{2}{\mathrm{MR}}(T_{1,0.25} - T_{0,0.25})$$

$$+ \frac{2}{\mathrm{MH}} [T_{0.25,1} - T_{0.25,0} - \mathrm{PH}(T_{0.25,0} - T_{\mathrm{ext}})]$$

$$- \mathrm{PR}_1(T_{0,0.25} - T_{\mathrm{ext}}) + \frac{\Delta Q_{0.25,0.25}}{C} \tag{8.90}$$

with the dimensionless numbers MR, MH, PH and PR_1. This last number is defined by

$$\mathrm{PR}_1 = \frac{R_i}{R_i + \frac{\Delta r}{4}} \frac{2k\Delta t}{\Delta r \rho C} \tag{8.19'''}$$

The temperature in position (0.25, 0.25) is approximated by one of the following relations:

$$T_{0.25,0.25} = T_{0,0} \tag{8.56'}$$

$$T_{0.25,0.25} = \tfrac{3}{4}T_{0,0} + \tfrac{1}{4}T_{1,1} \tag{8.57'}$$

$$T_{0.25,0.25} = \tfrac{1}{2}T_{0,0} + \tfrac{1}{4}T_{0,1} + \tfrac{1}{4}T_{1,0} \tag{8.58'}$$

8.9 Hollow Cylinder of Finite Length with or Without Reaction, with Temperature-Dependent Thermal Parameters

The new temperatures after an interval of time are given in various places without demonstration, as they have been already calculated, especially in the case of the solid cylinder of finite length with temperature-dependent thermal parameters.

8.9.1 Within the Cylinder Wall, $1 \leqslant n \leqslant \mathrm{NR} - 1$ $\quad 1 \leqslant h \leqslant 2\mathrm{NH} - 1$ (Fig. 8.5)

The new temperature is given by

$$\mathrm{TN}_{n,h} = T_{n,h} + \frac{\Delta t}{r\Delta r} \frac{1}{(\rho C)_{n,h}} [\Gamma_{n,h} - \Gamma_{n+1,h}]$$

$$+ \frac{\Delta t}{(\Delta z)^2} \frac{1}{(\rho C)_{n,h}} [G_{n,h} - G_{n,h+1}] + \left(\frac{\Delta Q}{C}\right)_{n,h} \tag{8.64'}$$

with $r = R_i + n\Delta r$ and with the functions G and Γ defined by

$$G_{n,h} = \lambda_{n,h-0.5}(T_{n,h-1} - T_{n,h}) \tag{8.62}$$

$$\Gamma_{n,h} = \left(\frac{R_i}{\Delta r} + n - 0.5\right) \lambda_{n-0.5,h}(T_{n-1,h} - T_{n,h}) \tag{8.63'}$$

The thermal conductivity at position $(n - 0.5,\ h)$ or $(n,\ h - 0.5)$ is obtained by using the temperature dependency of this parameter, after calculating the temperature at these positions by the relations

$$T_{n,h-0.5} = \tfrac{1}{2}(T_{n,h} + T_{n,h-1}) \tag{8.66}$$

$$T_{n-0.5,h} = \tfrac{1}{2}(T_{n,h} + T_{n-1,h}) \tag{8.66'}$$

8.9.2 Given Temperature on the Surface

The same conditions (8.87) of Sect. 8.8.2 are available.

8.9.3 Finite Coefficient of Convection at the Surface

As calculations have been made already with temperature-dependent thermal parameters, the new temperature after an interval of time Δt is given without demonstration at the following places:

1. External circular surface
2. Internal circular surface
3. Plane surface
4. Corner next to the external surface
5. Corner next to the internal surface

External Circular Surface: $n = \mathrm{NR} \qquad 1 \leqslant h \leqslant 2\mathrm{NH} - 1$ (*Fig. 8.7*)

The new temperature is given by

$$\begin{aligned}\mathrm{TN}_{\mathrm{NR}-0.25,h} = T_{\mathrm{NR}-0.25,h} &+ \frac{\Delta t}{(\Delta z)^2}\frac{1}{(\rho C)_{\mathrm{NR}-0.25,h}}[G_{\mathrm{NR}-0.25,h} - G_{\mathrm{NR}-0.25,h+1}]\\ &+ \frac{2\Delta t}{(R_e - \Delta r/4)\Delta r(\rho C)_{\mathrm{NR}-0.25,h}}\Gamma_{\mathrm{NR},h}\\ &- \frac{R_e}{R_e - \Delta r/4}\frac{2\Delta t k}{\Delta r(\rho C)_{\mathrm{NR}-0.25,h}}(T_{\mathrm{NR},h} - T_{\mathrm{ext}}) + \left(\frac{\Delta Q}{C}\right)_{\mathrm{NR}-0.25,h}\end{aligned} \tag{8.68'}$$

with the functions G and Γ defined in Sect. 8.9.1 by Eq. (8.62) and (8.63').

The temperature at position ($\mathrm{NR} - 0.25$, $h - 0.5$) is approximated by the relation

$$T_{\mathrm{NR}-0.25,h-0.5} = \tfrac{3}{8}(T_{\mathrm{NR},h-1} + T_{\mathrm{NR},h}) + \tfrac{1}{8}(T_{\mathrm{NR}-1,h-1} + T_{\mathrm{NR}-1,h}) \tag{8.91}$$

and the value of the thermal conductivity at this temperature is obtained by using the temperature dependency of this parameter.

Internal Circular Surface: $n = 0 \qquad 1 \leqslant h \leqslant 2\mathrm{NH} - 1$

The new temperature after an interval of time Δt on the internal circular surface is determined in a way similar to that used for the external circular surface (Fig. 8.11).

$$TN_{0.25,h} = T_{0.25,h} + \frac{\Delta t}{(\Delta z)^2} \frac{1}{(\rho C)_{0.25,h}} [G_{0.25,h} - G_{0.25,h+1}]$$
$$- \frac{2\Delta t}{(R_i + \Delta r/4)\Delta r(\rho C)_{0.25,h}} \Gamma_{1,h}$$
$$- \frac{R_i}{R_i + \Delta r/4} \frac{2\Delta tk}{\Delta r(\rho C)_{0.25,h}} (T_{0,h} - T_{ext}) + \left(\frac{\Delta Q}{C}\right)_{0.25,h} \quad (8.92)$$

The temperatures at positions (0.25, h) and (0.25, $h - 0.5$) are approximated by using the relations shown above.

Plane Surface, $h = 0$, $1 \leqslant n \leqslant NR - 1$ (Fig. 8.8)

The new temperature on the plane surface is given by the same relation as that obtained for the solid cylinder.

$$TN_{n,0.25} = T_{n,0.25} + \frac{\Delta t}{r\Delta r(\rho C)_{n,0.25}} [\Gamma_{n,0.25} - \Gamma_{n+1,0.25}]$$
$$+ \frac{2\Delta t}{\Delta z^2} \frac{1}{(\rho C)_{n,0.25}} [k\Delta z(T_{ext} - T_{n,0}) - G_{n,1}] + \left(\frac{\Delta Q}{C}\right)_{n,0.25} \quad (8.69')$$

The temperatures at positions (n, 0.25) and ($n - 0.5$, 0.25) are approximated by using the relations (8.70) and (8.71) for the case of the solid cylinder (Sect. 8.5.3).

Corner Next to the External Circular Surface (Fig. 8.9)

The new temperature is given by

$$TN_{NR-0.25,0.25} = T_{NR-0.25,0.25} + \frac{2\Delta t}{(R_3 - \Delta r/4)\Delta r} \frac{1}{(\rho C)_{NR-0.25,0.25}} \Gamma_{NR,0.25}$$
$$- \frac{2R_e}{R_e - \Delta r/4} \frac{\Delta t}{\Delta r} \frac{k}{(\rho C)_{NR-0.25,0.25}} (T_{NR,0.25} - T_{ext})$$
$$+ \frac{\Delta r}{(\Delta z)^2} \frac{2}{(\rho C)_{NR-0.25,0.25}} [k\Delta z(T_{ext} - T_{NR-0.25,0}) - G_{NR-0.25,1}]$$
$$+ \left(\frac{\Delta Q}{C}\right)_{NR-0.25,0.25} \quad (8.73')$$

where the parameter (ρC) is at position (NR − 0.25, 0.25).

The temperatures at positions (NR − 0.5, 0.25), (NR − 0.25, 0.5) and (NR − 0.25, 0.25) are approximated by Eqs (8.74), (8.75) and (8.58) of Sect.

8.5.3 and the thermal parameters are calculated with their temperature dependency relationships.

Corner Next to the Internal Circular Surface (Fig. 8.12)

The new temperature at the internal circular surface is obtained in a way similar to that used for the external circular surface.

$$\begin{aligned}
\mathrm{TN}_{0.25,0.25} = T_{0.25,0.25} &- \frac{2\Delta t}{(R_\mathrm{i} + \Delta r/4)\Delta r}\frac{1}{(\rho C)_{0.25,0.25}}\Gamma_{1,0.25} \\
&- \frac{2R_\mathrm{i}}{R_\mathrm{i} + \Delta r/4}\frac{\Delta t}{\Delta r}\frac{k}{(\rho C)_{0.25,0.25}}(T_{0,0.25} - T_\mathrm{ext}) \\
&+ \frac{\Delta t}{(\Delta z)^2}\frac{2}{(\rho C)_{0.25,0.25}}[k\Delta z(T_\mathrm{ext} - T_{0.25,0}) - G_{0.25,1}] \\
&+ \left(\frac{\Delta Q}{C}\right)_{0.25,0.25} \qquad (8.93)
\end{aligned}$$

while the temperatures at positions (0.5, 0.25), (0.25, 0.5) and (0.25, 0.25) are approximated by

$$T_{0.5,0.25} = \tfrac{3}{8}(T_{0,0} + T_{1,0}) + \tfrac{1}{8}(T_{0,1} + T_{1,1}) \qquad (8.74')$$

$$T_{0.25,0.5} = \tfrac{3}{4}(T_{0,0} + T_{0,1}) + \tfrac{1}{8}(T_{1,0} + T_{1,1}) \qquad (8.75')$$

$$T_{0.25,0.25} = \tfrac{1}{2}T_{0,0} + \tfrac{1}{4}T_{0,1} + \tfrac{1}{4}T_{1,0} \qquad (8.58')$$

and the thermal parameter (ρC) is at the position (0.25, 0.25).

Chapter 9

Numerical Analysis for a Sphere

9.1 Introduction

Sometimes, large spherical samples are prepared in industry, but more common is the production of small beads. The general problem must be applicable whatever the size may be, as the time of heating is proportional to the square of the radius of the spherical sample.

Spherical samples are generally solid, and only the case of a solid sphere is studied. The case of a hollow sphere can be easily considered, with the help of the studies which are made in the case of cylinder and hollow cylinder.

Very often, the solid spheres are heated by immersing them in heated surroundings, and heat transfer can be considered as radial only.

Two cases are of interest: whether the thermal conductivity is constant, or it is temperature-dependent. The interesting case where the thermal parameters depend not only on temperature but also on the value of the state of cure is treated in the same way.

9.2 Solid Sphere with Radial Heat Transfer, with or Without Internal Reaction, with Constant Thermal Parameters

Heat transfer is radial only. The circular cross-section passing through the centre of the sphere is considered (Fig. 9.1) for clarity.

The radius R is divided in N equal intervals of thickness Δr. The radial abscissa of each point limiting one of these intervals is defined by an integer n with the following relations:

$$r = n\Delta r, \qquad 0 \leqslant n \leqslant N, \qquad R = N\Delta r \tag{9.1}$$

Δr being the increment of space.

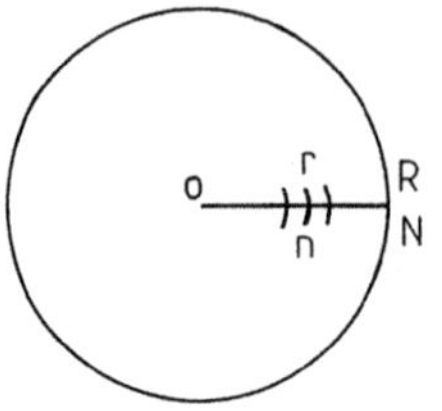

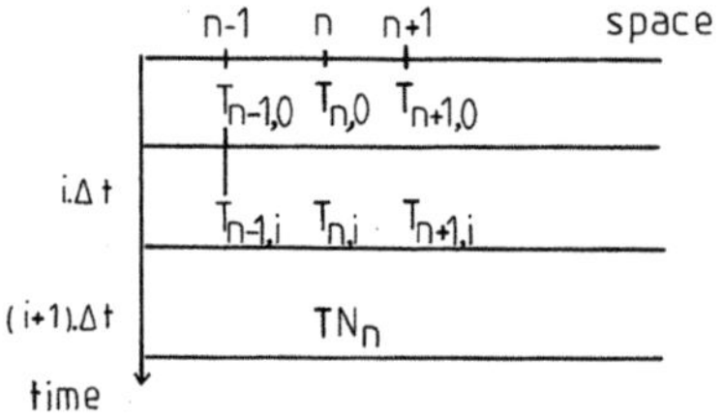

Fig. 9.1. Space–time diagram for numerical analysis. Cross-section of the sphere of radius R.

9.2.1 Within the Sphere

The spherical membrane of mean radius r, with the same centre as that of the sphere, and of thickness Δr is drawn (Fig. 9.2). The heat balance during the interval of time $[t, t + \Delta t]$ is evaluated within this spherical membrane:

$$[\underset{n-0.5}{A\phi} - \underset{n+0.5}{A\phi}]\,\Delta t + V\rho\left(\frac{dQ}{dt}\right)_n \Delta t = V\rho C(\mathrm{TN}_n - T_n) \tag{9.2}$$

where A is the area of each spherical surface limiting this membrane, ϕ is the radial heat flux passing through this surface, V is the volume of the membrane and $Q = Q(r, t)$ is the amount of heat generated per unit mass at the radial abscissa r up to time t.

The volume of the spherical membrane is given by

$$V = \frac{4\pi}{3}\left[\left(r + \frac{\Delta r}{2}\right)^3 - \left(r - \frac{\Delta r}{2}\right)^3\right] = \frac{4}{3}\pi\left[3r^2\Delta r + \frac{(\Delta r)^3}{4}\right] \tag{9.3}$$

$$V = 4\pi\left[r^2 + \frac{(\Delta r)^2}{12}\right]\Delta r = 4\pi(\Delta r)^3\left(n^2 + \frac{1}{12}\right) \tag{9.3'}$$

By replacing the area of each surface and the radial heat flux by their value, the heat balance within the membrane becomes

$$-4\pi(\Delta r)^2(n - 0.5)^2\lambda\left(\frac{\partial T}{\partial r}\right)_{n-0.5} + 4\pi(\Delta r)^2(n + 0.5)^2\lambda\left(\frac{\partial T}{\partial r}\right)_{n+0.5}$$

$$= 4\pi\left(n^2 + \frac{1}{12}\right)\frac{(\Delta r)^3}{\Delta t}\rho C\left(\mathrm{TN}_n - T_n - \frac{\Delta Q_n}{C}\right) \tag{9.4}$$

The gradient of temperature at each position $n - 0.5$ (or $n + 0.5$) is approximated by the chord slope taken between the positions $n - 1$ and n (or n and $n + 1$).

$$\left(\frac{\partial T}{\partial r}\right)_{n-0.5} = \frac{T_n - T_{n-1}}{\Delta r} \tag{9.5}$$

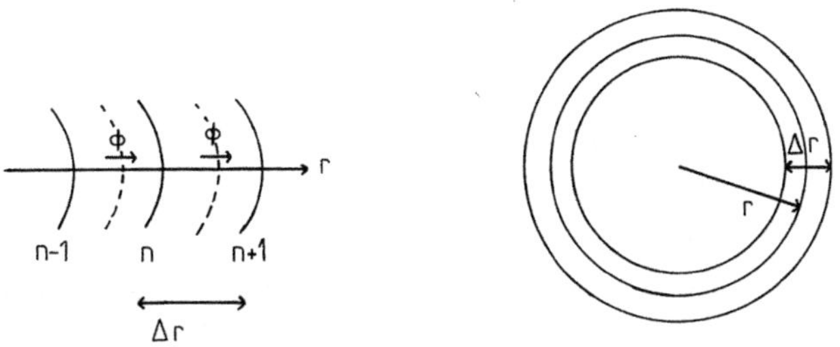

Fig. 9.2. Spherical membrane of mean radius r and thickness Δr. Heat balance during the increment of time Δt.

The equation expressing the heat balance within the membrane is thus

$$(\Delta r)^2(n-0.5)^2\lambda\,\frac{T_{n-1}-T_n}{\Delta r}+(\Delta r)^2(n+0.5)^2\lambda\,\frac{T_{n+1}-T_n}{\Delta r}$$
$$=\left(n^2+\frac{1}{12}\right)\frac{(\Delta r)^3}{\Delta t}\,\rho C\left(\mathrm{TN}_n-T_n-\frac{\Delta Q_n}{C}\right) \tag{9.6}$$

The new temperature within this spherical membrane after time Δt is thus given as a function of the temperatures obtained at the previous time in the same membrane and adjacent membranes.

$$\mathrm{TN}_n=T_n+\frac{1}{M\left(n^2+\frac{1}{12}\right)}[(n-0.5)^2(T_{n-1}-T_n)+(n+0.5)^2(T_{n+1}-T_n)]$$
$$+\frac{\Delta Q_n}{C} \tag{9.7}$$

where ΔQ, at each position n, is calculated as shown in Sect. 7.4, and M is the dimensionless number defined by

$$M=\frac{(\Delta r)^2}{\Delta t}\,\frac{\rho C}{\lambda} \tag{9.8}$$

Centre of the Sphere, $r = 0$ or $n = 0$

The equation obtained for calculating the new temperature at position n cannot be used for the centre of the sphere.

The small sphere of centre O and of radius $\frac{1}{2}\Delta r$ is considered (Fig. 9.3). The heat balance is calculated within this small sphere during the increment of time Δt.

$$-(A\phi)_{0.5}\Delta t+V\rho\Delta Q_0=V\rho C(\mathrm{TN}_0-T_0) \tag{9.9}$$

By replacing the area A and volume V of this small sphere, and by approximating the gradient of temperature at position 0.5 by the chord slope between the positions 0 and 1:

$$\phi_{0.5}=\lambda\,\frac{T_0-T_1}{\Delta r} \tag{9.10}$$

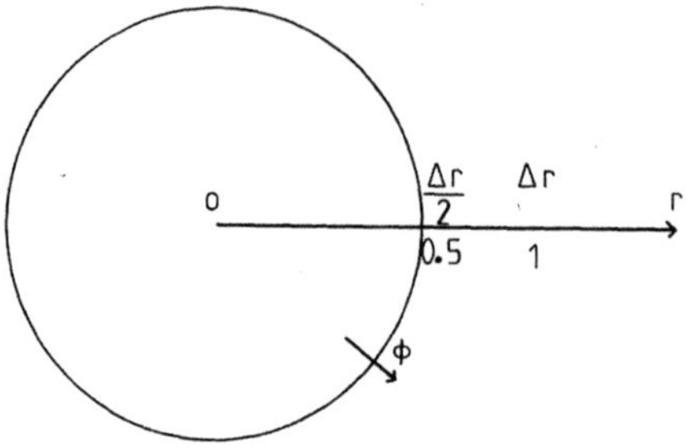

Fig. 9.3. Heat balance within the sphere of centre O and radius $\frac{1}{2}\Delta r$, during the interval of time $[t, t + \Delta t]$.

the heat balance becomes

$$-4\pi\left(\frac{\Delta r}{2}\right)^2 \lambda \frac{T_0 - T_1}{\Delta r} \Delta t = \frac{4}{3}\pi\left(\frac{\Delta r}{2}\right)^3 \rho C \left(\mathrm{TN}_0 - T_0 - \frac{\Delta Q_0}{C}\right) \tag{9.11}$$

The new temperature at the centre of the sphere after time Δt is thus given by

$$\mathrm{TN}_0 = T_0 + \frac{6}{M}(T_1 - T_0) + \frac{\Delta Q_0}{C} \tag{9.12}$$

9.2.2 On the Surface

Given Temperature on the Surface

In this simple case, the temperature on the surface is either constant, or given as a function of time

$$T_s = T(R, t) = T_N = g(t) \tag{9.13}$$

Finite Coefficient of Convection at the Surface

The boundary condition is

$$\phi_N = h(T_N - T_{\mathrm{ext}}) \tag{9.14}$$

The spherical membrane of mean radius $(N - 0.25)\Delta r$ and thickness $\frac{1}{2}\Delta r$, located next to the surface of the sphere is considered (Fig. 9.4). The heat balance during the increment of time Δt is evaluated within this spherical membrane by taking into account heat transferred by conduction through the spherical surface at the position $N - 0.5$, heat transferred by convection at the external surface and heat generated by the reaction.

$$[(A\phi)_{N-0.5} - (A\phi)_N]\Delta t + V\rho\Delta Q_{N-0.25} =$$

$$4\pi(N - 0.25)^2(\Delta r)^2 \frac{\Delta r}{2} \rho C(\mathrm{TN}_{N-0.25} - T_{N-0.25}) \tag{9.15}$$

By expressing the area A and heat flux ϕ at various abscissae by their value, the heat balance becomes

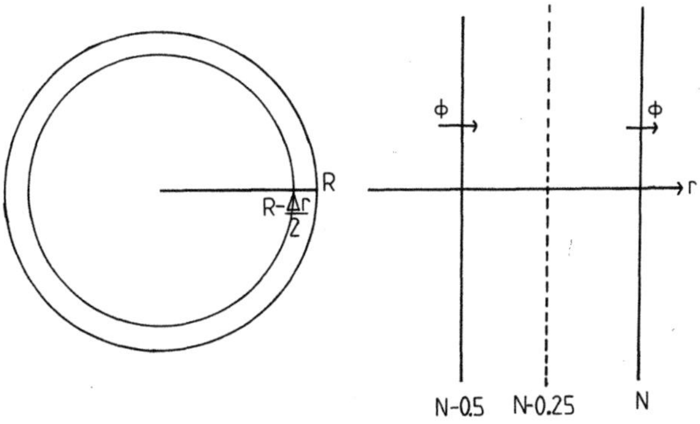

Fig. 9.4. Heat balance within the spherical membrane of mean radius $(N - 0.25)\Delta r$ and thickness $\frac{1}{2}\Delta r$, located next to the external surface.

$$\left[4\pi(N - 0.5)^2(\Delta r)^2\lambda \frac{T_{N-1} - T_N}{\Delta r} - 4\pi N^2(\Delta r)^2 h(T_N - T_{ext})\right]\Delta t$$

$$= 4\pi(N - 0.25)(\Delta r)^2 \frac{\Delta r}{2}\rho C\left(TN_{N-0.25} - T_{N-0.25} - \frac{\Delta Q_{N-0.25}}{C}\right) \quad (9.16)$$

The new temperature after time Δt within this membrane is given by

$$TN_{N-0.25} = T_{N-0.25} + \frac{2}{M}\left(\frac{N - 0.5}{N - 0.25}\right)^2(T_{N-1} - T_N)$$

$$+ 2P\left(\frac{N}{N - 0.25}\right)^2(T_{ext} - T_N) + \frac{\Delta Q_{N-0.25}}{C} \quad (9.17)$$

with the dimensionless number P

$$P = \frac{\Delta t}{\Delta r}\frac{h}{\rho C} \quad (9.18)$$

As the number $N - 0.25$ is not an integer, the temperature at this position is not calculated. The temperature at this position is approximated by one of the following two assumptions:

$$TN_{N-0.25} - T_{N-0.25} = TN_N - T_N \quad (9.19)$$

$$TN_{N-0.25} - T_{N-0.25} = \tfrac{3}{4}(TN_N - T_N) + \tfrac{1}{4}(TN_{N-1} - T_{N-1}) \quad (9.20)$$

When the second assumption is made, the new temperature at the surface TN_N is expressed in terms of the new temperature at the position $(N - 1)$. The new temperature TN_{N-1} must be calculated beforehand by using the equation obtained within the solid sphere (Eq. (9.7)).

9.2.3 Stability of Calculations

In the equation giving the new temperature TN_n, for each value of n $(0 \leq n \leq N)$, the whole coefficient of the temperature T_n must be positive. Thus Eqs (9.7),

(9.12) and (9.17) impose respectively

$$1 - \frac{(n-0.5)^2 + (n+0.5)^2}{\left(n^2 + \frac{1}{12}\right) M} \geqslant 0 \quad \text{for} \quad 1 \leqslant n \leqslant N-1 \tag{9.21}$$

$$1 - \frac{6}{M} \geqslant 0 \quad \text{or} \quad M \geqslant 6 \quad \text{or} \quad \Delta t \leqslant \frac{(\Delta r)^2}{6\alpha} \tag{9.22}$$

$$\frac{2}{M}\left(\frac{N-0.5}{N-0.25}\right)^2 + 2P\left(\frac{N}{N-0.25}\right)^2 \leqslant 1$$

$$\text{or} \quad \Delta t \leqslant \frac{(N-0.25)^2 \Delta r^2}{2\alpha(N-0.5)^2 + \frac{2h\Delta r}{\rho C} N^2} \tag{9.23}$$

Remark. If the parameters ρ, C and h are variable (with the temperature or the state of cure), but λ constant, the previous Eqs (9.7), (9.12) and (9.17) are still valid, with variable values for M, C and P.

9.3 Solid Sphere with Radial Heat Transfer, with or Without Reaction, with Temperature-Dependent Thermal Parameters

As shown above, all thermal parameters may vary with temperature, and also with the value of the state of cure. If the thermal conductivity λ is variable with the position, the previous equations (9.7) and (9.12) are not valid, because the derivative in the Eq. (1.26) (Sect. 1.3.3) is then expressed as follows:

$$\frac{\partial}{\partial r}\left(\lambda r^2 \frac{\partial T}{\partial r}\right) = \lambda \frac{\partial}{\partial r}\left(r^2 \frac{\partial T}{\partial r}\right) + \frac{\partial \lambda}{\partial r} r^2 \frac{\partial T}{\partial r} \tag{9.24}$$

The last term vanishes if λ is constant. Therefore Eq. (1.27) is not valid if λ is variable with the position.

9.3.1 Within the Sphere

The spherical membrane of mean radius r and thickness Δr, with the same centre as that of the solid sphere is considered (Fig. 9.2). The heat balance during the time increment Δt within this membrane leads to

$$[(A\phi)_{n-0.5} - (A\phi)_{n+0.5}]\Delta t + V\rho\Delta Q_n = 4\pi\left(n^2 + \frac{1}{12}\right)(\Delta r)^3 (\rho C)_n (\mathrm{TN}_n - T_n) \tag{9.25}$$

with the area A of each spherical surface of radius $(n-0.5)\Delta r$ and $(n+0.5)\Delta r$, and the heat flux ϕ given by

$$\phi_{n-0.5} = -\left(\lambda \frac{\partial T}{\partial r}\right)_{n-0.5} = \lambda_{n-0.5} \frac{T_{n-1} - T_n}{\Delta r} \tag{9.26}$$

Upon defining the function G

$$G_n = (n - 0.5)^2 \lambda_{n-0.5}(T_n - T_{n-1}) \quad \text{for} \quad 1 \leqslant n \leqslant N \tag{9.27}$$

the new temperature after time Δt, TN_n, is thus expressed by

$$TN_n = T_n + \frac{1}{n^2 + \frac{1}{12}} \frac{\Delta T}{\Delta r^2} \frac{1}{(\rho C)_n} (G_{n+1} - G_n) + \frac{\Delta Q_n}{C_n} \tag{9.28}$$

The thermal conductivity λ in the function G_n must be determined at the radial abscissa $(n - 0.5)\Delta r$. The temperature at this position is thus estimated by the linear interpolation:

$$T_{n-0.5} = \tfrac{1}{2}(T_{n-1} + T_n) \tag{9.29}$$

Centre of the Sphere, $r = 0$ and $n = 0$

The equation evaluating the temperature at radial abscissa different from zero cannot be used for the centre of the sphere. The small sphere of centre O and radius $\frac{1}{2}\Delta r$ is thus considered (Fig. 9.3). The heat balance during the time increment Δt is determined within this small sphere.

$$-(A\phi)_{0.5}\Delta t + V\rho_0 \Delta Q_0 = \frac{4}{3}\pi\left(\frac{\Delta r}{2}\right)^3 (\rho C)_0 (TN_0 - T_0) \tag{9.30}$$

This equation becomes

$$4\pi\left(\frac{\Delta r}{2}\right)^2 \left(\lambda \frac{\partial T}{\partial r}\right)_{0.5} \Delta t = \frac{4}{3}\pi\left(\frac{\Delta r}{2}\right)^3 (\rho C)_0 \left(TN_0 - T_0 - \frac{\Delta Q_0}{C_0}\right) \tag{9.31}$$

and the new temperature after time Δt at the centre of the sphere, TN_0, is thus given by

$$TN_0 = T_0 + 24 \frac{\Delta t}{\Delta r^2} \frac{1}{(\rho C)_0} G_1 + \frac{\Delta Q_0}{C_0} \tag{9.32}$$

with the function G_1 given by Eq. (9.27):

$$G_1 = (0.5)^2 \lambda_{0.5}(T_1 - T_0) \tag{9.27'}$$

Eq. (9.32) can also be written as

$$TN_0 = T_0 + 6 \frac{\Delta t}{\Delta r^2} \frac{\lambda_{0.5}}{(\rho C)_0} (T_1 - T_0) + \frac{\Delta Q_0}{C_0} \tag{9.32'}$$

9.3.2 On the Surface

Given Temperature on the Surface

The temperature on the surface is either constant or given as a function of time:

$$T_s = T(R, t) = T_N = g(t) \tag{9.13}$$

Finite Coefficient of Heat Convection at the Surface

In the same way as for constant thermal parameters, the spherical membrane of mean radius $(N-0.25)\Delta r$ and thickness $\frac{1}{2}\Delta r$, next to the surface is shown (Fig. 9.4). The heat balance within the increment of time Δt within this spherical membrane is evaluated by considering the heat transferred by conduction through the spherical surface at position $(N-0.5)\Delta r$ and heat transferred by convection at the external surface:

$$[(A\phi)_{N-0.5} - (A\phi)_{\mathrm{N}}]\Delta t + V(\rho\Delta Q)_{N-0.25} =$$

$$4\pi(N - 0.25)^2(\Delta r)^2 \frac{\Delta r}{2} (\rho C)_{N-0.25}(\mathrm{TN}_{N-0.25} - T_{N-0.25}) \qquad (9.33)$$

The new temperature after time Δt within this membrane, $\mathrm{TN}_{N-0.25}$, can thus be obtained in terms of the previous temperatures

$$\mathrm{TN}_{N-0.25} = T_{N-0.25} + \frac{2}{(N-0.25)^2} \frac{\Delta t}{\Delta r^2} \frac{1}{(\rho C)_{N-0.25}} G_N$$

$$+ 2\left(\frac{N}{N-0.25}\right)^2 \frac{\Delta t}{\Delta r} \frac{h}{(\rho C)_{N-0.25}} (T_{\mathrm{ext}} - T_{\mathrm{N}}) + \left(\frac{\Delta Q}{C}\right)_{N-0.25} \qquad (9.34)$$

The temperatures at the surface, can be approximated by one of the following two relations:

$$T_{N-0.25} = T_N \qquad (9.19')$$

$$T_{N-0.25} = \tfrac{3}{4}T_N + \tfrac{1}{4}T_{N-1} \qquad (9.20')$$

and likewise for TN.

The temperature at radial abscissa $(N-0.5)\Delta r$ is approximated by

$$T_{N-0.5} = \tfrac{1}{2}(T_N + T_{N-1}) \qquad (9.29)$$

enabling one to determine the thermal parameter $\lambda_{N-0.5}$ in the function G_N defined by Eq. (9.27).

Chapter 10

Numerical Analysis for a Parallelepiped with Three-Dimensional Heat Transfer and Cure Reaction

10.1 Introduction

The parallelepiped problem is frequently encountered in the cure of thermosets. An analytical solution exists only when no reaction occurs in the resin and the thermal parameters are constant, for particular initial and boundary conditions. The problem is thus treated in a general way by considering heat transfer by conduction through the resin and heat evolved from the cure reaction. In this general case, the thermal parameters are temperature-dependent.

Following these general results, it is easy to obtain the solutions in the following two basic cases:

1. When the thermal parameters are temperature-dependent and no reaction occurs (the general solutions can be used, when the heat of cure vanishes by making $k_0 = 0$)
2. When the thermal parameters are constant, and the cure reaction takes place

The problem is studied for various places:

1. Within the resin
2. On the surface of the resin with heat conduction, the resin being in perfect contact with the mould
3. On the surface of the resin (or mould) with heat convection, the resin (or mould) being in contact with a surrounding fluid

A corner of the parallelepiped is chosen as origin, 0, and the three corresponding edges as coordinate axes, $0x$, $0y$, $0z$. The parallelepiped of dimensions $2X$, $2Y$ and $2Z$ is divided into small parallelepipeds of dimensions Δx, Δy and Δz, and whose edges are parallel to the directions $0x$, $0y$ and $0z$ (Fig. 10.1). The corners of each small parallelepiped are defined by the three integers i, j, k:

$$x = i\Delta x, \qquad 0 \leq i \leq 2N_1, \qquad X = N_1 \Delta x \tag{10.1}$$

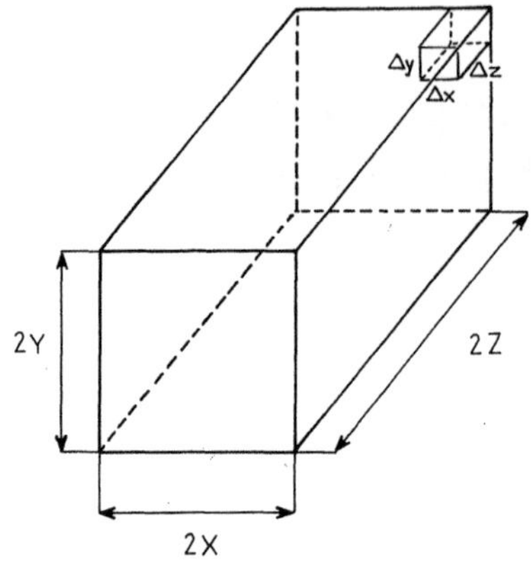

Fig. 10.1. Parallelepiped of dimensions $2X$, $2Y$, $2Z$ with a small parallelepiped of dimensions Δx, Δy, Δz, centred at position i, j, k.

$$y = j\Delta y, \qquad 0 \leqslant j \leqslant 2N_2, \qquad Y = N_2 \Delta y \tag{10.1$'$}$$

$$z = k\Delta z, \qquad 0 \leqslant k \leqslant 2N_3, \qquad Z = N_3 \Delta z \tag{10.1$''$}$$

It is of interest to take $N_3 = N_2 = N_1$.

10.2 Temperature-Dependent Thermal Parameters with Reaction

Three positions in the resin are considered:

1. Within the resin
2. On the surface of the resin, when it is in contact with a solid, such as a mould
3. On the surface of the mould or resin, when it is in contact with a surrounding fluid

10.2.1 Within the Resin (Fig. 10.2)

The small parallelepiped is considered, with dimensions Δx, Δy, Δz and $i\Delta x$, $j\Delta y$, $k\Delta z$, as coordinates of the centre. The heat balance during the interval of time $[t, t + \Delta t]$ is evaluated within this parallelepiped by considering heat conduction along the three axes, as well as heat evolved from the cure reaction.

$$\begin{aligned}\Delta x \Delta y \Delta z (\rho C)_{i,j,k}(\mathrm{TN}_{i,j,k} - T_{i,j,k}) &= \Delta y \Delta z (\phi^1_{i-0.5,j,k} - \phi^1_{i+0.5,j;k})\Delta t \\ &+ \Delta x \Delta z (\phi^2_{i,j-0.5,k} - \phi^2_{i,j+0.5,k})\Delta t \\ &+ \Delta x \Delta y (\phi^3_{i,j,k-0.5} - \phi^3_{i,j,k+0.5})\Delta t \\ &+ \Delta x \Delta y \Delta z \rho_{i,j,k} \Delta Q_{i,j,k}\end{aligned} \tag{10.2}$$

where the three components of heat flux by conduction, ϕ^1, ϕ^2, ϕ^3 are given by

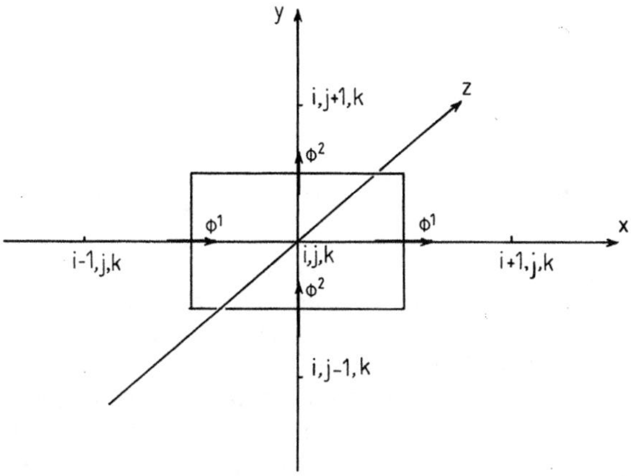

Fig. 10.2. Heat transfer within the small parallelepiped of dimensions Δx, Δy, Δz, centred at position i, j, k.

$$\phi^1_{i-0.5,j,k} = -\left(\lambda \frac{\partial T}{\partial x}\right)_{i-0.5,j,k} = \lambda_{i-0.5,j,k} \frac{T_{i-1,j,k} - T_{i,j,k}}{\Delta x} \tag{10.3}$$

and likewise for ϕ^2 and ϕ^3.

By dividing by $\Delta x \Delta y \Delta z (\rho C)_{i,j,k} \Delta t$, the new temperature $\mathrm{TN}_{i,j,k}$ can be expressed in terms of the previous temperatures at this place $T_{i,j,k}$ and the six adjacent places

$$\begin{aligned}\mathrm{TN}_{i,j,k} = T_{i,j,k} &+ \frac{\Delta t}{\Delta x (\rho C)_{i,j,k}} (\phi^1_{i-0.5,k} - \phi^1_{i+0.5,j,k}) \\ &+ \frac{\Delta t}{\Delta y (\rho C)_{i,j,k}} (\phi^2_{i,j-0.5,k} - \phi^2_{i,j+0.5,k}) \\ &+ \frac{\Delta t}{\Delta z (\rho C)_{i,j,k}} (\phi^3_{i,j,k-0.5} - \phi^3_{i,j,k+0.5}) \\ &+ \frac{\Delta Q_{i,j,k}}{C_{i,j,k}}\end{aligned} \tag{10.4}$$

where ΔQ, at each position i, j, k, is calculated as shown in Sect. 7.4. This equation can be used when no reaction occurs, by making the pre-exponential factor k_0 equal to zero in the rate of heat evolved from cure.

If the thermal conductivity λ is constant, Eq. (10.4) may be simplified, as follows:

$$\begin{aligned}\mathrm{TN}_{i,j,k} = T_{i,j,k} &+ \frac{\Delta Q_{i,j,k}}{C_{i,j,k}} + \frac{1}{\mathrm{MX}} (T_{i-1,j,k} - 2T_{i,j,k} + T_{i+1,j,k}) \\ &+ \frac{1}{\mathrm{MY}} (T_{i,j-1,k} - 2T_{i,j,k} + T_{i,j+1,k}) \\ &+ \frac{1}{\mathrm{MZ}} (T_{i,j,k-1} - 2T_{i,j,k} + T_{i,j,k+1})\end{aligned} \tag{10.4'}$$

where MX, MY, MZ are the dimensionless moduli defined as follows:

$$\mathrm{MX} = \frac{\rho C(\Delta x)^2}{\lambda \Delta t} \qquad \mathrm{MY} = \frac{\rho C(\Delta y)^2}{\lambda \Delta t} \qquad \mathrm{MZ} = \frac{\rho C(\Delta z)^2}{\lambda \Delta t} \tag{10.5}$$

The values of ρ and C may depend on the position and time.

10.2.2 Interface Between Resin and Mould (Fig. 10.3)

The parallelepipedic resin sample is assumed to be surrounded by a parallelepipedic mould having the same centre, and the internal and external surfaces of the mould are parallel. The contact is assumed to be perfect, i.e. the temperature is the same on both sides of the interface.

Three external edges of the mould are chosen as coordinate axes. The thickness of the mould along the $0x$ direction is divided into NX equal slices of thickness Δx_{m}. The thickness of the resin along the $0x$ direction is divided into $2N_1$ equal slices of thickness Δx. The same operation is made along the other directions $0y$ and $0z$, with the numbers of slices NY and NZ, $2N_2$ and $2N_3$, of respective thicknesses Δy_{m} and Δz_{m}, and Δy and Δz.

As shown in Fig. 10.3, the two small parallelepipeds on both sides of the resin–mould interface, are considered. Their dimensions are $\frac{1}{2}\Delta x_{\mathrm{m}}$, Δy and Δz for the mould and $\frac{1}{2}\Delta x$, Δy and Δz for the resin.

The heat balance during the increment of time Δt is evaluated by considering heat conduction within the resin and the mould, as well as heat evolved from cure reaction in the resin:

$$\begin{aligned}
&\Delta y \Delta z\left[\frac{\Delta x_{\mathrm{m}}}{2}(\rho_{\mathrm{m}} C_{\mathrm{m}})_{\mathrm{NX}-0.25,j,k} + \frac{\Delta x}{2}(\rho C)_{\mathrm{NX}+0.25,j,k}\right](\mathrm{TN}_{\mathrm{NX},j,k} - T_{\mathrm{NX},j,k}) \\
&= \Delta t \Delta y \Delta z\left[\left(\lambda_{\mathrm{m}} \frac{\partial T}{\partial x}\right)_{\mathrm{NX}-0.5,j,k} - \left(\lambda \frac{\partial T}{\partial x}\right)_{\mathrm{NX}+0.5,j,k}\right] \\
&+ \Delta t \frac{\Delta x_{\mathrm{m}}}{2} \Delta z\left[\left(\lambda_{\mathrm{m}} \frac{\partial T}{\partial y}\right)_{\mathrm{NX}-0.25,j-0.5,k} - \left(\lambda_{\mathrm{m}} \frac{\partial T}{\partial y}\right)_{\mathrm{NX}-0.25,j+0.5,k}\right] \\
&+ \Delta t \frac{\Delta x}{2} \Delta z\left[\left(\lambda \frac{\partial T}{\partial y}\right)_{\mathrm{NX}+0.25,j-0.5,k} - \left(\lambda \frac{\partial T}{\partial y}\right)_{\mathrm{NX}+0.25,j+0.5,k}\right] \\
&+ \Delta t \frac{\Delta x_{\mathrm{m}}}{2} \Delta y\left[\left(\lambda_{\mathrm{m}} \frac{\partial T}{\partial z}\right)_{\mathrm{NX}-0.25,j,k-0.5} - \left(\lambda_{\mathrm{m}} \frac{\partial T}{\partial z}\right)_{\mathrm{NX}-0.25,j,k+0.5}\right] \\
&+ \Delta t \frac{\Delta x}{2} \Delta y\left[\left(\lambda \frac{\partial T}{\partial z}\right)_{\mathrm{NX}+0.25,j,k-0.5} - \left(\lambda \frac{\partial T}{\partial z}\right)_{\mathrm{NX}+0.25,j,k+0.5}\right] \\
&+ \frac{\Delta x}{2} \Delta y \Delta z(\rho \Delta Q)_{\mathrm{NX}+0.25,j,k}
\end{aligned} \tag{10.6}$$

Upon defining

$$\mathrm{FX} = \Delta x_{\mathrm{m}}(\rho_{\mathrm{m}} C_{\mathrm{m}})_{\mathrm{NX}-0.25,j,k} + \Delta x(\rho C)_{\mathrm{NX}+0.25,j,k} \tag{10.7}$$

$$\mathrm{LX}_{u,v} = \Delta x_{\mathrm{m}}(\lambda_{\mathrm{m}})_{\mathrm{NX},u,v} + \Delta x \lambda_{\mathrm{NX},u,v} \tag{10.8}$$

and making the approximations

$$T_{\mathrm{NX}\pm 0.25,u,v} = T_{\mathrm{NX},u,v} \quad \text{for} \quad u = j \text{ or } j \pm 0.5 \quad \text{and} \quad v = k \text{ or } k \pm 0.5 \tag{10.9}$$

the new temperature at the interface, $TN_{\mathrm{NX},j,k}$, after time Δt is expressed in terms of the previous temperatures T at the same place and the adjacent six places, as follows:

$$\begin{aligned} \mathrm{TN}_{\mathrm{NX},j,k} &= T_{\mathrm{NX},j,k} \\ &+ \frac{2\Delta t}{\mathrm{FX}}\left[\frac{(\lambda_{\mathrm{m}})_{\mathrm{NX}-0.5,j,k}}{\Delta x_{\mathrm{m}}}(T_{\mathrm{NX}-1,j,k} - T_{\mathrm{NX},j,k}) + \frac{\lambda_{\mathrm{NX}+0.5,j,k}}{\Delta x}(T_{\mathrm{NX}+1,j,k} - T_{\mathrm{NX},j,k})\right] \\ &+ \frac{\Delta t}{\mathrm{FX}\Delta y^2}[\mathrm{LX}_{j-0.5,k}(T_{\mathrm{NX},j-1,k} - T_{\mathrm{NX},j,k}) + \mathrm{LX}_{j+0.5,k}(T_{\mathrm{NX},j+1,k} - T_{\mathrm{NX},j,k})] \\ &+ \frac{\Delta t}{\mathrm{FX}\Delta z^2}[\mathrm{LX}_{j,k-0.5}(T_{\mathrm{NX},j,k-1} - T_{\mathrm{NX},j,k}) + \mathrm{LX}_{j,k+0.5}(T_{\mathrm{NX},j,k+1} - T_{\mathrm{NX},j,k})] \\ &+ \frac{\Delta x}{\mathrm{FX}}(\rho\Delta Q)_{\mathrm{NX}+0.25,j,k} \end{aligned} \tag{10.10}$$

If λ_{m} and λ are constant, this equation may be simplified, as follows:

$$\begin{aligned} \mathrm{TN}_{\mathrm{NX},j,k} &= T_{\mathrm{NX},j,k} \\ &+ \frac{2\Delta t}{\mathrm{FX}}\left[\frac{\lambda_{\mathrm{m}}}{\Delta x_{\mathrm{m}}}(T_{\mathrm{NX}-1,j,k} - T_{\mathrm{NX},j,k}) + \frac{\lambda}{\Delta x}(T_{\mathrm{NX}+1,j,k} - T_{\mathrm{NX},j,k})\right] \\ &+ \frac{\mathrm{LX}\Delta t}{\mathrm{FX}\Delta y^2}[T_{\mathrm{NX},j-1,k} - 2T_{\mathrm{NX},j,k} + T_{\mathrm{NX},j+1,k}] \\ &+ \frac{\mathrm{LX}\Delta t}{\mathrm{FX}\Delta z^2}(T_{\mathrm{NX},j,k-1} - 2T_{\mathrm{NX},j,k} + T_{\mathrm{NX},j,k+1}) \\ &+ \frac{\Delta x}{\mathrm{FX}}(\rho\Delta Q)_{\mathrm{NX}+0.25,j,k} \end{aligned} \tag{10.11}$$

where FX may also depend on j and k, at each step, if ρ or C are not constant.

10.2.3 Edge and Corner Between Resin and Mould

For the edge and corner between the resin and mould, only the principle of calculation is given, as very long equations are obtained.

Edge (Fig. 10.4)

The edge parallel with the x axis is considered, with the small parallelepiped in the resin of dimensions Δx, $\frac{1}{2}\Delta y$ and $\frac{1}{2}\Delta z$ located next to this edge and the other two small parallelepipeds in the mould. The heat balance during the interval of time $[t, t + \Delta t]$ is evaluated within these small parallelepipeds by considering heat conduction along the x, y and z axes as well as heat evolved from the cure reaction in the resin, as shown in Fig. 10.4.

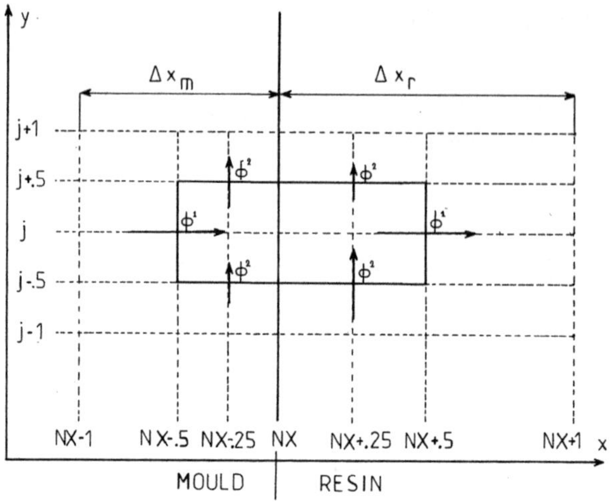

Fig. 10.3. Heat transfer at the resin–mould interface.

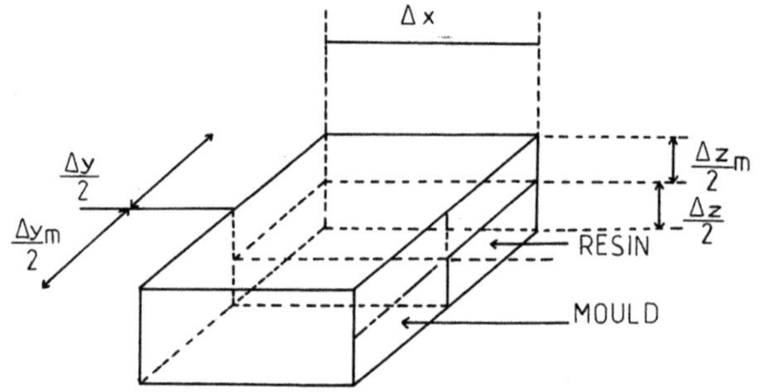

Fig. 10.4. Heat transfer at the edge of a resin and a mould.

Corner (Fig. 10.5)

The small parallelpiped in the resin of dimensions $\frac{1}{2}\Delta x$, $\frac{1}{2}\Delta y$ and $\frac{1}{2}\Delta z$ is considered as well as the other three small parallelepipeds in the mould (Fig. 10.5). The heat balance during the increment of time Δt is evaluated by considering heat conduction along the three axes, and heat evolved from the reaction.

10.2.4 Surface of the Mould or Resin in Contact with a Fluid (Fig. 10.6)

The small parallelepiped of dimensions $\frac{1}{2}\Delta x$, Δy and Δz next to the surface is considered. The rectangular cross-section perpendicular to the $0z$ axis is only drawn in Fig. 10.6. The heat balance during the increment of time Δt is evaluated

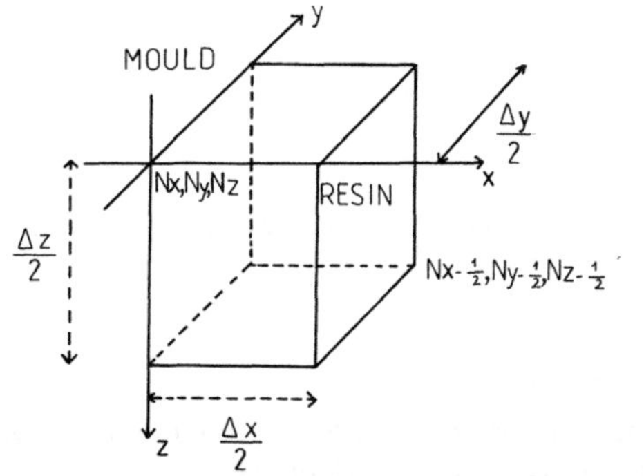

Fig. 10.5. Heat transfer at the corner of a resin and a mould.

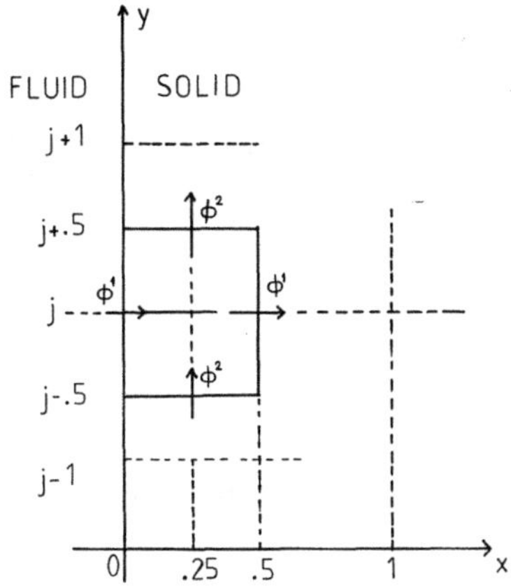

Fig. 10.6. Heat transfer at the surface of a parallelepipedic solid in contact with a surrounding fluid.

within this small parallelepiped by considering the heat by conduction through the solid (resin or mould), the heat by convection at the surface, and the heat evolved from cure reaction in the resin. For instance, the value of the integer i is zero on this surface.

$$\frac{\Delta x}{2}\,\Delta y \Delta z (\rho C)_{0.25,j,k}[\mathrm{TN}_{0.25,j,k} - T_{0.25,j,k}]$$

$$= \Delta y \Delta z[\phi^1_{0.5,j,k} - \phi^1_{0,j,k}]\Delta t + \frac{\Delta x}{2}\,\Delta y[\phi^3_{0.25,j,k-0.5} - \phi^3_{0.25,j,k+0.5}]\Delta t$$

$$+ \frac{\Delta x}{2}\,\Delta z[\phi^2_{0.5,j-0.5,k} - \phi^2_{0.5,j+0.5,k}]\Delta t + \frac{\Delta x \Delta y \Delta z}{2}\,(\rho \Delta Q)_{0.25,j,k} \qquad (10.12)$$

By making the simple assumption

$$\mathrm{TN}_{0.25,j,k} - T_{0.25,j,k} = \mathrm{TN}_{0,j,k} - T_{0,j,k} \qquad (10.13)$$

the new temperature on the surface after time Δt is thus obtained as follows:

$$\mathrm{TN}_{0,j,k} = T_{0,j,k} + \frac{2\Delta t}{\Delta x(\rho C)_{0.25,j,k}} [\phi^1_{0.5,j,k} - \phi^1_{0,j,k}]$$

$$+ \frac{\Delta t}{\Delta z(\rho C)_{0.25,j,k}} [\phi^3_{0.25,j,k-0.5} - \phi^3_{0.25,j,k+0.5}]$$

$$+ \frac{\Delta t}{\Delta y(\rho C)_{0.25,j,k}} [\phi^2_{0.25,j-0.5,k} - \phi^2_{0.25,j+0.5,k}] + \left(\frac{\Delta Q}{C}\right)_{0.25,j,k} \qquad (10.14)$$

In this equation, the various heat fluxes by conduction are evaluated as shown above. The heat flux by convection is given by

$$\phi^1_{0,j,k} = h(T_{0,j,k} - T_{\text{ext}}) \qquad (10.15)$$

where T_{ext} is the temperature of the surrounding atmosphere which is far enough away from the solid.

The heat evolved from the cure reaction can be approximated by the following relation:

$$\left(\frac{\Delta Q}{C}\right)_{0.25,j,k} = \frac{3}{4}\left(\frac{\Delta Q}{C}\right)_{0,j,k} + \frac{1}{4}\left(\frac{\Delta Q}{C}\right)_{1,j,k} \qquad (10.16)$$

10.3 Constant Thermal Parameters with Reaction

10.3.1 Within the Resin (Fig. 10.2)

From the heat balance evaluated during the increment of time Δt within the small parallelepiped of dimensions Δx, Δy and Δz, the new temperature after time Δt, $\mathrm{TN}_{i,j,k}$ can be expressed in terms of the previous temperatures obtained at the same place and size adjacent places.

The heat flux by conduction, with constant thermal conductivity, is given by

$$\phi^1_{i-0.5,j,k} = \lambda \frac{T_{i-1,j,k} - T_{i,j,k}}{\Delta x} \qquad (10.17)$$

According to Eq. (10.4′), the new temperature $\mathrm{TN}_{i,j,k}$ is as follows:

$$\mathrm{TN}_{i,j,k} = T_{i,j,k} + \frac{1}{\mathrm{MX}} [T_{i-1,j,k} - 2T_{i,j,k} + T_{i+1,j,k}]$$

$$+ \frac{1}{\mathrm{MY}} [T_{i,j-1,k} - 2T_{i,j,k} + T_{i,j+1,k}]$$

$$+ \frac{1}{\mathrm{MZ}} [T_{i,j,k-1} - 2T_{i,j,k} + T_{i,j,k+1}] + \frac{\Delta Q_{i,j,k}}{C} \qquad (10.18)$$

10.3.2 Surface of the Mould or Resin in Contact with a Fluid

The temperature on the surface of a mould or of a resin in contact with a fluid can be expressed in a rather simple equation when the thermal parameters are constant.

From the heat balance evaluated during the time increment Δt within the small parallelepiped of dimensions $\frac{1}{2}\Delta x$, Δy and Δz, located next to the surface (Fig. 10.6), the new temperature on the surface can be obtained as follows, the value of the integer i being zero on the surface:

$$\begin{aligned} \mathrm{TN}_{0,j,k} = T_{0,j,k} &+ \frac{2}{\mathrm{MX}}[T_{1,j,k} - T_{0,j,k}] \\ &- \frac{2\Delta th}{\Delta x \rho C}(T_{0,j,k} - T_{\text{ext}}) \\ &+ \frac{1}{\mathrm{MZ}}[T_{0,j,k-1} - 2T_{0,j,k} + T_{0,j,k+1}] \\ &+ \frac{1}{\mathrm{MY}}[T_{0,j-1,k} - 2T_{0,j,k} + T_{0,j+1,k}] \\ &+ \frac{1}{4C}[3\Delta Q_{0,j,k} + \Delta Q_{1,j,k}] \end{aligned} \tag{10.19}$$

by making the following assumption:

$$T_{0.25,j,k} = T_{0,j,k} \tag{10.20}$$

Of course, the calculation can also be made by using other assumptions, but the equations obtained are more complex.

In the case of a mould in contact with a fluid, the term ΔQ vanishes, by putting $k_0 = 0$.

10.3.3 Edge and Corner of the Mould or Resin in Contact with a Fluid

At the edge of a resin or a mould in contact with a fluid, the new temperature after time Δt is expressed in terms of the previous temperatures in various places. For instance, for the edge along the z axis, where the corresponding integers have the values $i = 0$ and $j = 0$, the following is obtained:

$$\begin{aligned} \mathrm{TN}_{0,0,k} = T_{0,0,k} &+ \frac{2}{\mathrm{MX}}[T_{1,0,k} - T_{0,0,k}] - \frac{2\Delta th}{\Delta x \rho C}[T_{0,0,k} - T_{\text{ext}}] \\ &+ \frac{2}{\mathrm{MY}}[T_{0,1,k} - T_{0,0,k}] - \frac{2\Delta th}{\Delta y \rho C}[T_{0,0,k} - T_{\text{ext}}] \\ &+ \frac{1}{\mathrm{MZ}}[T_{0,0,k+1} - 2T_{0,0,k} + T_{0,0,k-1}] \\ &+ \frac{1}{4C}[3\Delta Q_{0,0,k} + \Delta Q_{1,1,k}] \end{aligned} \tag{10.21}$$

At the corner, the following equation can be obtained, with the integers $i = 0$, $j = 0$ and $k = 0$:

$$\mathrm{TN}_{0,0,0} = T_{0,0,0} + \frac{2}{\mathrm{MX}}[T_{1,0,0} - T_{0,0,0}] - \frac{2\Delta th}{\Delta x\rho C}[T_{0,0,0} - T_{\mathrm{ext}}]$$

$$+ \frac{2}{\mathrm{MY}}[T_{0,1,0} - T_{0,0,0}] - \frac{2\Delta th}{\Delta y\rho C}[T_{0,0,0} - T_{\mathrm{ext}}]$$

$$+ \frac{2}{\mathrm{MZ}}[T_{0,0,1} - T_{0,0,0}] - \frac{2\Delta th}{\Delta z\rho C}[T_{0,0,0} - T_{\mathrm{ext}}]$$

$$+ \frac{1}{4C}[3\Delta Q_{0,0,0} + \Delta Q_{1,1,1}] \tag{10.22}$$

by making various approximations

$$T_{0.25,0.25,0.25} = T_{0,0,0} \tag{10.23}$$

$$T_{0.25,0.25,k} = T_{0,0,k} \tag{10.23'}$$

and analogous relations.

10.3.4 Stability of Calculations. Symmetry

For each possible system of integers i, j, k, in the equations which give the new temperature $\mathrm{TN}_{i,j,k}$, the whole coefficient of the corresponding temperature $T_{i,j,k}$ must be positive. Thus Eqs (10.18), (10.19), (10.21) and (10.22) give the conditions

$$\frac{2}{\mathrm{MX}} + \frac{2}{\mathrm{MY}} + \frac{2}{\mathrm{MZ}} < 1 \tag{10.24}$$

$$\frac{2}{\mathrm{MX}} + \frac{2h\Delta t}{\Delta x\rho C} + \frac{2}{\mathrm{MY}} + \frac{2}{\mathrm{MZ}} < 1 \text{ and the same conditions for } \Delta y \text{ and } \Delta z \tag{10.25}$$

$$\frac{2}{\mathrm{MX}} + \frac{2h\Delta t}{\Delta x\rho C} + \frac{2}{\mathrm{MY}} + \frac{2h\Delta t}{\Delta y\rho C} + \frac{2}{\mathrm{MZ}} < 1 \text{ and the same conditions} \tag{10.26}$$

$$\frac{2}{\mathrm{MX}} + \frac{2}{\mathrm{MY}} + \frac{2}{\mathrm{MZ}} + \frac{2h\Delta t}{\rho C}\left(\frac{1}{\Delta x} + \frac{1}{\Delta y} + \frac{1}{\Delta z}\right) < 1 \tag{10.27}$$

If the temperature at the surface is given as a function of time and position, only the condition (10.24) intervenes, and it gives

$$\left(\frac{\alpha}{(\Delta x)^2} + \frac{\alpha}{(\Delta y)^2} + \frac{\alpha}{(\Delta z)^2}\right)\Delta t < \frac{1}{2} \quad \text{with} \quad \alpha = \frac{\lambda}{\rho C} \tag{10.24'}$$

If there is at the surface a finite coefficient of heat convection, the condition (10.27) is sufficient and gives

$$\left[\frac{\alpha}{(\Delta x)^2} + \frac{\alpha}{(\Delta y)^2} + \frac{\alpha}{(\Delta z)^2} + \frac{h}{\rho C}\left(\frac{1}{\Delta x} + \frac{1}{\Delta y} + \frac{1}{\Delta z}\right)\right]\Delta t < \frac{1}{2} \tag{10.27'}$$

Symmetry

If the initial and boundary conditions are symmetric with respect to the three planes $i = N1$, $j = N2$, $k = N3$, the time of calculations may be reduced by taking at each step

$$\text{for} \quad 0 \leqslant j \leqslant N2 \qquad 0 \leqslant k \leqslant N3 \qquad T(N1+1, j, k) = T(N1-1, j, k) \tag{10.28}$$

$$\text{for} \quad 0 \leqslant k \leqslant N3 \qquad 0 \leqslant i \leqslant N1 \qquad T(i, N2+1, k) = T(i, N2-1, k) \tag{10.29}$$

$$\text{for} \quad 0 \leqslant i \leqslant N1 \qquad 0 \leqslant j \leqslant N2 \qquad T(i, j, N3+1) = T(i, j, N3-1) \tag{10.30}$$

Chapter 11

Determination of the Kinetics of Cure and Its Problems

11.1 Techniques Used for Measuring the Change in Properties During Cure

For many years, various techniques have been developed and tested for the cure of rubber. A comparison between these techniques used for rubber and the techniques which have been developed more recently for studying the cure of thermosets may be of interest, because the problems are very similar.

11.1.1 Cure of Rubbers

Tests for the rate of cure and scorch during the cure process should be distinguished from tests for determining the state of cure on vulcanised material. The second type measures the physical or mechanical properties of test pieces vulcanised for various times: tensile properties, dynamic stress and strain properties, and swelling being the properties commonly chosen.

Tests for Measuring the Physical Properties of Vulcanised Rubbers

Among the various tests used for measuring the quality of the final rubber sample [1], some of them are especially worth noting: hardness, tensile stress–strain properties, dynamic stress–strain properties and swelling in a solvent.

Hardness. Hardness measurement is the simplest way of measuring the elastic modulus of a rubber by determining its resistance to a rigid indenter to which a force is applied. Indentation involves deformation not only in compression but also in tension and shear. But in the case of a perfectly elastic rubber, the three moduli characterising these deformations are closely related, and hardness can be regarded as depending only on Young's modulus.

Static Tensile Stress–Strain Properties. After indentation hardness, the most common kind of stress–strain determination is that made in tension. The main characteristic of rubber is its ability to stretch to several times its original length, in spite of the fact that rubber is very often used in compression or shear. Since the early days of the rubber industry, tensile stress–strain properties have been considered as a general guide to the quality of a rubber, because it is sensitive not only to the state of cure of the rubber but also to the presence of a filler and plasticiser content. Generally, dumb-bell shaped test pieces are strained at a constant rate, and force and corresponding extension are simultaneously recorded. The force readings are expressed as stresses by reference to the original cross-section area of the test piece, but corrections can be made in order to obtain the actual cross-section area during the stretching. As shown in Fig. 11.1, where a typical tensile stress–strain curve is shown, there is no linear elastic portion as is usual with metals. Instead of a modulus, the stress at various percentage elongations, usually at 100% and 200%, is quoted.

Dynamic Stress and Strain Properties. The rubber sample is subjected to cyclic deformation and the stress and strain are monitored. Besides free vibration in which the test piece is set into oscillation and the amplitude decay due to damping in the sample is followed, the most interesting method is with forced vibration in which the oscillation is maintained by an apparatus. These forced vibration machines may operate either at resonance or away from resonance. As rubber is not elastic only, but rather viscoelastic, its response to dynamic stress is thus a combination of an elastic and a viscous response, and energy is absorbed in each cycle. With a sinusoidal strain

$$\gamma = \gamma_0 \sin \omega t \tag{11.1}$$

the stress is not in phase with the strain because of the viscoelasticity of the rubber, and the deformation lags behind the force by the angle δ.

$$\tau = \tau_0 \sin (\omega t + \delta) \tag{11.2}$$

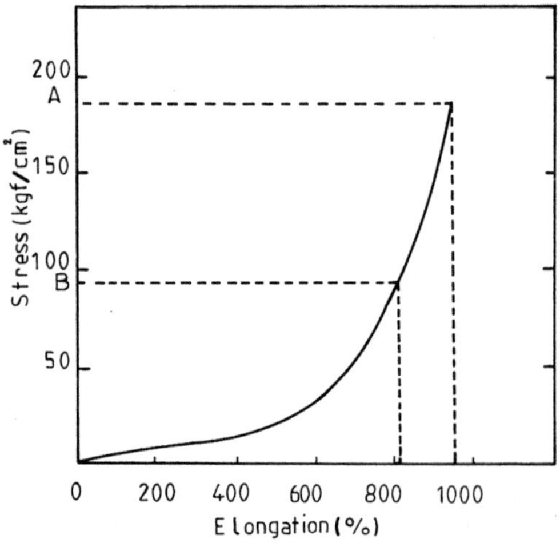

Fig. 11.1. Stress–strain curve for vulcanised rubber.

where γ is the strain, γ_0 the strain amplitude, ω is the angular frequency, τ is the stress and τ_0 the stress amplitude.

The stress is considered as a complex vector with two components: one in phase with the displacement (τ'), and one 90° out of phase (τ'') (Fig. 11.2). The loss tangent (or loss factor) is given by

$$\tan\delta = \frac{\tau''}{\tau'} \tag{11.3}$$

Dynamic properties are important in various industrial applications including springs and dampers, and they are more useful than the properties measured with the simpler static test. The results of dynamic tests are dependent on the test conditions: shape of the test piece, mode of deformation, strain amplitude and history, frequency and temperature. Moreover, the repetition of strain cycles may cause changes in the dynamic properties for various reasons, one being that a stress softening occurs at the beginning of the test, and the dynamic properties reach a steady equilibrium level after a few cycles. Another reason is the continuous change in property levels due to the generation of heat within the rubber sample.

Dynamic properties are dependent on both temperature and frequency. The effect of temperature on the complex modulus E^x and loss factor $\tan\delta$ is generally described by curves shown in Fig. 11.3. Increasing frequency shifts the curves to the right.

The Young modulus in tension or compression E^x is given by

$$E^x = \sqrt{E'^2 + E''^2} \tag{11.4}$$

E' being the in-phase component, and E'' the out-of phase component. Thus

$$E' = T_0 \cos\delta, \qquad E'' = T_0 \sin\delta, \qquad E^x = T_0$$

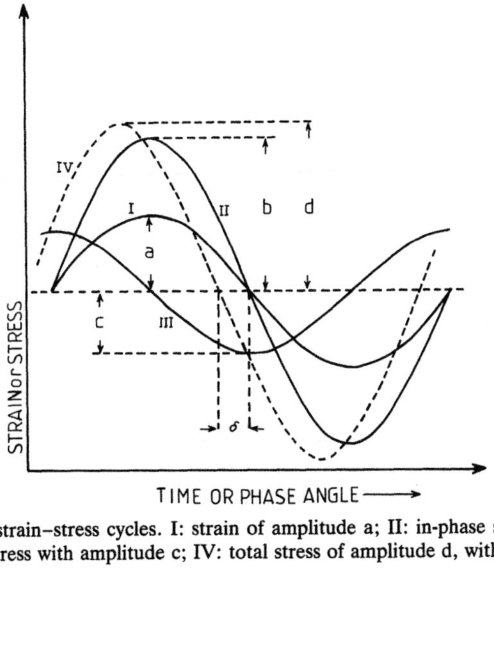

Fig. 11.2. Sinusoidal strain–stress cycles. I: strain of amplitude a; II: in-phase stress with amplitude b; III: out-of-phase stress with amplitude c; IV: total stress of amplitude d, with the phase angle δ.

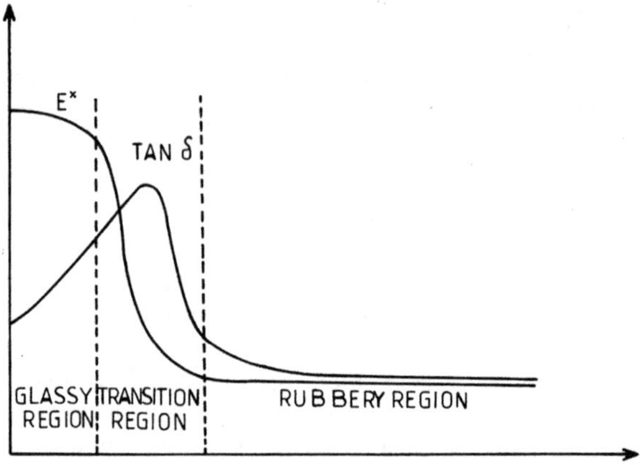

Fig. 11.3. Effect of temperature on the complex modulus E^x in tension–compression and the loss factor tan δ.

Swelling in a Liquid. Tests in which rubber is immersed in liquids are called swelling tests, because the liquid may enter the rubber with a resulting change in volume of the rubber. Volume change is a very good measure of the resistance of a rubber sample to a given liquid. A high degree of swelling indicates that the rubber is not suitable for use in contact with this liquid, because the action of a liquid on rubber results not only in the absorption of liquid by the rubber, but also in the extraction of soluble constituents from the rubber. Moreover, the degree of swelling can be related to the state of cure of the rubber, and especially to the crosslink density.

The test piece of rubber is immersed in the test liquid for the chosen time at the chosen temperature, following the International Standard Methods (e.g. ISO 1817). The preferred method for volume change is the gravimetric method, and the test piece is weighed before and after immersion.

In fact, the degree of volume change with time generally follows the shape illustrated in Fig. 11.4. It is easy to see that a single time of exposure in the liquid can give only a rough or even a false idea of the process of liquid absorption. As shown in earlier studies [2, 3], the process of liquid absorption is controlled by diffusion with constant or concentration dependent diffusivity, and two parameters are of interest, the diffusivity and maximum amount of liquid absorbed after infinite time. The following step of desorption for the presaturated sample is controlled by diffusion of the liquid within the solid and evaporation at the surface, and two additional parameters are of importance, the rate of evaporation and the concentration of liquid which is in equilibrium with the surrounding atmosphere.

Methods for Measuring the Physical Properties of Rubber During Cure

Various methods have been developed in order to follow a physical property of the rubber during its cure. Some of the most common techniques are briefly described below.

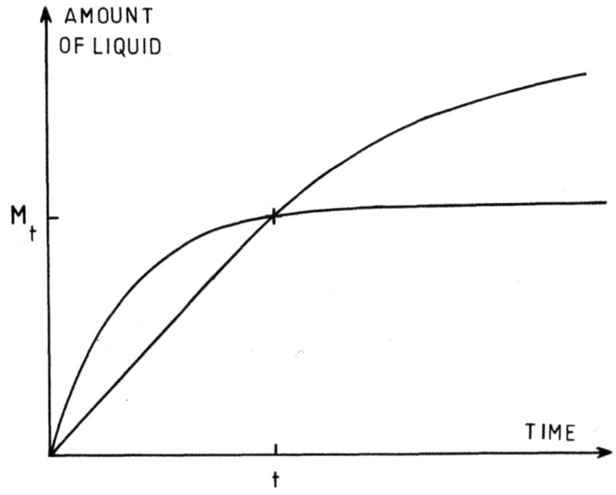

Fig. 11.4. Kinetics of absorption of liquid which can be supposed with a simple test and one reading of absorption.

Mooney Viscometer. This consists of a rotor and stator made of a cavity wall in which the rubber is placed. Anti-slip grooves are on both the rotor and cavity walls. The rotor is driven at constant speed, and the torque is measured. When the compounded rubber is subjected to plasticity measurement at a high enough temperature and for a long time, the cure takes place, and the plasticity test becomes also a test for scorch or rate of cure. The Mooney viscometer is thus also used to measure scorch, i.e. the onset of vulcanisation, and even the rate of cure. The most important change in a rubber when vulcanisation starts is an increase in stiffness and an increase in the elastic component of its viscoelastic deformation.

Wallace–Shawbury Curometer. In this curometer, a small paddle is embedded in the rubber placed in a heated cavity, and is reciprocated. In order to measure the change in stiffness, either the change in amplitude of oscillation at constant force or the change in force necessary to produce constant amplitude is monitored.

Oscillating Disc Rheometer. In the oscillating disc type, a bi-conical disc is embedded in the rubber located in a closed cavity in about the same way as for the Mooney. The disc is oscillated through constant angular displacement and the torque required is monitored. Various results such as minimum torque, cure induction time, optimum cure time for 90%, and maximum torque, are given with the help of a microprocessor.

Moving Die Rheometer. This rheometer was introduced in 1985. It is a rotorless curometer in which one half of the die enclosing the test rubber oscillates or reciprocates. There is no paddle or disc within the test rubber. The two dies are heated directly, with lower die oscillated and reaction torque/pressure transducer positioned above the upper die.

Conclusions on Results Given by Rheometers and Curometers. Comparisons between cure times predicted by the oscillating disc rheometer were made [1, p. 96], showing that temperature distribution, and hence state of cure, is not uniform throughout the test rubber.

Moreover, the moving die rheometer gives shorter times of cure than the oscillating disc rheometer due to faster heat transfer to the sample, and different torque values due to the die design [4] (Fig. 11.5).

Temperature Measurements During Cure [5]. The sample with a thickness of 0.5 mm is placed in an isothermally heated press for a given time, and is then quickly removed, quench cooled in water. The degree of cure is measured by swelling in toluene [5]. The process is repeated for different times to build up the complete cure curve at the same temperature. The same process is then repeated at a different temperature. It takes a few weeks to determine a complete family of cure–time curves for a rubber by using this technique. Another disadvantage with this method is that the scatter in the data is too high, especially at low levels of cure. Measurements are also made on a Monsanto rheometer and a Wallace–Shawbury curometer. As shown in Fig. 11.6, both these instruments overestimate the isothermal cure times at 200 °C and in fact at all temperatures. The rheometer overestimates the cure time because the rotor acts as a heat sink; the curometer overestimates the cure time because of a lack of pressure.

A mathematical model has been developed [5] for predicting cure profiles. The model is based on the whole of each isothermal cure time curve, and not just a single point. The temperature history at the centre can be calculated.

An effective method of calculating the time dependent temperature distribution has also been described elsewhere [6], by considering heat conduction within the rubber and internal heat due to the vulcanisation reaction. Although no simple kinetic equation is obeyed during the entire vulcanisation period, a first-order expression describes the reaction rate, i.e., the rate is proportional to the

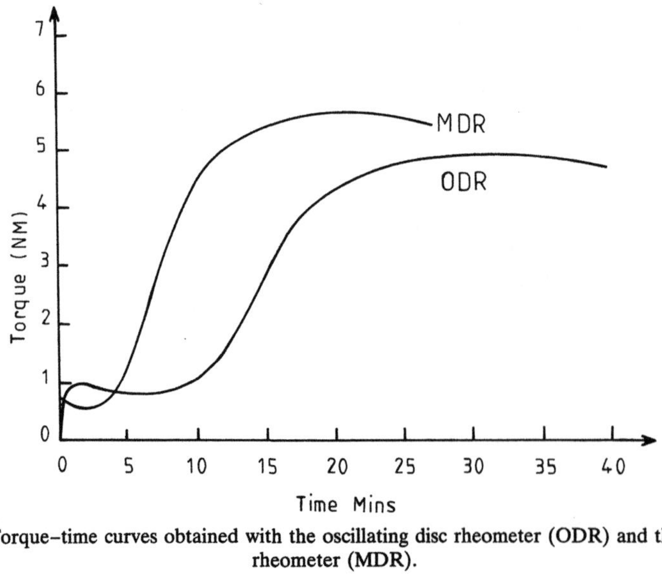

Fig. 11.5. Torque–time curves obtained with the oscillating disc rheometer (ODR) and the moving die rheometer (MDR).

remaining sulphur. The state of cure is calculated from the temperature history at any location

$$\mathrm{SOC} = \int_0^t r(T)\,dt \qquad (11.5)$$

where $r(T)$ expresses the instantaneous rate of vulcanisation, which is referred to the rate of a reference temperature of 149 °C. A temperature coefficient of vulcanisation of 1.85, is also chosen meaning that the reaction rate increases by a factor of 1.85 per 10 °C.

In another study [7], the kinetic parameters of the overall reaction of rubber vulcanisation by sulphur are determined by isothermal calorimetry. Numerical models are built and tested for calculating the profiles of temperature and state of cure developed within the rubber mass by simulating the injection moulding process [8]. The rate of heat evolved from the cure reaction is described by a first-order reaction with a single activation energy, and all parameters are determined by isothermal calorimetry. The main assumption which is made for calculating the state of cure is that the partial enthalpy evolved up to time t, as a fraction of the total enthalpy evolved, represents the conversion up to time t as well as the state of cure at time t. The validity of the model is determined by comparing the temperature–time histories at the midplane of the rubber sheet obtained by experiments and calculation.

11.1.2 Cure of Thermosetting Resins

Several techniques have been developed recently in order to determine the physical properties of the resin after cure, and some of them are also able to follow the process of cure.

Thermomechanical Analyser (TMA)

The most common is the thermomechanical analyser (developed by various firms). This apparatus is essentially made up of a furnace (or a cryostat) and of a

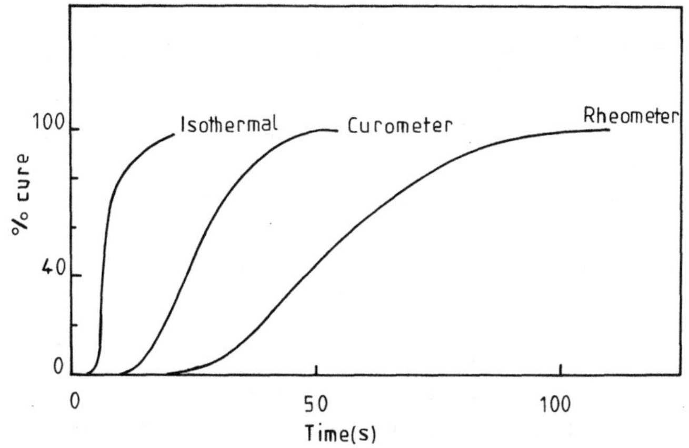

Fig. 11.6. Comparison of isothermal, curometer and rheometer results at 200 °C for cure of rubber.

transducer for measuring the variations in length of the sample. The transducer provides continuous measurement of the probe's position as it rests on the sample. This technique is used to measure the change in dimension of the sample material as a function of temperature and stress. It provides a sensitive measure of temperature dependent properties, which are extremely important in practice. Using thermomechanical analysis, the degree of crosslinking in a resin can be controlled through detecting the glass transition. Moreover, the coefficient of expansion of the pre-impregnated material can be measured before and after the glass transition temperature T_g.

Applying strain to a sample (compression, penetration, traction, flexion) can modify its behaviour. Thermomechanical analysers are able to study how material samples behave under strain.

The value of the temperature of glass transition T_g can also be determined by using calorimetry techniques working in scanning mode.

The thermomechanical system can thus measure the static properties of a material, and its changes with temperature.

Dynamic Mechanical Analysis (DMA)

Dynamic mechanical analysis is the measurement of the mechanical response of a material as it is deformed under periodic stress. It provides users with the information necessary to predict the performance of a material over a wide range of conditions. Measurements of primary interest include modulus (in phase and out of phase), $\tan \delta$, compliance, stress relaxation and creep. These properties characterise the viscoelastic performance of the material as a function of time, temperature and frequency.

Generally, the sample is clamped between two parallel arms, and these arms are held by low-stiffness spring pivots, which allow movement. An electromagnetic motor attached to one arm drives the arms and sample at an amplitude and frequency selected by the operator, and displacement and frequency are measured by a linear variable differential transformer.

Four modes of operation can be used, depending on the apparatus:

With a fixed frequency, a sinusoidal applied stress forces the sample to undergo oscillatory motion at a fixed deformation. When the sample is viscoelastic, maximum strain does not occur at the same time as maximum stress and the phase lag is defined by phase angle δ.

Working at resonant frequency provides higher damping sensitivity than fixed frequency, making possible detection of subtle transitions in a material.

The stress relaxation profile of a viscoelastic material can be determined by measuring stress decay at a constant sample deformation for selected times and temperatures.

For creep measurement, a fixed load stress is applied to the sample and the resulting strain deformation is measured.

When the apparatus does not require a clamp, the sample clamping problems are eliminated. Materials such as rubber, which get very soft over their entire temperature range, can be studied. Samples which transform from a viscous liquid to a solid, such as uncured resins, can be easily studied during the entire curing cycle.

Apparatus Measuring the Resistance to Torsion

With this apparatus, the sample is heated by thermal convection through air with the help of an oven kept at constant temperature. The sample in the shape of a strip is held with two clamps. One of these clamps rotates alternatively through a given small angle, and the resistance to torsion is measured with the other clamp.

As it is possible to make temperature measurements not only in the surrounding atmosphere in the oven and also in the sample itself during the process, the mechanical property, i.e., essentially the resistance to torsion, is correlated with the value of the state of cure obtained by considering heat transferred by convection and conduction as well as internal heat generated by the cure reaction [9]. This correlation is described in Chap. 17.

Temperature Measurement During Cure

Various studies on the cure of different resins have been made by monitoring the temperature. Thermosets are generally polymerised and processed in a single operation which involves the irreversible transformation of a low molecular weight resin into a network polymer. The curing reaction of most thermosets is very exothermic. Since polymers exhibit low thermal conductivity, the temperature and rate of reaction can vary considerably within the curing mass, particularly for thick castings. Therefore, the degree of cure as well as the temperature is a function of position and time.

Of course, it is possible to measure the temperature at various places. However, in almost all cases, the temperature is measured at the midplane of a sheet [9–13], or along the longitudinal axis of a long cylinder [14, 15]. These positions for the sheet and cylinder are selected for various reasons:

1. It is easier to place a thermocouple at the midplane of a thin sheet or at the centre of a cylinder than elsewhere, without modifying the shape of the sample.
2. This position represents a plane of symmetry for the sheet and an axis of symmetry for the cylinder. As a result, the first derivative of temperature with respect to the abscissa is always equal to zero. A slight change in position of the thermocouple around the right position may thus have a smaller effect than elsewhere in the solid.
3. At this position, the increase in temperature due to the exothermic reaction is at its highest, and a more sensitive measurement is obtained.

These temperature measurements are generally made on samples with a simple shape, in order to test the validity of a numerical model describing the process by considering heat transfer and heat evolved from the cure reactions [9–15].

Various examples of this technique are given in Chap. 17.

Swelling of the Cured Resin

Some studies have been made on the behaviour of cured resins towards various kinds of liquids. As cured resins are often used for high performance coatings, it is of interest to know the resistance of the coating to given liquids. The problem is rather more complex for cured resins than for elastomers, because of the glassy state of the resin at the temperature at which it is used. In a few cases, for instance when the liquid is water and the selected temperature is around 80°–100 °C, the process of liquid transport within the resin is controlled by transient diffusion with a constant diffusivity. The following stage of desorption of the water which has entered the resin is more complex, and two characteristics are worth noting:

The stage of desorption is controlled by diffusion within the resin and evaporation of the liquid from the surface.

Only a part of the water located in the resin evaporates at 20 °C, and after drying steps at higher temperatures, e.g. 80° and 100 °C, complete dryness of the coating is not achieved [16].

In another case, when the liquid is toluene, the process of liquid absorption is particular, with a constant rate of absorption until the resin dissolves. This process is under investigation [17].

11.1.3 Comparison Between the Cure of Rubbers and Thermosets

Two main differences between the behaviour of rubbers and thermosets during their cure are worth noting:

1. While viscous rubber is transformed into a viscoelastic material during cure, remaining soft, viscous liquid resin turns into a solid material.
2. Another difference of fundamental importance is the exothermic nature of the curing reaction. It is rather low for rubber, being proportional to the sulphur percentage. For thermosets, the cure enthalpy is very high, in the 40–100 cal g^{-1} range, depending on the type of the resin and especially on the amount of filler and other inert additives.

Following these main characteristics of the cure of rubbers and thermosets, it is easy to understand that the techniques used for studying cure can be of the same basic nature, with some differences.

Thermomechanical Properties

The thermomechanical properties are more widely used for the cure of rubber than for thermosets. These properties are easy to follow for rubber, either under static or dynamic conditions. The conditions are more difficult for the thermosets, as the viscous liquid resin turns into a solid hard material during cure.

Another fact is worth pointing out: rubber has been vulcanised throughout the past hundred years by rule of thumb. Then the techniques based on the change of mechanical properties, and especially the viscosity and viscoelasticity, have been widely developed over many decades. Thermosets are quite new, and the techniques used have only been developed in recent years.

Calorimetry Studies

As the curé enthalpy is rather low for rubber, it is more difficult to obtain the kinetics of heat evolved from the cure of this material, because of a lack of sensitivity. This drawback does not appear with thermosets, as the enthalpy of cure is very high. Sometimes, this high exothermic reaction is responsible for a large increase in temperature in the sample. This point is specifically considered in Chaps 12 and 13.

Another fact worth mentioning is that uncured rubbers and thermosets result from a mixture of various components. Some problems arise when taking such a small sample from the batch in that it may be not representative of the homogeneity of the batch. This problem is examined in Sect. 11.2.3.

Swelling Properties

As rubbers are very often in contact with various liquids, their swelling properties have been widely studied in spite of the facts that the problems are not so simple [1–3]. Only a few studies have been made for thermosets, and the problems are very complex because thermosets initially in glassy state may become more and more elastomeric as the liquid penetrates the solid [16, 17].

11.2 Calorimetry, Its Principles and Problems

In calorimetry techniques, enthalpy changes accompanying a physical or chemical event in the sample are monitored either as a function of temperature or time. The result is a profile of the rate of enthalpy change, either as a function of temperature as the sample is heated at a known linear rate in differential scanning calorimetry, or as a function of time, the calorimeter being held at constant temperature in isothermal differential calorimetry.

11.2.1 Isothermal Differential Calorimetry (DC)

For constant temperature operation in the DC mode, the calorimeter is stabilised at the desired temperature, the sample, previously at room temperature, is introduced, and time base recording started. The sample is then strongly heated, and the exothermic reaction takes place. At a selected time, the sample is removed from the calorimeter and cooled to room temperature. The cured sample is then reintroduced into the calorimeter, kept at the same temperature, in order to obtain the reference baseline.

Various problems or even drawbacks appear in the DC technique.

1. Two experimental limitations on the temperature range over which meaningful data can be obtained result from the principle of the technique. On the one hand, the lowest temperature for cure studies is determined by the calorimeter sensitivity. If the rate of cure is very low, the rate of enthalpy change is so

small that it cannot be measured. On the other hand, the highest temperature at which measurements can be made is determined by the induction period for the cure reactions to initiate. If the induction period is shorter than the time required for the calorimeter to establish thermal equilibrium after the sample is introduced, this first part of the cure exotherm is not obtained. In general, for thermosetting resins as well as for accelerated sulphur cure systems, a temperature window of around 20°–30 °C is found over which meaningful data can be obtained.

2. In the DC experiment, in principle, the enthalpy of all reactions occurring at a constant applied temperature up to a chosen time is determined. But, in fact, the applied temperature is the set calorimeter temperature and not the true sample temperature during cure, for various reasons. Because of the rather low rate of heat transfer through the sample, steep gradients of temperature are developed within the resin when the sample is introduced into the heated calorimeter. Because of local heating arising from the exothermic reactions, other gradients of temperature are developed within the sample with a higher temperature at the centre. Because of these exothermic reactions, the temperature in the sample is higher than that of the calorimeter (this fact appears during the exothermic part of the curve, when heat flux is emitted from the sample to the calorimeter sensors). Briefly, it can be said that the sample is only twice at a constant temperature: when the heat flux is zero at the end of the heating stage following the introduction of the sample, and at the end of the process when the cure approaches completion (Fig. 11.7).
3. Another drawback appears for the value of the cure enthalpy which is measured. The enthalpy associated with the overall exotherm is determined by measuring the area under the exotherm after the selection of a baseline. This baseline can be obtained in two ways: by considering only the exothermic part, when the heat flux is emitted from the sample (Fig. 11.8), or by considering

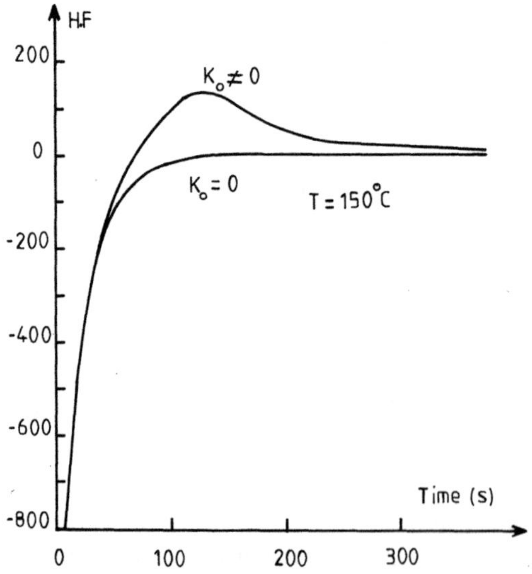

Fig. 11.7. Heat flux–time curves obtained with cure ($k_0 \neq 0$) and without cure ($k_0 = 0$) in DC for epoxy resin at 150 °C, registered with a computer.

the reference baseline obtained when the precured sample at room temperature is reintroduced into the calorimeter (Fig. 11.7). Choosing the second reference baseline is a better way of working, but it necessitates coupling the calorimeter with a computer capable of quickly registering a high change in the heat flux [18]. This is the reason why the value of enthalpy of cure may differ with the temperatures at which they have been measured.

4. In order to facilitate the operations of introduction and removal of the sample in the calorimeter, the contact between the sample and calorimeter thus becomes imperfect, introducing the presence of a thin layer of air between the sample and calorimeter. This presence of the air layer is responsible for a change in the heat flux response which modifies the kinetics.

11.2.2 Differential Scanning Calorimetry (DSC)

In the DSC experiment, starting with fresh uncured resin, the sample is scanned at a selected heating rate from room temperature up to a temperature at which the reaction isotherm is complete. The cure exotherm is usually observed in the 90–140 °C range, depending on the type of resin. After the resin has been given a precure, it is quench-cooled and then scanned under conditions identical to the initial scan, in order to give the reference baseline. In general, the rescan curve can be superimposed on the initial curve to give the baseline which would be observed in the absence of cure (Fig. 11.9). A difference between these curves can express the difference in the temperature dependence of the specific heat of the cured product as formed after the initial DSC experiment with the original resin.

Of course, the parameter of concern in DSC experiments is the heating rate selected for the oven. As it takes some time for heat to be transferred by conduction through the holder and sample, there is always a difference in the temperature of the oven and sample. As a result, an endothermic heat flux is observed, clearly shown in Fig. 11.9 between 40 and 80 °C. Moreover, a high

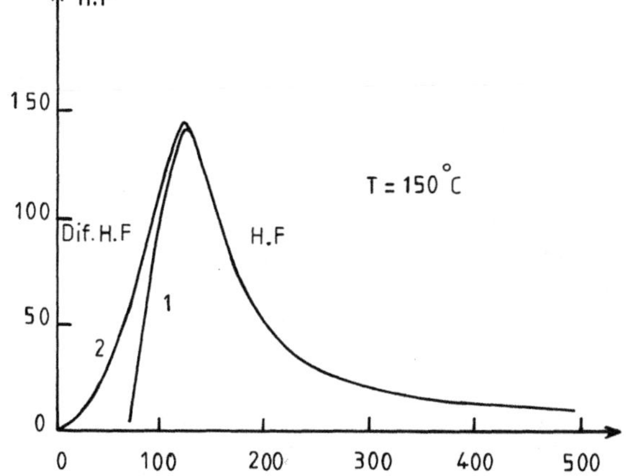

Fig. 11.8. Exothermic part of heat flux–time curves for epoxy resin, at 150 °C (1). Difference in heat flux with cure reaction and without cure (2).

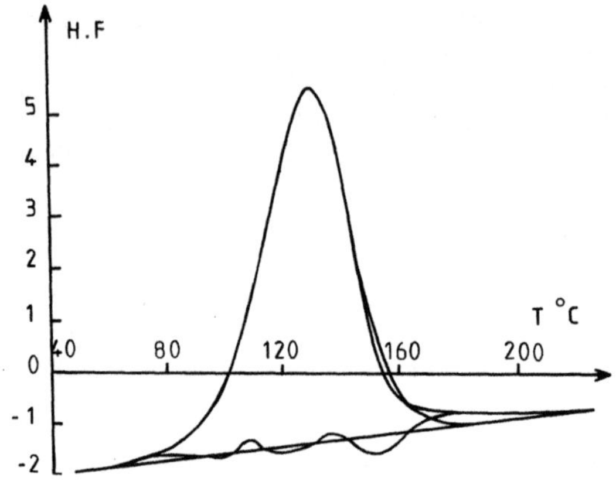

Fig. 11.9. Heat flux–temperature histories for epoxy resin in DSC, with fresh uncured resin and with precured resin.

heating rate is responsible for the development of steep gradients of temperature within the sample when no reaction occurs, with a higher temperature on the surface. Other gradients of temperature are expanded during the reaction, their values depending on the enthalpy of cure and on the temperature dependence of the rate of cure. For instance, with a high cure enthalpy, the temperature at the middle of the sample may become higher than in other places when reaction takes place, and especially at the maximum of the heat flux–temperature curve. With a low value of the cure enthalpy, with a rubber for instance, the temperature at the centre of the sample may be always lower than on the surface.

The size of the sample is also of great interest, especially in the case of mixtures.

How to select the best value for the heating rate is an important question. This is specifically studied in Chap. 13. However, it can be said now that the heating rate must be chosen by considering the size of the sample and the cure enthalpy of the material, with the following statement: the higher the enthalpy of cure and the size of the sample, the lower the heating rate in DSC.

11.2.3 Size of the Sample in Calorimetry; Good Contact Between Sample and Holder

In sample preparation it is imperative that a good contact is made between the sample and the sample holder, in order to improve the heat transfer efficiency at this interface. Elastomers have nerve, and in order to produce a flat surface they must be cut without a deforming force. The problem is even more complex with thermosets and composites. But an obvious statement may be made: the smaller the sample, the more difficult it is to obtain a good contact between the sample and the holder.

A good contact must also be obtained between the sample holder and calorimeter. As specifically described in Chap. 12, the presence of a thin air layer is of great importance in DC experiments. This effect is less important in DSC as shown in Chap. 13.

Sample Size and Consequences

The small sample size used in calorimetry requires that all ingredients in the sample be very well dispersed. Dispersion problems do arise on occasion and are generally observed by a lack of reproducibility in the DSC or DC curves. The question which always arises is how representative is a very small sample (5 mg) taken from a large batch (from 100–1000 g as mixed with laboratory equipment) of the homogeneity of the batch. In the case of a rubber, occasions have arisen when sulphur is poorly dispersed either because of large particles of sulphur failing to be broken or because of poor milling operations [19]. It was even said that, under these conditions, a 5 mg sample could be virtually all sulphur! As for thermosets, the problems of mixing are even more sensitive, for the following reasons:

1. In the case of a binary system obtained by mixing the two components, it is well known that mixing becomes more and more difficult when the size of the sample is reduced. The distribution of the two components in the calorimetry sample is of great importance, as found for epoxy resin, [20] either for the enthalpy and kinetic parameters or for the mechanical properties of the final material.
2. In the case of a composite with a viscous liquid resin and solid filler or additives, the problems are also difficult. Very often, fibres are so long that it is not possible to insert them into a small holder.
3. In the case of woven fibres impregnated with resin, it is even more difficult to have a sample located in the holder.
4. The quality of contact between the sample holder and calorimeter is better achieved when the sample is larger.

As shown in these examples, there are some advantages in using a large sample, e.g. 100–200 mg in a cylinder with a diameter of 5 mm, or even 2–10 g in a cylinder with a radius of more than 1 cm. In our calorimetric studies, cylindrical samples of 100–200 mg with a diameter of 5 mm for the resin are used (Chaps 12 and 13).

Resulting from this increase in size of the sample, some inconveniences occur. The main drawback is due to the rate of heat transfer by conduction through the sample which is rather low with thermosets. For this reason, lower values have to be used for the heating rate in DSC. As shown in Chap. 13, the heating rate must be inversely proportional to the square of the main dimension of the sample, e.g., radius for a cylinder with radial heat, thickness of a plane sheet.

11.2.4 Making a Choice Between DC and DSC

It is perhaps too early to make a choice between DC and DSC, as the following two chapters are devoted to these techniques. However, it may be of interest at this time to point out in a few lines the results already obtained in Sect. 11.2.

Many people prefer the DC technique, thinking this is the easier technique, capable of working under isothermal conditions. However, three experiments within a narrow temperature range are needed. Moreover, as has been shown, the temperature in the sample is not always constant, and it is not true to speak of a process run under isothermal conditions.

The DSC technique already appears as perhaps the best way of working. As the problem of homogeneity of the sample necessitates a rather large sample, the selection of the best value for the heating rate is of great concern. Values ranging from 0.5° to 2 °C per minute can be used for a cylindrical sample with a diameter of 5 mm.

References

1. Brown RP. Physical testing of rubbers, Applied Science Publishers, London, 1979
2. Vergnaud JM. Liquid transport process in polymeric materials. Modeling and industrial application, Prentice Hall, Englewood Cliffs NJ, 1990
3. Khatir Y, Bouzon J, Vergnaud JM. Liquid sorption by rubber sheets and evaporation; models and experiments. Polym Test 1986; 6:253
4. Di Mauro P. Recent developments in testing, analysis and equipment for the rubber industry. Int Rubber Conf, Kyoto, 1985, p 50
5. Hands D, Horsfall F. A new method for simulating industrial cure process. Rapra Report 44, 1980
6. Prentice GA, Williams MC. Numerical evaluation of the state of cure in a vulcanizing rubber article. Rubber Chem Technol 1980; 53:1023
7. Accetta A, Le Parlouer P, Vergnaud JM. Kinetic parameters of the overall reaction of scrap rubber vulcanization by 2% sulfur. Thermochim Acta 1982; 59:149
8. Sadr A, Granger R, Vergnaud JM. Calculation of profiles of temperature and state of cure developed within the rubber mass in injection molding systems. J Appl Polym Sci 1984; 29:955
9. El Brouzi A, Dursapt M, Vergnaud JM. Simultaneous determination of torque and state of cure of a thermoset composite by experiment and modeling of the process. Plast Rubber Comp Process Applic 1991; 15:13
10. Pusiatcioglu SY, Hassler JC, Fricke AL, McGee HA. Effect of the temperature gradients of cure and stress gradients in thick thermoset castings. J Appl Polym Sci 1980; 25:381
11. Nixon JA, Hutchinson JM. Analysis of the cure of sheet moulding compound. I. Development of the model. Plast Rubber Process Applic 1985; 5:349
12. Hutchinson JM, Nixon JA. Analysis of the cure of sheet moulding compound. II. Application of the model to various moulding geometries. Plast Rubber Process Applic 1985; 5:359
13. Khouider A, Vergnaud JM. Effect of temperature of motionless air on the cure of vulcanizates after removal from the mold. J Appl Polym Sci 1986; 32:5301
14. Chater M, Vergnaud JM. Effect of kinetic and thermal parameters on the process of cure of epoxy resin in a cylindrical mould. Eur Polym J 1987; 23:563
15. Chater M, Vergnaud JM. Study of the process of cure of epoxide resin heated by liquid oil at constant temperature. Eur Polym J 1988; 24:245
16. Laoubi S, Vergnaud JM. Modelling the process of absorption and desorption of water by coatings made of epoxy resin. Eur Polym J 1990; 26:1359
17. Laoubi S, Bouzon J, Vergnaud JM. Modelling the process of absorption of toluene by coatings made of epoxy resin. Eur Polym J, in press
18. Vergnaud JM. Modelling in calorimetry working under isothermal conditions or in scanning mode. Thermochim Acta 1987; 114:15
19. Brazier DW. Calorimetric studies of rubber vulcanisation and vulcanisates. In: Grassie N (ed.) Developments in polymer degradation, vol 3. Applied Science Publishers, London, 1981, pp 27–61
20. Chater M, Lalart D, Michel-Dansac F, Vergnaud JM. Variation of enthalpy and kinetic parameters of the cure of epoxide resin with the composition of the binary system. Eur Polym J 1986; 22:805

Chapter 12

Isothermal Differential Calorimetry

12.1 Introduction

Some information on isothermal differential calorimetry (DC) has already been given in the previous chapter (Sect. 11.2.1).

The DC technique may attract researchers, because it is simple, and working under isothermal conditions is of interest. But in fact there are many drawbacks in this technique: one is due to the principle of this method, with the sample temperature which varies with time; the other results from the quality of contact between the sample and calorimeter.

12.1.1 Change of Sample Temperature with Time

Two stages can be considered during the cure process with the DC technique: a heating stage, and a reaction stage.

Heating Stage

Following the principle of this technique, the sample at room temperature is introduced into the calorimeter, which has been previously stabilised at the selected temperature. The sample is strongly heated, and the sample temperature increases up to the calorimeter temperature. During this fast heating period of the sample, gradients of temperature are developed within the sample, with a high temperature on the surface. The reaction may start on the surface of the sample during this heating period. Resulting from this strong heating, a high endothermic heat flux (i.e. from the calorimeter to the sample) can be observed when a computer is coupled with the calorimeter.

Reaction Stage

When the temperature of the sample is the same as that of the calorimeter, the heat flux is equal to zero. Because of the internal heat generated by the exothermic cure reaction, the temperature in the sample is increased and becomes higher than that of the calorimeter; thus an exothermic heat flux is observed. During this reaction stage, other gradients of temperature can be developed within the sample, with a higher temperature at its centre. Because of the consumption of reactants in the sample, the rate of heat evolved from the cure decreases. Thus the heat flux decreases, after having passed through a maximum value.

12.1.2 Quality of Contact Between the Sample and Calorimeter

The quality of the contact between two materials at different temperatures is an important factor, and worth considering. If the contact is perfect, the temperature on each side of the interface is the same. The contact between a solid and a fluid is generally perfect. But between two solids there is often a thin interstice containing air or a fluid. Then the heat flux across the interface, F, is proportional to the temperature difference between the two sides of this interface, because it is proportional to the temperature gradient in the interstice:

$$F = k(T1_s - T2_s)$$

where $T1_s$ and $T2_s$ are the temperatures of the solids 1 and 2, respectively, at the surface separating the two solids. If the coefficient k is infinite, the difference $T1_s - T2_s$ is equal to zero, then the contact is perfect.

Perfect quality of contact is difficult to achieve generally. In the case of a small sample, the quality of contact is far from perfect, not only between the holder and calorimeter but also between the sample itself and the sample holder. And perhaps worse, it is difficult to appreciate how bad these contacts are when the sample is small. Another thing which must be considered is that heating takes some time, and using a small sample cannot reduce this time to zero.

In the case of a large sample, it is possible to determine the quality of the contact between the calorimeter and sample holder and between the holder and sample. This quality between the holder and sample can thus be easily determined by sawing the holder and sample at the end of the experiments, and examining the contact.

The quality of the contact between the holder and calorimeter is particularly considered in this chapter.

12.1.3 Modelling of the Process in DC

Modelling the process is of great help in gaining a fuller insight into the process itself. Mathematical treatment of heat transfer is not effective in this case where a reaction takes place. The numerical model must take into account all the facts, especially the heat transferred by conduction through the sample and the heat generated by the reaction. It must also enable the heat flux through the holder–calorimeter interface to be obtained. This result is of great interest, because a comparison between heat fluxes obtained by experiments and by calculation enables one to test the validity of the model.

For many numerical models, the following data can be calculated: the heat flux between the calorimeter and the external surface of the holder, and the profile of temperature as well as the profile of state of cure developed within the sample.

Modelling of the process is achieved in DC by considering either the quality of the contact, or the value of the enthalpy of reaction.

12.2 DC with Direct Contact Between the Sample and Calorimeter

12.2.1 Numerical Model

Mathematical Treatment of the Problem [1, 2]

Several basic assumptions are made for the heat transfer process and kinetics of reaction in the sample.

1. The sample is cylindrical in shape with a circular cross-section diameter of 6.8 mm for the holder and 5 mm for the sample.
2. The flow of heat conduction is radial, the longitudinal heat transfer being neglected.
3. There is no flow and no molecular diffusion in the sample.
4. The contact between the sample and calorimeter is direct: the holder is neglected. But the contact between the sample and calorimeter is imperfect.
5. The kinetics of heat evolved from the cure reaction is expressed in terms of temperature by an Arrhenius equation.
6. The thermal parameters are constant.

The equation of transient heat conduction through the cylindrical cross-section of the sample is expressed as follows:

$$\rho C \frac{\partial T}{\partial t} = \frac{1}{r}\frac{\partial}{\partial r}\left(r\lambda \frac{\partial T}{\partial r}\right) + \rho \frac{\partial Q}{\partial t} \tag{12.1}$$

where ρ is the density of the material, C the specific heat of material, and λ the thermal conductivity of the material, with the heat transferred by radial conduction, and the heat evolved from the cure reaction on the right-hand side, respectively.

The rate of heat evolved from the overall cure reaction is given by the simple expression with a constant activation energy, and a first-order reaction:

$$\frac{dQ(t, r)}{dt} = k_0(Q_\infty - Q(t, r)) \exp\left(-\frac{E}{RT(t, r)}\right) \tag{12.2}$$

where k_0 is the pre-exponential factor, E is the activation energy, R the ideal gas constant, $Q = Q(t, r)$ is the heat evolved per mass unit at radial abscissa r, from the reaction during the interval of time $[0, t]$, and Q_∞ is the corresponding quantity after infinite time.

Initial and boundary conditions obey the above assumptions.

$$t = 0, \quad 0 \leqslant r \leqslant R, \quad T = T_s \quad \text{(sample space)} \tag{12.3}$$

$$t > 0, \quad -\lambda_s \frac{\partial T}{\partial r}(t, R) = k\,(T(t, R) - T_m) \quad \text{(interface with oven)} \tag{12.4}$$

Numerical Analysis

Because of the internal heat generated by the reaction, Eqs (12.1) and (12.2) cannot be solved by mathematical treatment. The problem must be solved by using a numerical method based on finite differences [1].

Let the circular cross-section be divided into N annuli with the following radii, $r - \Delta r$, r, $r + \Delta r$, of constant thickness Δr (Fig. 12.1). Let us consider the integers n for space and i for time, as

$$r = n\Delta r, \quad 0 \leqslant n \leqslant N, \quad R = N\Delta r \tag{12.5}$$

$$t = i\Delta t \tag{12.5'}$$

The heat balance is evaluated within the annulus, of height the unit length, located between the radii $r - \frac{1}{2}\Delta r$ and $r + \frac{1}{2}\Delta r$ during the interval of time $[t, t + \Delta t]$, by considering the heat transferred by conduction through the cylindrical surfaces of radii $r - \frac{1}{2}\Delta r$ and $r + \frac{1}{2}\Delta r$, and heat evolved from the cure reaction. The new temperature after time Δt can thus be expressed as a function of the previous temperatures obtained in the annuli of radii $r - \Delta r$, r and $r + \Delta r$.

$$\mathrm{TN}_n = \frac{1}{M}[T_{n+1} + (M - 2)T_n + T_{n-1}] + \frac{\Delta r}{2Mr}[T_{n+1} - T_{n-1}] + \frac{\Delta Q_{i,n}}{C} \tag{12.6}$$

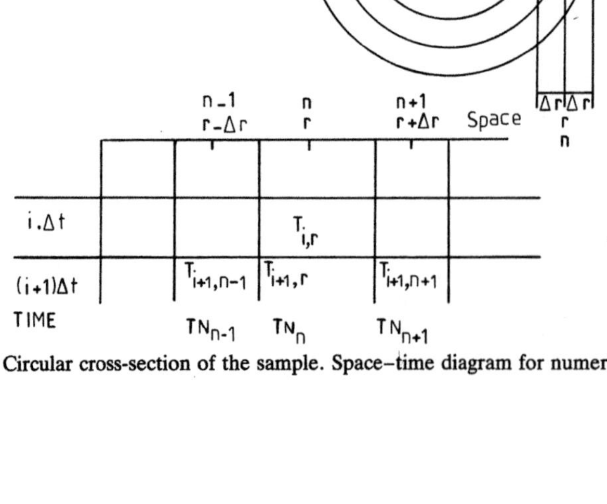

Fig. 12.1. Circular cross-section of the sample. Space–time diagram for numerical analysis.

where TN_n is the new temperature at radial abscissa $r = n\Delta r$ after time Δt, at time $(i+1)\Delta t$, T_n is the temperature in the same position at time $i\Delta t$ and M is a dimensionless number given by

$$M = \frac{(\Delta r)^2}{\Delta t}\frac{1}{\alpha_s} \tag{12.7}$$

where α_s is the thermal diffusivity of the sample.

As the above equation is not applicable for $r = 0$, the temperature in the centre of the sample must be evaluated in another way. A cylinder of radius $\frac{1}{2}\Delta r$ and unit height is considered. From the heat balance determined in this cylinder during the interval of time $[t, t + \Delta t]$, by taking into account heat transferred by conduction and heat evolved from the reaction, the new temperature at the centre of the cylinder is thus obtained:

$$TN_0 = T_0 + \frac{4}{M}(T_1 - T_0) + \frac{\Delta Q_0}{C} \tag{12.8}$$

where TN_0 is the new temperature after time Δt at the centre of the cylinder, at time $(i+1)\Delta t$, and T_0 is the temperature at the centre of the cylinder at time $i\Delta t$.

At the interface between the sample and oven ($r = R$ and $n = N$), there is the equation obtained in Sect. 8.2.2:

$$TN_N = T_N + \frac{2}{M}\frac{N - 0.5}{N - 0.25}(T_{N-1} - T_N) + \frac{2P}{M}(T_m - T_N) \tag{12.9}$$

where

$$P = \frac{k\Delta r}{\lambda_s}\frac{N}{N - 0.25} \tag{12.10}$$

The heat flux HF between the cylindrical sample and calorimeter is obtained from the simple equation:

$$HF = \lambda_s \frac{3T_N - 4T_{N-1} + T_{N-2}}{2\Delta r} \tag{12.11}$$

where the gradient of temperature has been evaluated by parabolic approximation.

The heat generated by the cure reaction during the increment of time Δt, at any place per mass unit, is calculated with the help of the following relation:

$$\Delta Q_{i,r} = Q_{i+1,r} - Q_{i,r} = k_0(Q_\infty - Q_{i,r})\exp\left(-\frac{E}{RT_{i,r}}\right)\Delta t \tag{12.12}$$

where

$$Q_{0,r} = 0 \tag{12.12'}$$

The amount of heat generated by the reaction at time $i\Delta t$ and radial abscissa r, per mass unit, is then

$$Q_{i,r} = \sum_{j=0}^{i-1} \Delta Q_{j,r} \tag{12.13}$$

12.2.2 Results for Cure of Low Enthalpy

Experiment

The main characteristic of the rubber compound used for this study is the sulphur percentage: 3%.

The values taken for the parameters concerned with the kinetics of reaction, and for the thermal parameters are shown in Table 12.1. These parameters remain constant and non-temperature dependent. Good contact was made between the holder and calorimeter with the help of sheets of aluminium, as well as between the holder and rubber using pressure.

The value obtained for the cure enthalpy is about the same at that given by some other authors [4]. The values selected [1] for the thermal parameters of the rubbers (thermal conductivity λ and heat capacity C) are of the same order of magnitude as those given in various papers [5, 6].

A calorimeter DSC 111 (Setaram, France) working under isothermal conditions was used in this study. Two holders (sample and reference) were introduced simultaneously into the sensitive zone of the calorimeter. The heat flux detector, consisting of thermocouples, connected the vessel to a block so that the temperature of the vessel was always close to that of the block. The heat flux transferred is proportional to the signal emitted.

Rather large samples of material are used in this type of calorimeter, between 100 and 200 mg. The sample holder is a cylinder of external diameter 6.8 mm and height of 7.7 mm, with a thickness of 0.9 mm, so that the sample is also cylindrical in shape with a diameter of 5 mm and a height of 7.5 mm.

After curing in the DC vessel, the sample was cooled at room temperature and replaced in the DC vessel under the same conditions in order to obtain the reference baseline.

Results for Heat Flux–Time Curves

The whole heat flux–time curve obtained by calculation for the cure of the rubber whose characteristics are described in Table 12.1 is shown in Fig. 12.2. The temperature is 20 °C for the sample when placed in the calorimeter, and 170 °C for the calorimeter. The largest part of the heat flux–time history is concerned with the endothermic stage corresponding to the heating of the sample. Thus one minute or so is necessary for the sample temperature to reach that of the calorimeter. Only a very small exothermic effect due to the cure reaction is obtained, in sharp contrast with the preceding high endothermic effect.

Table 12.1. Parameters for the rubber sample

Heat transfer	$\lambda = 5.5 \times 10^{-4}$ cal^{-1} s cm^{-1} deg^{-1}, $C = 0.37$ cal g^{-1} deg^{-1}, $\rho = 1.1$ g cm^{-3}, $\alpha = 1.3 \times 10^{-3}$ cm^2 s^{-1}		
Cure enthalpy	$Q = 10$ cal g^{-1}		
Kinetic reaction	Order $n = 1$,	$k_0 = 3 \times 10^9$ s^{-1},	$E = 25\,400$ cal mol^{-1}
Sample	$R = 3.4$ mm,	$T_{in} = 20$ °C,	$T_m = 160, 170, 180$ °C

Calculation is done by taking a value for the coefficient k, such that the curves calculated with this value are in best agreement with the experimental curves.

Fig. 12.2 also illustrates the development of heat transfer through the sample during the first minute of the sample history in the calorimeter. Very steep gradients of temperature are developed over the sample radius, especially in the first period of heating, with a high temperature on the surface of the sample. It is difficult to appreciate the other gradients of temperature resulting from the cure reaction, for instance at 90 s.

Results for Comparison between Heat Flux–Time Curves and Profiles of Temperature and State of Cure

The exothermic part of the heat flux is expressed as a function of time in Fig. 12.3, as obtained by calculation at 180 °C. Two types of curves are shown, one corresponding to the first cure reaction and calculated with $k_0 = 3 \times 10^9\ \mathrm{s}^{-1}$, and the other to the reheated sample with no reaction which is obtained by putting $k_0 = 0$ in the same model. These curves are about the same as those obtained through experiments.

Fig. 12.3 also illustrates the increase in the state of cure obtained at different places in the sample ($r = 0$, centre $r = \frac{1}{2}R - r = R$, surface).

The increase in the temperature obtained at the same places in the sample during this period is shown in Fig. 12.4, as well as the profiles of temperature developed through the sample in Fig. 12.5.

A comparison between the results illustrated in Figs 12.3–12.5 is of great interest. We can simultaneously follow the profiles of the heat flux emitted

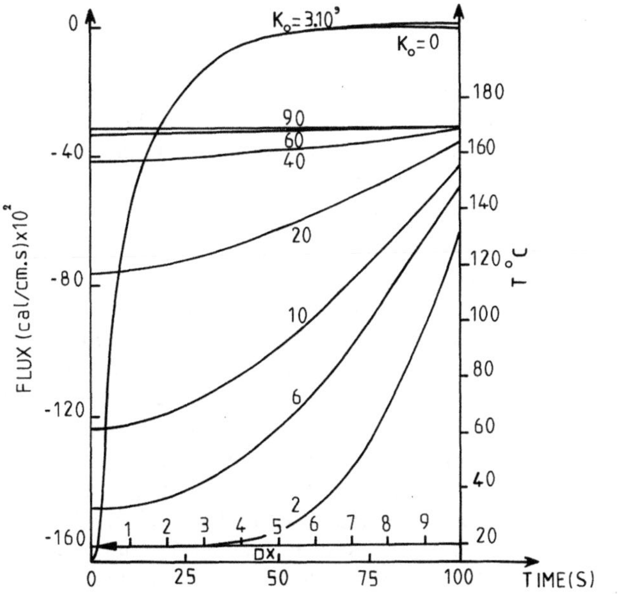

Fig. 12.2. Heat flux vs time at 170 °C in DC. Profiles of temperature over the radius of the sample at various times.

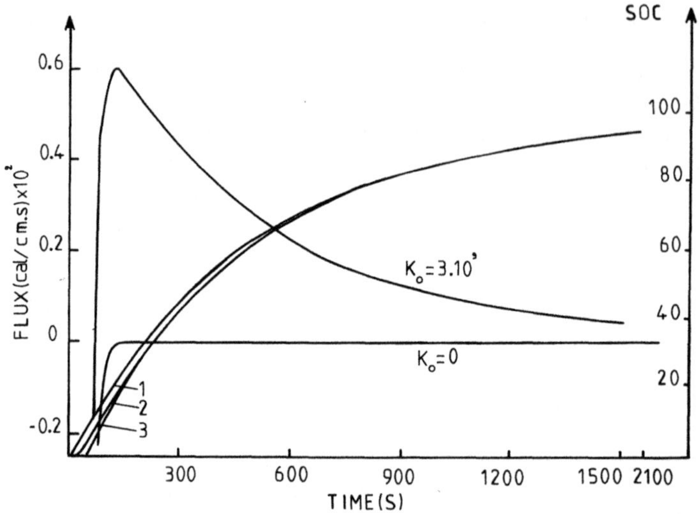

Fig. 12.3. Heat flux vs time calculated for the heated uncured rubber (with $k_0 = 3 \times 10^9\ s^{-1}$) and for the reheated cured rubber (with $k_0 = 0$), at 180 °C. SOC vs time at different places throughout the sample: (1) $r = R$; (2) $r = \frac{1}{2}R$; (3) $r = 0$.

through the calorimeter–sample interface, and those of the temperature and state of cure determined at different places through the sample. A result worth noting, concerns the fact that the maximum heat flux is reached at about the same time at which the maximum temperature in the sample is attained, as can be appreciated from Figs 12.3 and 12.4.

Effect of the Calorimeter Temperature

The effect of the calorimeter temperature on heat flux–time curves can be appreciated by comparing Fig. 12.3 at 180 °C, 12.6 at 170 °C and 12.7 at 160 °C. The profiles of temperature developed through the sample can also be compared in Figs 12.4 (180 °C), 12.8 (170 °C) and 12.9 (160 °C).

The following conclusions are worth noting:

1. There is a narrow window for the temperature range over which experiments can yield interesting results. On the one hand, the low temperature limit for cure studies is defined by the calorimeter detector sensitivity, as well as by the time-consuming experiment. On the other hand, too high a temperature is associated with a higher sensitivity for heat flux measurements, but the induction period of the cure reaction, if there is one, cannot be obtained.
2. Associated with the heat flux emitted through the calorimeter–sample interface is an increase in the temperature in the sample, with the following statement: the higher the heat flux, the higher the increase in temperature in the sample. These results are quantitatively expressed in Table 12.2, where some characteristics of the maximum value of the heat flux can be compared: the value of the maximum heat flux ($cal\,cm^{-2}\,s^{-1}$); the value ΔT at which the temperature is raised in excess of that of the calorimeter, by considering the centre of the sample ($r = 0$), and the position at which $r = \frac{1}{2}R$.

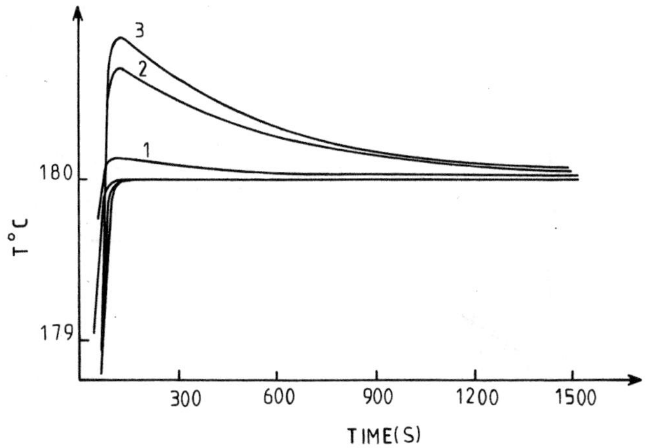

Fig. 12.4. Temperature vs time curves at different places for the heated rubber and reheated rubber, at 180 °C: (1) $r = R$; (2) $r = \frac{1}{2}R$; (3) $r = 0$.

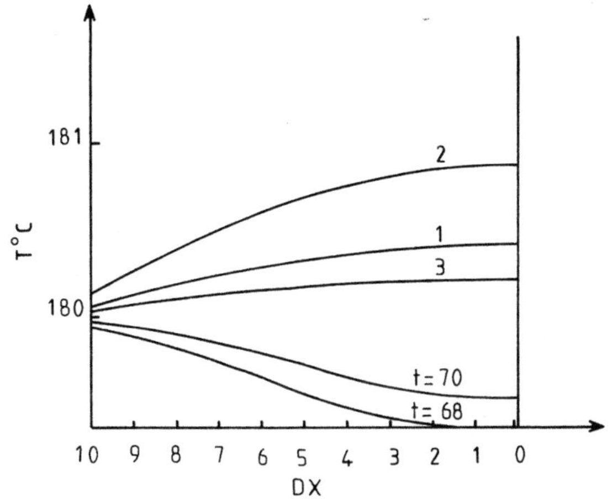

Fig. 12.5. Profiles of temperature developed over the radius of the cross-section of the sample, at various times: (1) when heat flux is half of the maximum; (2) when heat flux is maximum. (1) $r = R$; (2) $r = \frac{1}{2}R$; (3) $r = 0$.

Conclusions for the Cure of Low Enthalpy in DC

A few conclusions are worth noting:

1. Numerical models can describe the process of cure in DC, and they are capable of providing one more piece of information on the process. In fact, a further insight into the nature of the process is obtained. The following results are obtained simultaneously: the heat flux–time curves, which are also obtained through experiments; the temperature–time histories at various places, as well as the profiles of temperature developed through the sample; the state of cure–time histories and their profiles.

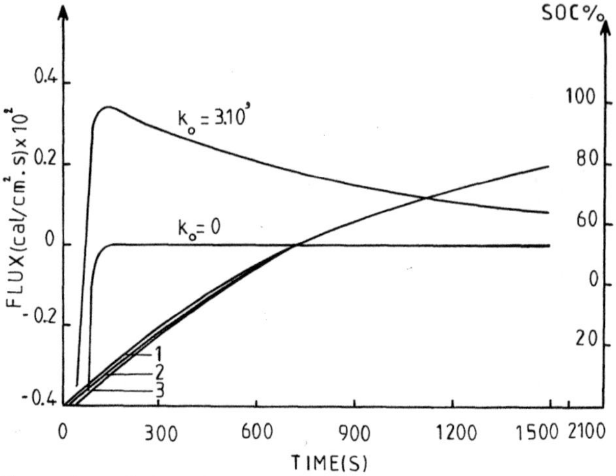

Fig. 12.6. Heat flux vs time calculated for the heated uncured rubber ($k_0 = 3 \times 10^9\ s^{-1}$) at 170 °C, and reheated precured rubber ($k_0 = 0$). SOC vs time at various places throughout the sample: (1) $r = R$; (2) $r = \frac{1}{2}R$; (3) $r = 0$.

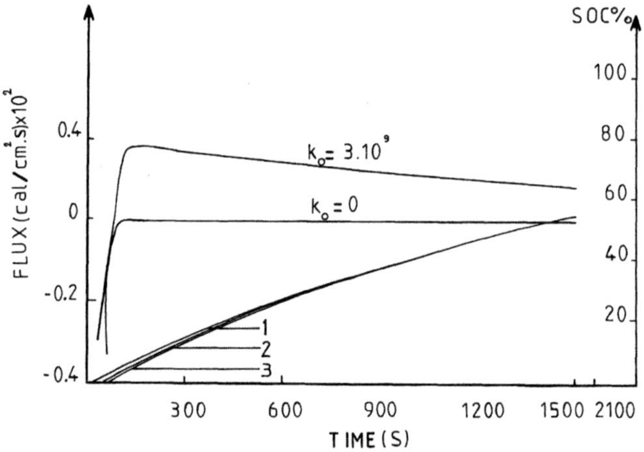

Fig. 12.7. Heat flux vs time for the heated uncured rubber ($k_0 = 3.10^9\ s^{-1}$) and reheated cured rubber ($k_0 = 0$) at 160 °C. SOC vs time at different places throughout the sample: (1) $r = R$; (2) $r = \frac{1}{2}R$; (3) $r = 0$.

2. Comparison between the heat flux–time curves obtained by calculation and by experiments enables one to test the validity of the model.
3. The temperature–time histories at various places within the sample have about the same shape as those of the heat flux–time histories.
4. There are two stages in the process: a first stage with a strong heating of the sample when it is placed in the calorimeter; another stage with the cure reaction period.
5. Steep gradients of temperature are developed within the sample with a higher temperature on the surface during the first stage of heating.

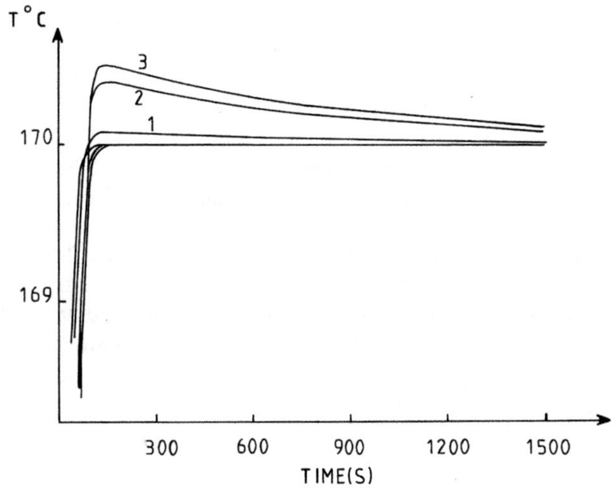

Fig. 12.8. Temperature–time histories at different places in the sample, for the heated uncured rubber at 170 °C: (1) $r = R$; (2) $r = \frac{1}{2}R$; (3) $r = 0$, and for the reheated cured rubber.

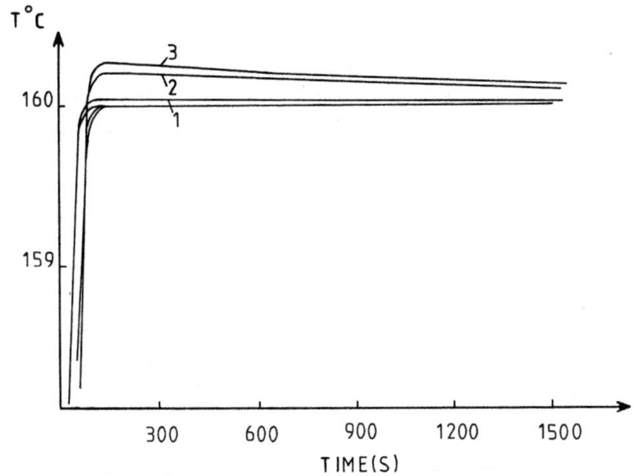

Fig. 12.9. Temperature–time histories at different places in the sample, for the heated uncured rubber at 160 °C: (1) $r = R$; (2) $r = \frac{1}{2}R$; (3) $r = 0$, and for the reheated cured rubber.

Table 12.2. Characteristics of maximum of heat flux

	Calorimeter temperature T(°C)		
	160	170	180
Heat flux max. (cal cm^{-2} s^{-1})	0.18	0.34	0.60
ΔT at $r = 0$ (°C)	0.26	0.50	0.88
ΔT at $r = \frac{R}{2}$ (°C)	0.20	0.39	0.60

6. During the reaction stage, other gradients of temperature are developed within the sample, with a higher temperature at the centre. The heat generated by the cure reaction is thus responsible for an increase in temperature in the sample, this increase being the highest at the centre.
7. It is difficult to separate the two stages exactly. It is thus clear that a part of the reaction is already completed during the first stage of heating. This fact is important for the measurement of the cure enthalpy. The total cure enthalpy can be obtained by the DC technique only when the data are collected by a computer capable of registering a wide range for the heat flux. For the simple recorders considering only the exothermic part of the curves, a significant portion of the cure enthalpy is lost with the following statement: the higher the calorimeter temperature, the lower the cure enthalpy obtained.
8. There is a narrow temperature window of about 30 °C over which data of interest can be obtained.

12.2.3 Results for Cure of High Enthalpy [2]

Experiment

An epoxy resin with a high enthalpy of cure was used in this experiment; the high enthalpy is the characteristic of interest here [2].

The values taken for the kinetic and thermal parameters are shown in Table 12.3. These parameters were kept constant in order to simplify the study. The rate of heat evolved from the overall cure reaction is expressed in terms of temperature by way of a simple Arrhenius equation with a single pre-exponential factor and a single activation energy, in spite of the fact that more complex equations have been presented in earlier literature [7–10].

The values of the kinetic parameters were determined from the heat flux–temperature curve obtained in scanning mode with a heating rate of 2 °C min^{-1}, by using the Freeman–Carroll method [11].

In sample preparation, great attention was given to ensure that good contact was made between the epoxide sample and the holder, as well as between the holder and calorimeter. About 150 mg of the compound was encapsulated in the holder.

The calorimeter (DSC 111, Setaram, France) can work either in isothermal or in scanning mode. In the present study concerned with isothermal conditions, the calorimeter was stabilised at the selected temperature, and both holders (for measurement and reference) were simultaneously introduced into the sensitive zone.

After cure in the calorimeter, the sample was extracted, cooled at room

Table 12.3. Parameters for cure of epoxy resin in DC

Kinetic reaction	Order $n = 1$, $E = 28000$ cal mol^{-1}, $k_0 = 3.6 \times 10^{12}$ s^{-1}, $Q_\infty = 65$ cal g^{-1}
Thermal parameters	$\lambda = 10^{-3}$ cal cm^{-1} deg^{-1} s^{-1}, $C = 0.32$ cal g^{-1} deg^{-1} $\rho = 1.25$ g cm^{-3}, $\alpha = 25 \times 10^{-4}$ cm^2 s^{-1}
Sample	$R = 3.4$ mm, $T_{in} = 20$ °C, $T_m = 130, 140, 150$ °C

temperature, and replaced in the calorimeter under the same conditions, to obtain the reference baseline.

Results for Heat Flux–Time Curves

Fig. 12.10 illustrates a typical calorimeter curve obtained for the cure of epoxy resin with the DSC 111 working under isothermal conditions at 150 °C. Several heat flux curves were calculated by using different values for the coefficient k shown in Eq. (12.10).

After the sample had been given a precure, it was quench-cooled to room temperature, and then heated in the calorimeter under conditions identical to the initial heating. This second curve was calculated with the help of the model by taking the value zero for the constant rate k_0.

The effect of the value of the temperature selected for the calorimeter on the profiles of the heat flux–time curves is clearly shown in Fig. 12.11, within the 130–150 °C range.

The following results, concerning the experimental limitation on the temperature range over which meaningful data can be obtained, are obvious:

1. The lowest possible temperature for cure studies of an epoxy resin is limited by the calorimeter sensitivity. When the cure rate is too low, the rate of enthalpy change is too small to be recorded.
2. The highest temperatures at which measurements can be obtained is determined by the length of the induction period for the cure reaction to initiate.

Results for Comparison between Heat Flux–Time Curves and Profiles of Temperature

Because of the low heat transfer rate through the resin and the high value of the cure enthalpy, a significant increase in temperature is observed in the resin during the cure in DC mode.

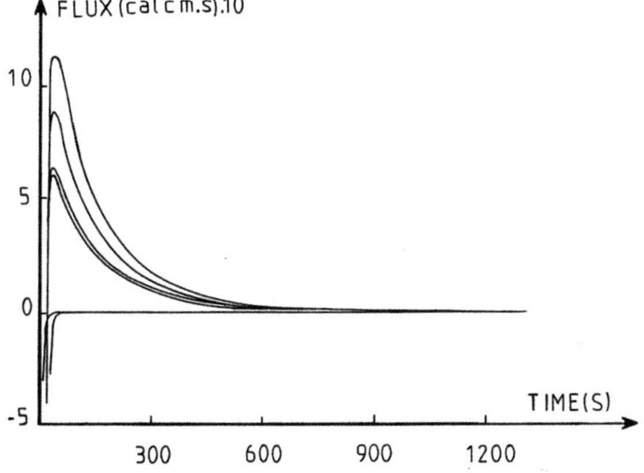

Fig. 12.10. Heat flux–time curves calculated at 150 °C, for different values of the coefficient k. Cure of epoxy resin in DC mode. The highest heat flux is the same as that obtained through experiments.

The profiles of temperature developed through the sample can be appreciated in Fig. 12.12 at 130 °C, where the temperature–time history is plotted for different places in the sample. These temperatures have been calculated with the help of the model by giving to the coefficient k a value such that the heat flux calculated with this value is in agreement with the experiments.

The effect of the temperature chosen for the calorimeter on the temperature profiles is illustrated in Fig. 12.13 at 140 °C, and Fig. 12.14 at 150 °C.

The following conclusions can be drawn:

1. The temperature within the resin during the cure in DC mode is far from constant. There is a significant increase in temperature in the resin, with a higher temperature at the centre.
2. The temperature–time history in the sample varies with the position in the

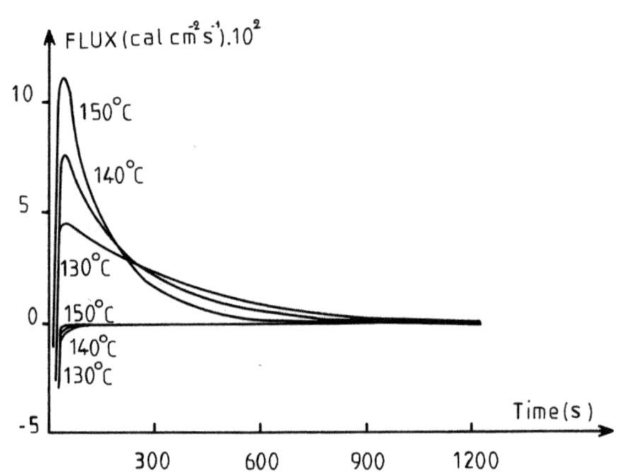

Fig. 12.11. Effect of temperature on the cure of epoxy resin in DC mode. Heat flux vs time curves obtained at various temperatures ranging from 130° to 150 °C.

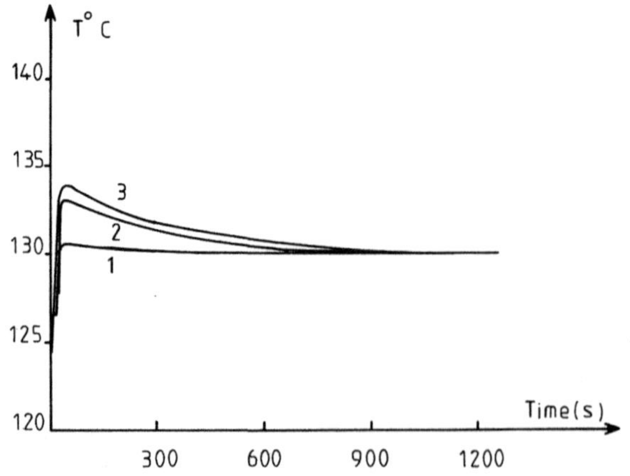

Fig. 12.12. Effect of temperature on the cure of epoxy resin in DC mode. Temperature–time histories at various places in the sample, with calorimeter temperature of 130 °C. (1) surface; (2) $r = \frac{1}{2}R$; (3) centre.

resin. These curves are of about the same shapes as those of the heat flux–time curves.

3. The effect of the temperature of the calorimeter on the increase in temperature within the sample is very important as shown in Table 12.4. For instance, a difference of about 7 °C between the centre and the surface of the sample can be seen for the maximum value reached by the temperature, when the calorimeter temperature is 150 °C.

Results for SOC–Time Histories in the Sample

As there are different temperature–time histories in various places within the sample, various SOC–time histories may also result.

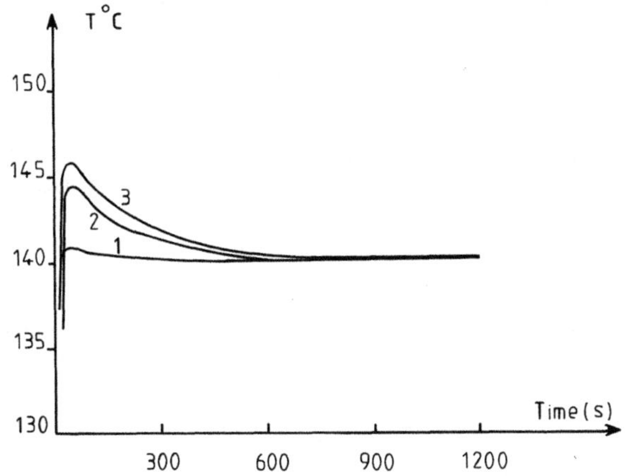

Fig. 12.13. Temperature–time histories in various places in the sample, with calorimeter temperature of 140 °C. (1) surface; (2) $r = \frac{1}{2}R$; (3) centre.

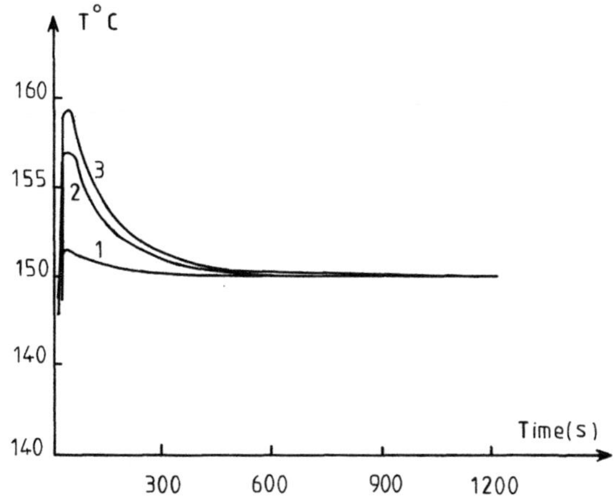

Fig. 12.14. Temperature–time histories in various places in the sample, with calorimeter temperature of 150 °C. (1) surface; (2) $r = \frac{1}{2}R$; (3) centre.

The different values for the SOC across the cross-section of the sample are illustrated in Fig. 12.15, when the calorimeter temperature is 150 °C. These results show that gradients of SOC through the section are developed as soon as the maximum temperature in the sample is reached, as well as the heat flux–time curves. These differences in the values of the SOC are maintained for about 500 s. Of course, the higher value for the SOC is obtained in the middle of the sample where the temperature is higher during the cure period. The difference between the extreme SOC values across the section does not exceed 7%, but it is significant.

Other SOC–time curves are shown for the calorimeter temperature kept at 130 °C (Fig. 12.16) and at 140 °C (Fig. 12.17), and the same conclusions can be drawn.

Conclusions for the Cure of High Enthalpy in DC

The same general conclusions are obtained for the cure of high and low enthalpy, with some differences in detail.

1. Numerical models are capable of describing the process of cure, by taking into account not only the kinetics of cure but also heat transferred through the

Table 12.4. Effect of calorimetry temperature on homogeneity across the sample section. Difference between the maximum temperature at some positions and the calorimeter temperature (°C)

Calorimeter temperature (°C)	Surface ($r = R$)	$r = R/2$	Centre ($r = 0$)
130	0.05	1.5	2.0
140	1.15	4.5	5.75
150	1.5	6.9	9.25

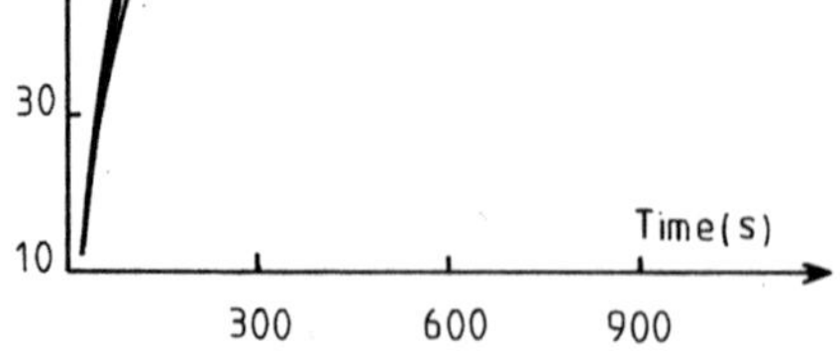

Fig. 12.15. SOC–time histories at different places for the cure of epoxy resin with calorimeter temperature of 150 °C. (1) surface; (2) $r = \frac{1}{2}R$; (3) centre.

sample. Of course, more information is given by the models than by experiments, with the heat flux, and the temperature and SOC within the sample.

2. The temperature–time histories in various positions within the sample are of about the same shape as those of the heat flux–time histories. The effect of the value of the cure enthalpy on these temperature–time histories appears very clearly. In the case of epoxy resin, a high increase in temperature is observed in the sample during the cure period, and this increase is especially high at the centre of the sample.
3. Following these temperature–time histories, various values of the state of cure are observed, depending on the position in the sample.
4. There is also a narrow temperature window of about 30 °C over which meaningful data are obtained.

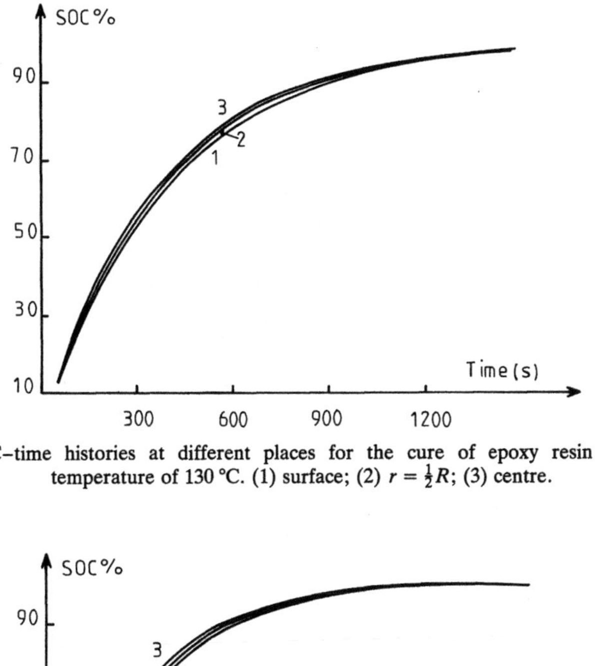

Fig. 12.16. SOC–time histories at different places for the cure of epoxy resin with calorimeter temperature of 130 °C. (1) surface; (2) $r = \frac{1}{2}R$; (3) centre.

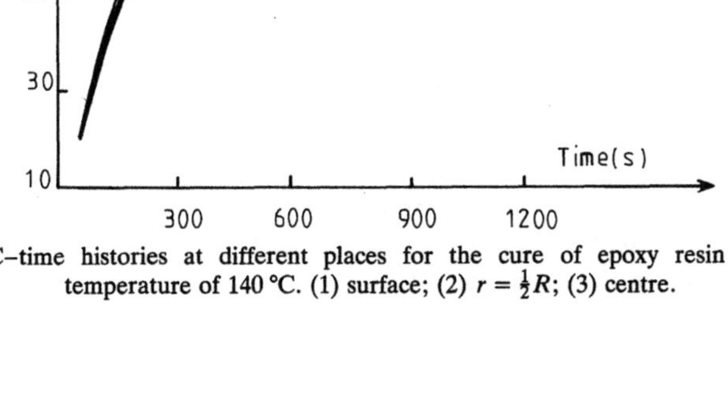

Fig. 12.17. SOC–time histories at different places for the cure of epoxy resin with calorimeter temperature of 140 °C. (1) surface; (2) $r = \frac{1}{2}R$; (3) centre.

12.3 DC with Indirect Contact Between the Sample and Calorimeter

Very often, in order to facilitate the insertion of the sample in the calorimeter as well as its extraction, the sample size is a little smaller than that of the calorimeter. As a result, a thin air layer appears between the sample and calorimeter. This fact is necessary for cylindrical samples, but also appears in the case of a small plane deposited on a plane sensor, because these two surfaces are not perfectly plane. Moreover, in this second case, the sample is heated through one surface only, while the cylindrical surface is radially heated.

In this section some emphasis is placed on this air layer located between the calorimeter and sample because it intervenes in the process of heat transfer in spite of its small thickness. This thin air layer is assumed to have a constant thickness around the sample.

12.3.1 Numerical Model [12, 13]

Assumptions

The following assumptions are made in order to clarify the problem and to simplify the process.

1. Because of the cylindrical shape of the sample and of the disposition of the sensitive heat cell, only radial heat through the circular cross-section is considered.
2. The thermal conductivity is taken to be constant.
3. The kinetics of the cure reaction of the resin is described by a simple equation with a constant activation energy.
4. The temperature of the heated oven does not vary when the sample, which was previously at room temperature, is introduced.
5. The thin layer of air located between the oven and the holder is cylindrical in shape, with constant thickness around the holder. This is obtained by placing the sample carefully.
6. Heat is transferred through this very thin air layer by conduction.

Mathematical Treatment

Following the above assumptions, the rate of heat transferred in or out of the sample is given by the equation of radial transient heat conduction.

$$\rho C \frac{\partial T}{\partial t} = \frac{1}{r} \frac{\partial}{\partial r}\left[r\lambda \frac{\partial T}{\partial r}\right] + \rho \frac{\partial Q}{\partial t} \tag{12.14}$$

where the contribution of the radial heat conduction and of the internal heat generated by the exothermic reaction can be seen on the right-hand side. $Q = Q(r, t)$ is the heat evolved, per mass unit at radial abscissa r, from the reaction during the interval of time $[0, t]$.

The initial and boundary conditions can be written as follows:

$$t = 0, \qquad 0 \leqslant r \leqslant R_s \ (T \text{ sample})$$
$$R_s < r < R_h \ (T \text{ holder})$$
$$R_h < r < R_m \ (T \text{ air layer}) \tag{12.15}$$
$$T > 0, \qquad r = R_m, \qquad T = T_m \ (\text{mould}) \tag{12.16}$$

where

R_s is the radius of the sample.
R_h is the external radius of the holder.
R_m is the internal radius of the oven of the calorimeter.
The air thickness is thus $R_m - R_h$.

At each position the rate of heat evolved from the cure reaction up to time t is expressed by

$$\frac{1}{Q_\infty}\frac{\partial Q}{\partial t}(r, t) = k_0\left(1 - \frac{Q(r, t)}{Q_\infty}\right)^n \exp\left(\frac{-E}{RT(r, t)}\right) \tag{12.17}$$

where

k_0 is the pre-exponential factor.
n is the order of the overall reaction.
E is the activation energy.
$Q(r, t)$ and Q_∞ are the heat evolved per mass unit from the reaction up to time t and after infinite time, respectively.

Numerical Analysis

As no analytical solution can be found to the problem, a numerical method with finite differences is built and tested.

The circular cross-section of the sample, holder and air layer is shown in Fig. 12.18. By considering the annuli of radii $r - \Delta r$, r and $r + \Delta r$, of thickness Δr, the heat balance during the time increment Δt is calculated in the annulus located between $r - \frac{1}{2}\Delta r$ and $r + \frac{1}{2}\Delta r$. As a result, the new temperature within this annulus after time Δt can be obtained as a function of temperature at the preceding time in the same annulus and in the two adjacent annuli.

The calculations are made in the resin, in the holder and in the air layer. The radius of the sample, R_s, the thickness of the holder, $R_h - R_s$, and the thickness of the air layer, $R_m - R_h$, are divided in equal parts, with a number of slices N_s, K_h and K_a, respectively. These slices have the following thicknesses, respectively:

$$\Delta r = \frac{R_s}{N_s}, \qquad \Delta r_h = \frac{R_R - R_s}{K_h}, \qquad \Delta r_a = \frac{R_m - R_h}{K_a}$$

The total number of slices is

$$N_m = N_s + K_h + K_a$$

Also

$$N_h = N_s + K_h$$

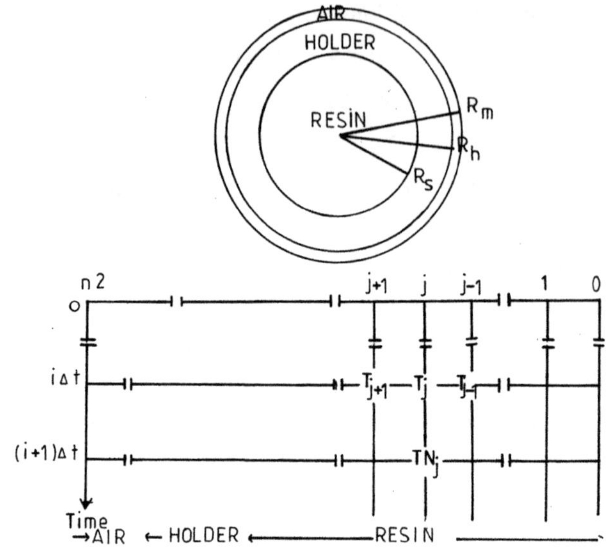

Fig. 12.18. Circular cross-section of the sample, with resin, holder and air layer. Temperature–space diagram for numerical analysis.

With each integer j, with $0 \leqslant j \leqslant N_m$, is associated a radial abscissa r. The temperatures at the position j at time t and time $t + \Delta t$, are denoted respectively T_j and TN_j.

Within the Resin, with $1 \leqslant j \leqslant N_s - 1$ $(r = \Delta r)$.

$$TN_j = \frac{1}{M_s}[T_{j-1} + (M_s - 2)T_j + T_{j+1}] + \frac{1}{2M_s j}[T_{j+1} - T_{j-1}] + \frac{\Delta Q_j}{C} \quad (12.18)$$

Centre of the Resin, with $j = 0$ $(r = 0)$. At the centre of the resin, a simple equation must be used:

$$TN_0 = T_0 + \frac{4}{M_s}(T_1 - T_0) + \frac{\Delta Q_0}{C} \quad (12.19)$$

The dimensionless number M_s is given by

$$M_s = \frac{(\Delta r)^2}{\Delta t}\frac{\rho C}{\lambda} \quad (12.20)$$

where ρ, C and λ are the density, heat capacity and thermal conductivity of the resin, respectively.

Within the Metal Holder, $N_s + 1 \leqslant j \leqslant N_h - 1$ $\quad r = R_s + (j - N_s)\Delta r_h$.

$$TN_j = \frac{1}{M_h}[T_{j-1} + (M_h - 2)T_j + T_{j+1}] + \frac{\Delta r_h}{2M_h(R_s + (j - N_s)\Delta r_h)}(T_{j+1} - T_{j-1}) \quad (12.21)$$

The dimensionless number M_h is given by

$$M_h = \frac{(\Delta r_h)^2}{\Delta t}\left(\frac{\rho C}{\lambda}\right)_h \tag{12.22}$$

where ρ, C and λ are the thermal parameters of the holder.

Within the Air Layer, $(N_h + 1 \leq j \leq N_m - 1)$ $\quad r = R_h + (j - N_h)\Delta r_a$.

$$\begin{aligned} TN_j = {} & \frac{1}{M_a}[T_{j-1} + (M_a - 2)T_j + T_{j+1}] \\ & + \frac{\Delta r_a}{2M_a[R_h + (j - N_h)\Delta r_a]}(T_{j+1} - T_{j-1}) \end{aligned} \tag{12.23}$$

The dimensionless number M_a is given by

$$M_a = \frac{(\Delta r_a)^2}{\Delta t}\left(\frac{\rho C}{\lambda}\right)_a \tag{12.24}$$

ρ, C, λ being the thermal parameters of the air layer.

Interface between the Resin and Holder, $j = N_s$.

$$\begin{aligned} TN_{N_s} = T_{N_s} + {} & \frac{1}{3(U + U_1)}[V(T_{N_s-1} - T_{N_s}) - V_1(T_{N_s} - T_{N_s+1}) \\ & - U(TN_{N_s-1} - T_{N_s-1})] \\ & - U_1(TN_{N_s+1} - T_{N_s+1}) + W[\tfrac{3}{4}\Delta Q_{N_s} + \tfrac{1}{4}\Delta Q_{N_s-1}] \end{aligned} \tag{12.25}$$

where the coefficients V, V_1, U and U_1 are:

$$V = \frac{8\lambda\Delta t}{\Delta r}\left(R_s - \frac{\Delta r}{2}\right), \qquad V_1 = \frac{8\lambda_h\Delta t}{\Delta r_h}\left(R_s + \frac{\Delta r_h}{2}\right)$$

$$U = \frac{\rho C\Delta r}{2}\left(2R_s - \frac{\Delta r}{2}\right), \qquad U_1 = \frac{(\rho C\Delta r)_h}{2}\left(2R_s + \frac{\Delta r_h}{2}\right)$$

$$W = \frac{4r\Delta r}{3(U + U_1)}\left(R_s - \frac{\Delta r}{2}\right) \tag{12.26}$$

Interface between the Holder and Air, $j = N_h$.

$$\begin{aligned} TN_{N_h} = T_{N_h} + {} & \frac{1}{3(U_2 + U_3)}[V_2(T_{N_h-1} - T_{N_h}) - V_3(T_{N_h} - T_{N_h+1}) \\ & - U_2(TN_{N_h-1} - T_{N_h-1}) - U_3(TN_{N_h+1} - T_{N_h+1})] \end{aligned} \tag{12.27}$$

where the coefficients V_2, V_3, U_2 and U_3 are

$$V_2 = \frac{8\lambda_h\Delta t}{\Delta r_h}\left[R_h - \frac{\Delta r_h}{2}\right], \qquad V_3 = \frac{8\lambda_a\Delta t}{\Delta r_a}\left[R_h + \frac{\Delta r_a}{2}\right]$$

$$U_2 = \frac{(\rho C\Delta r)_h}{2}\left[2R_h - \frac{\Delta r_h}{2}\right], \quad U_3 = \frac{(\rho C\Delta r)_a}{2}\left[2R_h + \frac{\Delta r_a}{2}\right] \tag{12.28}$$

Interface between the Air Layer and Calorimeter, $j = N_a$. The contact between a solid and a fluid being perfect, the following simple equation is used:

$$T_{N_a} = T_m \tag{12.29}$$

where T_m is the temperature of the mould.

Conditions for Stability of Calculations. The time increment Δt must be sufficiently small so that the following conditions for stability of calculations are verified:

$$\left.\begin{array}{l} M_s > 4, \quad M_h > 2, \quad M_a > 2 \\ 3(U + U_1) - (V + V_1) > 0 \\ 3(U_2 + U_3) - (V_2 + V_3) > 0 \end{array}\right\} \tag{12.30}$$

Heat Evolved from the Cure Reaction. At each position, the heat evolved from the cure reaction up to time $t = i\Delta t$, is calculated, as in Sect. 7.4.1, with the relations

$$Y = \frac{Q_t}{Q_\infty} = 1 - [1 + (n - 1)S_t]^{1/(1-n)} \tag{12.31}$$

where the function S is defined by:

$$S_t = \int_0^t k_0 \exp\left(-\frac{E}{RT_\theta}\right) d\theta \quad \text{with} \quad 0 \leqslant \theta \leqslant t \tag{12.32}$$

The function S at time $t + \Delta t$ is obtained by the following recurrent relation:

$$S_{t+\Delta t} = S_t + k_0 \Delta t \exp\left(-\frac{E}{RT_t}\right) \quad \text{with } S_0 = 0 \tag{12.33}$$

The quantity ΔQ in Eqs (12.18), (12.19) and (12.25) is thus obtained:

$$\Delta Q = Q_\infty (Y_{t+\Delta t} - Y_t) \tag{12.34}$$

The SOC is obtained from the ratio

$$\text{SOC}_t = \left(\frac{Qt}{Q_\infty}\right) 100 \tag{12.35}$$

Heat Flux. The heat flux which enters the resin at time t is determined next to the surface of the resin by the following equation:

$$HF = \lambda_s \frac{3T_R - 4T_{R-\Delta r} + T_{R-2\Delta r}}{2\Delta r} \tag{12.36}$$

where the gradient of temperature is evaluated by parabolic approximation.

12.3.2 Results

Experiment

A calorimeter DSC 111 (Setaram, France) working under isothermal conditions was used. Two holders with the sample and reference are simultaneously introduced into the calorimeter. The calorimeter is coupled with a microcomputer which collects and then processes the data (heat flux vs time).

A fairly large sample can be used in this type of calorimeter, with 100–200 mg of resin. The dimensions for the cylindrical holder are: external diameter 6.8 mm, internal diameter 5 mm and height 7.5 mm. The thickness of the air layer is around 0.25 mm.

An epoxy resin (150 mg) was encapsulated in the stainless steel holder. The kinetic properties are previously determined by calorimetry with the same apparatus working in scanning mode at the heating rate of 2 °C min^{-1}. The values obtained for this resin are shown in Table 12.5.

The heat evolved from the cure reaction is assumed to be described by the Arrhenius equation, and the activation energy and the order of reaction are obtained from the experimental curves by using the Freeman and Carroll technique. The pre-exponential constant k_0 is calculated by comparing the values of the heat flux obtained at the maximum either by experiments or by calculation [14]. The validity of these kinetic parameters is tested by superimposing the calculated and experimental heat flux–temperature histories. The main assumption made in this calculation is that the temperature throughout the sample is the same as that of the calorimeter [15]. As shown in Fig. 12.19, a good correlation between the theoretical and experimental curves is obtained, proving the validity of these kinetic parameters.

Results for Heat Flux–Time Curves [12]

Experiments were performed under isothermal conditions at 150 °C using the same calorimeter (DSC 111, Setaram), samples and material [12]. Two kinds of results are shown:

1. The heat flux–time profile obtained when the fresh uncured resin, previously at room temperature, is placed into the calorimeter maintained at a constant temperature of 150 °C (Fig. 12.20).
2. The heat flux–time profile with the reference baseline, which is obtained by dropping the cured sample previously cooled at room temperature into the calorimeter at 150 °C (Fig. 12.21).

The numerical model described above is used for calculating both these heat flux–time histories. The first, with cure reaction (Fig. 12.20), is calculated by considering not only the heat transfer by conduction through the sample, holder

Table 12.5. Kinetic and thermal parameters of the resin

Kinetic reaction	$n = 1.56$ $Q_\infty = 38$ cal g^{-1}	$k_0 = 2 \times 10^{23}$ s^{-1}	$E = 50018$ cal mol^{-1}
	Thermal parameters		
	λ (cal cm^{-1} s^{-1} deg^{-1})	ρ (g cm^{-3})	C (cal g^{-1} deg^{-1})
Sample (resin)	8×10^4	1.47	$0.3 + 5 \times 10^{-4} \times T$ with T (°C)
Holder (metal)	0.107	7.86	0.108
Air at 125 °C	7.6×10^{-5}	9.44×10^{-4}	0.24

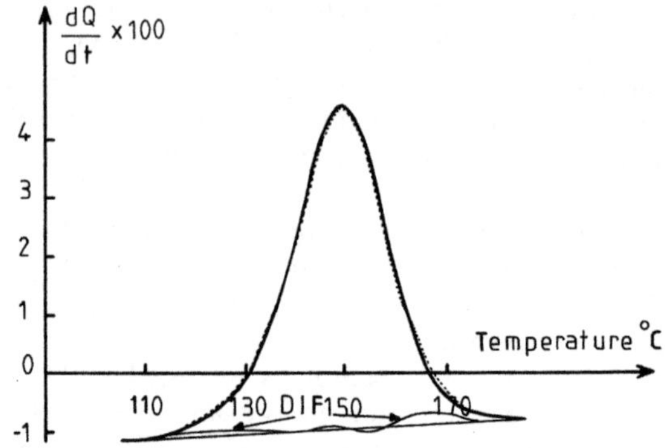

Fig. 12.19. Heat flux–temperature histories obtained in DSC technique, with and without reaction for an epoxy resin heating rate = 2 °C min^{-1}. —, experiment; – – –, calculation.

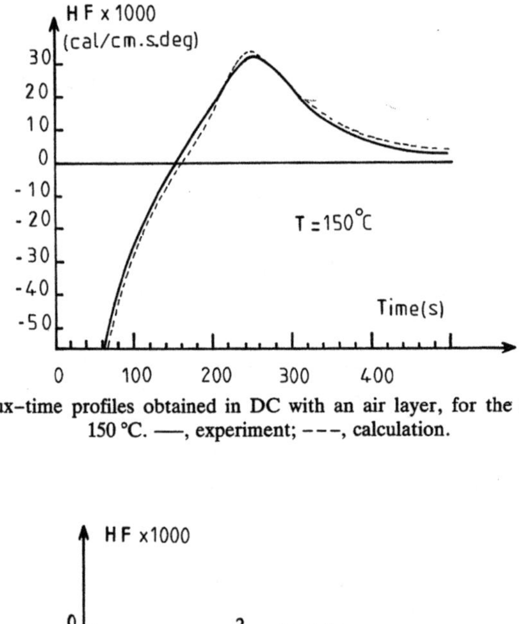

Fig. 12.20. Heat flux–time profiles obtained in DC with an air layer, for the cure of epoxy resin at 150 °C. —, experiment; – – –, calculation.

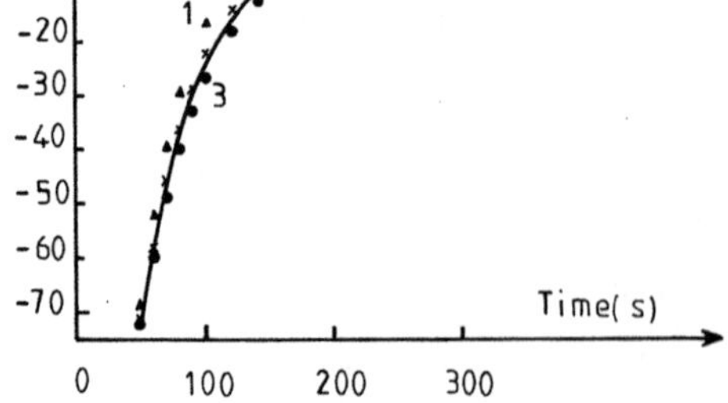

Fig. 12.21. Heat flux–time profiles obtained in DC with an air layer, for a cured epoxy resin at 150 °C. Thickness of the air layer: (1) 0.2 mm; (2) 0.25 mm; (3) 0.3 mm.

and air layer, but also the heat evolved from the cure reaction. The second, without cure reaction (Fig. 12.21), is obtained with the same model, and a value of zero for the pre-exponential term k_0 which eliminates the effect of heat reaction.

In Fig. 12.21, where no reaction occurs, the effect of the value attributed to the thickness of the air layer is considered when it is varied within the 0.20–0.30 mm range. The best agreement is obtained between the experimental and calculated curves when this value is 0.27 mm.

A good agreement is also shown in Fig. 12.20, where the process consists of heat transfer by conduction and internal heat generated by the reaction.

The effect of the air layer can be appreciated by comparing the heat flux–time curves calculated either by taking into account the air layer (Fig. 12.20) or by assuming that the contact between the holder and calorimeter is perfect (Fig. 12.22). The main effect of the air layer on the heat flux–time profile is to increase the time necessary for the cure. Thus, a period of time of at least 150 s is necessary for the heat fux to reach the value of zero and 250 s to reach the maximum value, when the air layer is considered. In the case of perfect contact, the time necessary to reach each of these objectives is considerably lower, around 50 to 80 s. The air layer is also responsible for a lower rate of heating of the sample.

Results for Profiles of Temperature and SOC

As in many other cases, the model is able to provide more information than experiments: the profiles of temperature and SOC can be determined by calculation (Fig. 12.23).

Some gradients of temperature are developed within the sample. A higher temperature is observed on the surface of the sample at the beginning of the process, during the heating period. When the reaction takes place, for instance after 200 s other temperature gradients are developed, with higher temperatures

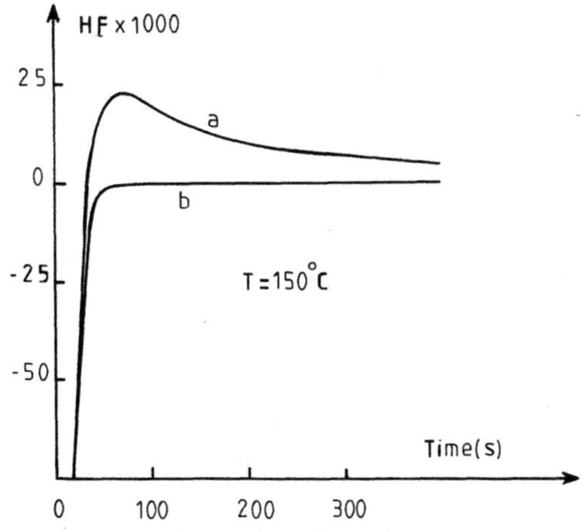

Fig. 12.22. Heat flux–time profiles obtained in DC without air layer, for an epoxy resin at 150 °C, **a** with and **b** without cure reaction.

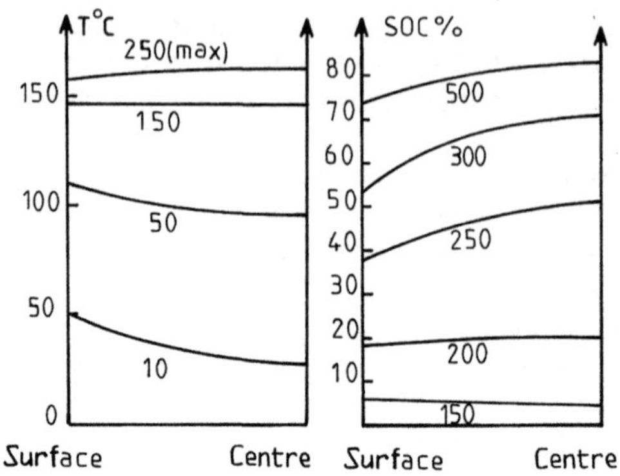

Fig. 12.23. Distribution of temperature (left) and SOC (right) through the sample with an air layer 0.27 mm thick for various times.

at the centre of the sample (Fig. 12.23 left). The SOC profiles developed within the sample are also of interest (Fig. 12.23 right). The following conclusions can be drawn from these curves:

1. The layer of air located between the calorimeter and holder must be taken into consideration when it exists, because of the heat transferred by conduction through this layer.
2. A period of time of around 150 s is necessary for the heat flux to reach a value of zero as shown in Figs 12.20, 12.21 and 12.23.
3. During this heating period, high gradients of temperature develop through the sample.
4. The cure reaction starts during this heating period. For instance, at the end of this period, an SOC between 4% and 5% is attained throughout the sample, following a low gradient through the cross-section of the sample.
5. During the period of reaction, when the heat flux emitted from the sample is positive, other gradients of temperature develop within the sample with a higher value at the centre of the sample. The same conclusions can also be drawn for the SOC.

The distribution of temperature and SOC through the sample is also calculated when the air layer is neglected and the contact between the holder and calorimeter is perfect (Fig. 12.24).

A comparison between the profiles of temperature and SOC developed within the sample with (Fig. 12.23) and without an air layer (Fig. 12.24) can lead to interesting conclusions about the effect of the air layer on the process. Some facts of interest are worth noting:

1. High temperature profiles develop in the case of a sample in good contact with the calorimeter. These temperature profiles are lower when the layer air is present.
2. Higher profiles of state of cure within the sample are found when no layer of air is present.

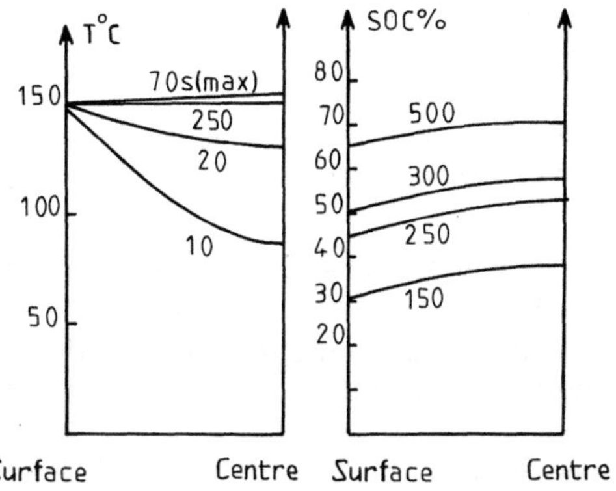

Fig. 12.24. Distribution of temperature (left) and SOC (right) through the sample without air layer, for various times.

3. The presence of the air layer is also responsible for a retardation in the cure reaction, especially at the beginning of the reaction when the heat flux has not reached its maximum value. In this case, lower profiles of temperature and SOC develop through the sample, the reaction proceeding at about the same rate everywhere in the sample. Of course, as already said above by considering the heat flux–time curves, it takes a longer period of time for the heat flux to reach its maximum value when the air layer is present (250 s instead of 75 s).
4. The presence of the air layer is also responsible for a higher increase in the temperature and state of cure after the maximum heat flux has passed. This fact can be seen precisely in Fig. 12.23 right and Fig. 12.24 for times of 300 and 500 s. In other words, the air layer decreases the rate of heat transfer whether the temperature in the sample is lower or higher than that of the calorimeter.

Conclusions on the Presence of an Air Layer

It has been shown above that in calorimetry it is difficult for a large cylindrical sample to have a perfect contact between the holder and calorimeter. An air layer thus occurs, located between the calorimeter and holder. This air layer can be eliminated by introducing a thin sheet of metal, but it is dangerous for the calorimeter to be in direct contact with the metal holder. The thermal dilatation of the metal holder can be responsible for breaking the sensitive part of the calorimeter.

A numerical model, taking into account all the known facts, and especially the presence of the air layer, is capable of describing the process, and thus providing more information.

Of course, the air layer modifies the process, because of the low thermal conductivity of air. The heat flux–time curves are thus modified considerably during the whole process: at the beginning, during the heating period of the sample, with a lower rate of heating, and after the heat flux has passed through its maximum, with a lower rate of cooling.

12.4 Conclusions

Some conclusions on the experiments and calculations made in the DC mode with a rubber of low cure enthalpy and with an epoxy resin of high cure enthalpy are as follows.

1. It is difficult to say that the process of cure is carried out under isothermal conditions, when the DC mode is used in calorimetry.
2. Two stages can be distinguished during the process: the first stage with strong heating of the sample after its introduction into the calorimeter, followed by the cure stage.
3. A temperature–time history is obtained by calculation with the help of a numerical model. These histories are of about the same shape as those of the heat flux–time curves. This increase in temperature within the sample during the cure period varies with the value of the cure enthalpy.
4. Because of the presence of the exothermic cure reaction and of the low heating rate through the sample, steep gradients of temperature are developed within the sample.
5. The results have been obtained using a fairly large sample, but the same conclusions can be drawn with a smaller sample.
6. Emphasis has been placed on the effect of a thin air layer located between the sample and calorimeter. This air layer is of help in inserting and removing the sample. Of course, it is possible to detect the presence of this air layer and even to measure its thickness in the case of a calorimeter with a large sample. In the case of a calorimeter with a small sample, this air layer still exists, but it is difficult to characterise it, and what is worse, it may vary from one sample to another.
7. The best heat transfer is obtained in the case of a cylindrical sample which is radially heated. In the case of a sample heated on one face only, being in contact with a heating plate, there is less efficient heat transfer.
8. The problem of the size of the sample is of great concern, especially when mixtures are studied. As shown in the case of rubbers or of epoxy resins, a fairly large sample taken from a batch is more representative of the homogeneity of the batch, than a small sample.

References

1. Khouider A, Rochette JB, Vergnaud JM. Study of heat flux and profiles of temperature and state of cure obtained in an isothermal calorimeter with rubber cure of low enthalpy. Thermochim Acta 1985; 89: 81
2. Chataing G, Chater M, Vergnaud JM. Determination of heat flux and profiles of temperature and state of cure in isothermal calorimetry with reaction of high enthalpy. Thermochim Acta 1985; 89: 107
3. McAdams WH. Heat transmission, McGraw-Hill Kogakusha, Tokyo, 3rd edn, 1954, p 50
4. Brazier DW, Nickel GH. Thermoanalytical methods in vulcanizate analysis. I. Differential scanning calorimetry and the heat of vulcanization. Rubber Chem Technol 1975; 48: 26
5. Hands D. The thermal transport properties of polymers. Rubber Chem Technol 1977; 50: 480

6. Gehman SD. Heat transfer in processing and use of rubber. Rubber Chem Technol 1967; 40: 36
7. Kamal MR. Thermoset characterization for moldability analysis. Poly Engng Sci 1974; 14: 231
8. Sickfield J, Mielke W. Application of thermal analysis for the investigation of epoxy resins. Progr Organ-Coat 1984; 12: 27
9. Pusatcioglu SY, Hassler JC, Fricke ALM, McGee HA. Effect of temperature gradients on cure and stress gradient in thick thermoset castings. J Appl Polym Sci 1980; 25: 381
10. Ryan ME, Dutta A. Kinetics of epoxy cure: a rapid technique for kinetic parameter estimation. Polymer 1979; 20: 203
11. Chataing G, Vergnaud JM. Effect of heating rate on the separation of various phenomena in the DSC technique. Thermochim Acta 1985; 94: 379
12. Liu H, Armand JY, Bouzon J, Vergnaud JM. Effect of the layer of air between the calorimeter and cylindrical samples in DC experiments. Thermochim Acta 1988; 130: 355
13. Armand JY, Bouzon J, Vergnaud JM. Modelling of the process in DC experiments by considering the layer of air between the calorimeter and holder. Thermochim Acta 1989; 147: 291
14. Kretzschmar K, Hoffmann KW. Kinetic aspects arising during the differential scanning calorimetric investigation of curing epoxy resins with anhydrides. Thermochim Acta 1985; 83: 139
15. Armand JY, Gonnet R, Jones R, Bouzon J, Touchard M, Vergnaud JM. Effect of heating rate and sample size on heat transfer through the sample in DSC. Thermochim Acta 1986; 103: 341

Chapter 13

Differential Scanning Calorimetry

13.1 Introduction

In spite of the fact that there are so many differences between the gas chromatography technique and calorimetry, it may be of interest to compare these two techniques, at least for a while.

13.1.1 Comparison between DSC and Gas Chromatography

Temperature plays an important role in gas chromatography (GC), as well as in calorimetry. In the 1950s, when GC was only used under isothermal conditions, with heavy loaded columns, the following statement applied: the best conditions for the chromatography of a liquid are when the working temperature is around the value of the boiling point of the liquid under atmospheric pressure. Following this simple statement, it is easy to understand why there was only a narrow range of GC applications: when a mixture of components is separated by GC, the more volatile components may experience too high a temperature and their retention times be too short, making it difficult to separate their peaks; however, this column temperature is too low for the less volatile components, necessitating a very long time and provoking the formation of wide peaks with low height which are difficult to detect. With GC with programming of temperature, which was developed in the early 1960s, this drawback disappeared. Starting with a low temperature for the column, the more volatile components can pass through the column at a low temperature and their retention times are not too short, while the less volatile components, having a rate of transport which increases with temperature, can thus leave the column at a temperature which corresponds to their boiling point, enabling the formation of narrow and high peaks. Of course, as far as temperature is considered in the GC technique, three parameters are of interest: the temperatures at which the process starts and is finished, and the rate of heating.

Of course, the comparison between these two techniques could be prolonged by considering the role played by the stationary liquid in the column. More sensitive detectors made possible a reduction in the amount of the liquid injected into the column. Not only was the column able to work under better conditions, but also a decrease in the percentage of stationary liquid used in the column made possible either a decrease in the retention time if temperature was kept at the above value, or a decrease in the column temperature.

Following the first part only of the comparison between the GC and calorimetry techniques, by considering the temperature of the oven, it has already been seen in Chaps 11 and 12 that for the DC technique working at constant temperature, a rather narrow window is found over which meaningful data can be obtained. The programming of temperature in calorimetry, in the same way as for GC, is able to enlarge the possibilities of the technique. The following three parameters are thus also of interest: the temperatures selected for the beginning and the end of the process, as well as the heating rate.

13.1.2 Principles of DSC

In the differential scanning calorimetry (DSC) experiment, the fresh uncured sample is scanned at a selected heating rate up to a final temperature at which the exotherm reaction is complete. The result is a profile of the rate of enthalpy change (called heat flux) as a function of temperature. After the sample has been given a complete cure, it is quench-cooled and then scanned under conditions identical to the initial scan, giving the reference baseline.

Heating Rate

Some parameters are of interest in the DSC technique. With the value selected for the heating rate, another important parameter is the dimensions of the sample. In fact, as considered in detail in Sects 13.3.1 and 13.3.2, the right value of the heating rate is related to the size of the sample, with the following statement: the smaller the sample size, the higher the heating rate. More accurately, the right value of the heating rate must be inversely proportional to the square of the main dimension of the sample, i.e., the radius for a cylinder radially heated, or for a sphere; the thickness for a sheet; the edge for a cube.

Treatment of Kinetic Curves Obtained

Another difference between the DSC and DC techniques exists with the way of treatment of the kinetic curves obtained. Most people are accustomed to treating the kinetic relationship under isothermal conditions. Special treatment is necessary to deal with the kinetics of heat obtained with the scanning mode and constant heating rate.

Modelling of the process in DSC studies can also be made by following two methods: when the heating rate is low enough, the temperature within the sample can be considered as about the same as that of the calorimeter, and simple models are obtained. When the heating rate is high, temperature gradients are developed within the sample, and the rate of heat transferred by conduction through the sample must be taken into account in the model.

13.2 Theoretical Aspects of DSC

Two numerical treatments are considered:

1. The treatment of the kinetic curves for the heat evolved from the cure reactions which are obtained in DSC.
2. The modelling of the process developed in the sample, with the DSC technique.

13.2.1 Treatment of the Kinetic Curves Obtained when the Heating Rate is Low

The main assumptions which are made in this case are that the sample is so small and the heating rate used in DSC is so low that the temperature throughout the sample is always uniform and about the same as that of the calorimeter. This means that the rate of heat transfer by conduction through the sample is very high compared to the rate of heating selected in the DSC.

Various methods can be used for the treatment of the kinetic curves obtained in DSC, these kinetic curves expressing the rate of change of the enthalpy as a function of the temperature [1]. The most well-known method is surely the Freeman–Carroll method of calculation presented in 1958.

For a given reaction, the rate of enthalpy change at time t is assumed to be expressed as a function of the enthalpy evolved up to time t and of the absolute temperature, according to the equation

$$\frac{dy}{dt} = k_0(1 - y)^n \exp\left(-\frac{E}{RT}\right) \tag{13.1}$$

where

$y = Q_t/Q_\infty$ is the ratio of the enthalpy evolved up to time t by the total enthalpy evolved from the reaction.

k_0 is the pre-exponential factor.

E is the activation energy.

n is the order of reaction.

T is the temperature (K).

R is the constant of ideal gas.

As the heating rate for the sample is the same as that of the calorimeter

$$\frac{dT}{dt} = b \tag{13.2}$$

and Eq. (13.1) is thus expressed in terms of temperature:

$$\frac{dy}{dT} = \frac{k_0}{b}(1 - y)^n \exp\left(-\frac{E}{RT}\right) \tag{13.3}$$

Freeman–Carroll Technique (1958)

The above equation, written in the logarithmic form,

$$\ln \frac{dy}{dT} = \ln \frac{k_0}{b} + n \ln (1 - y) - \frac{E}{RT} \tag{13.4}$$

by using the finite differences becomes

$$\Delta \ln \frac{dy}{dT} = n \Delta \ln (1 - y) - \frac{E}{R} \Delta \left(\frac{1}{T}\right) \tag{13.5}$$

This equation can be rewritten as follows:

$$\frac{\Delta \ln \dfrac{dy}{dT}}{\Delta \ln (1 - y)} = n - \frac{E}{R} \frac{\Delta \left(\dfrac{1}{T}\right)}{\Delta \ln (1 - y)} \tag{13.6}$$

Thus by plotting the values of the left-hand side of the above equation as a function of $\Delta(1/T)/(\Delta \ln(1 - y))$, a straight line can be obtained. The activation energy can be calculated from the value of the slope of this straight line, and the order of the reaction is given by the intercept with the ordinate axis.

Finite increments of temperature are considered, as well as the corresponding values of y and dy/dT which must be evaluated at each temperature.

The value of the constant pre-exponential k_0 can be determined easily by writing that the rate of enthalpy change obtained at a given temperature, either by experiment or calculation with the kinetic parameters, is the same.

Superposition of Experimental and Calculated Curves

A good way of testing the validity of the kinetic parameters determined by any method consists of comparing the experimental curve with the curve obtained by calculation with the kinetic parameters [1, 2].

13.2.2 Modelling of the Process with a High Heating Rate

When the rate of heating chosen for DSC is high, the rate of heat transfer by conduction through the sample may be a limiting factor and thus it must be taken into account.

The numerical model is the same as that described in the case of the DC technique with the air layer located between the external surface of the holder and the calorimeter (Sect. 12.3.1) or without it (Sect. 12.2.1). The best model is, of course, the one taking into account the air layer, if there is any.

Assumptions

The following assumptions are made:

1. Radial heat by conduction through the circular cross-section of the holder and sample is considered.

2. The heat generated by the overall cure reaction can be described by any type of equation, and especially by a kinetic equation with an apparent order, and an energy activation according to the Arrhenius expression.
3. The air layer may or may not exist.
4. The properties (density, heat capacity, thermal conductivity) may be constant or not.
5. The increase in temperature of the calorimeter may follow a constant rate or not.

Mathematical Treatment

With the above assumptions, the equation for heat conduction through the circular cross-section is

$$\rho C \frac{\partial T}{\partial t} = \frac{1}{r}\frac{\partial}{\partial r}\left[r\lambda \frac{\partial T}{\partial r}\right] + \rho \frac{\partial Q}{\partial t} \tag{13.7}$$

where ρ, C, λ are the density, heat capacity and thermal conductivity of the material, respectively, and $\partial Q/\partial t$ is the rate of heat evolved from the cure reaction, per unit mass.

The initial and boundary conditions are:

$$t = 0, \quad 0 \leqslant r \leqslant R_s \quad (T \text{ sample}) \tag{13.8}$$

$$R_s < r \leqslant R_h \quad (T \text{ holder}) \tag{13.9}$$

$$R_h < r \leqslant R_a \quad (T \text{ air layer}) \tag{13.10}$$

$$t > 0, \quad r = R_a, \quad T = T_m(t)(\text{calorimeter}) \tag{13.11}$$

The rate of heat evolved from the overall cure reaction up to time t is expressed by

$$\frac{dy}{dt} = k_0(1 - y)^n \exp\left(-\frac{E}{RT}\right) \tag{13.1}$$

where $y = Q_t/Q_\infty$ is the extent of heat evolved up to time t.

Numerical Analysis

The problem has to be solved by a numerical method with finite differences. Let us consider the following circles of radii: $r - \Delta r$, r and $r + \Delta r$, taken on the circular cross-section (Fig. 13.1). The heat balance during the increment of time Δt is evaluated within the annulus of radius r and thickness Δr, and enables one to obtain the new temperature within this annulus after time Δt. If λ is constant, the following is obtained:

$$\mathrm{TN}_r = \frac{1}{M}[T_{r+\Delta r} + (M-2)T_r + T_{r-\Delta r}] + \frac{\Delta r}{2Mr}[T_{r+\Delta r} - T_{r-\Delta r}] + \frac{1}{C}\Delta Q_r \tag{13.12}$$

where TN_r and T_r are the new temperature after time Δt, at time $t + \Delta t$, and the previous temperature at time t, in the annulus of radius r, respectively.

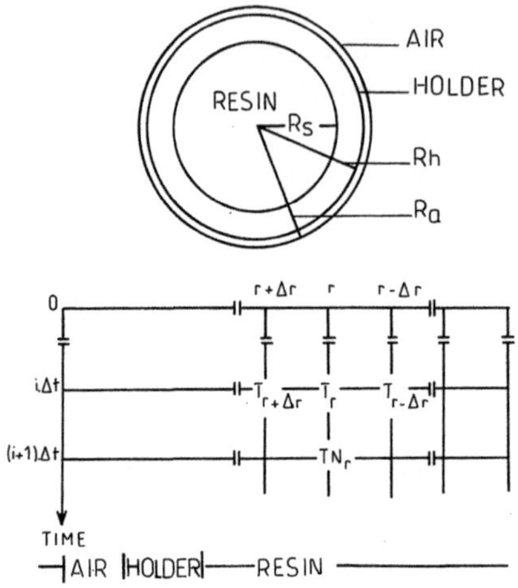

Fig. 13.1. Temperature–space diagram for numerical analysis. Circular cross-section of the sample.

For the centre of the cylinder ($r = 0$) the following equation is suitable for calculating the temperature:

$$\mathrm{TN}_0 = \frac{4}{M}\,[T_{\Delta r} - T_0] + T_0 + \frac{1}{C}\,\Delta Q_0 \tag{13.13}$$

where the dimensionless number M is given by

$$M = \frac{(\Delta r)^2}{\Delta t}\,\frac{\rho C}{\lambda} \tag{13.14}$$

When the air layer is considered, the numerical model described in Sect. 12.3.1 is used.

Eq. 13.12, without the contribution of the internal heat from the cure reaction, is also used for the holder, with another value of Δr.

Using a simpler model without an air layer, the following equation obtained in Sect. 8.2.2 can be used for the calorimeter–holder interface:

$$\mathrm{TN}_{R_\mathrm{h}} = T_{R_\mathrm{h}} + Z_2(T_{R_\mathrm{h}-\Delta r} - T_{R_\mathrm{h}}) + \frac{2P}{M}\,(T_\mathrm{m} - T_{R_\mathrm{h}}) \tag{13.15}$$

where

$$Z_2 = \frac{2}{M}\,\frac{R_\mathrm{h} - (\Delta r/2)}{R_\mathrm{h} - (\Delta r/4)} \quad \text{and} \quad P = \frac{k\Delta r}{\lambda}\,\frac{R_\mathrm{h}}{R_\mathrm{h} - (\Delta r/4)}$$

Z_2 and P being two dimensionless numbers, and k the coefficient of heat convection.

The heat flux transferred through the calorimeter–holder interface is obtained by the following expression which uses a parabolic approximation for the gradient:

$$\mathrm{HF} = \lambda\,\frac{3T_{R_\mathrm{h}} - 4T_{R_\mathrm{h}-\Delta r} + T_{R_\mathrm{h}-2\Delta r}}{2\Delta r} \tag{13.16}$$

13.3 Results Obtained with DSC: Effect of Heating Rate

The main parameter in the DSC technique is the heating rate, and some emphasis is placed in this section on the effect of the heating rate in various cases. In the same way as in GC with programming of temperature, the value of the heating rate has an important effect on the separation of the heat flux–temperature peaks. With a very low heating rate, the apparatus is thus capable of giving more information, especially when several reactions take place and some of these reactions can be obtained separately.

The sample size is also an important factor, as shown in Chap. 12, and the value chosen for the heating rate depends largely on the size of the sample, with the simple statement: the larger the size of the sample, the lower the heating rate. In fact, more precise relationships can be obtained (Sect. 13.3.2).

The enthalpy of the overall reaction is the main characteristic of a resin. The relationship between the heating rate in DSC and the enthalpy of cure is thus of interest.

The kinetic parameters also play an important role in the process of cure. The effect of the value given to the heating rate on the shape of the heat flux–temperature curves, as well as on the kinetic parameters, is studied specifically.

13.3.1 Heating Rate and Separation of Various Phenomena in the Sample [2]

The purpose is to determine all the kinetic parameters of various reactions when several peaks can be distinguished in the DSC curve. Very often, these peaks are very close, and it is only possible to distinguish a disturbance in the shape of the peak representing the heat flux–temperature curve. The heating rate is the important parameter when the shape of the heat flux–temperature curve is studied in order to find the best conditions able to determine separately two phenomena occurring in the resin.

Experiment

Materials. The formulations used were composed of 2608 epoxy resin (CDF Chimie) which basically contains DGEBA epoxy molecules, an anhydride hardener with tertiary amine as catalyst. The formulation investigated contained 80 parts of hardener to 100 parts of resin, by weight.

The epoxy amine mixture was prepared at room temperature, by adding the curing agent with continuous stirring until a clear mixture was obtained.

Apparatus. Samples of about 150 mg were placed in hermetically sealed iron–steel holders. Calorimetric measurements were made with a DSC 111 (Setaram), by using various values for the heating rate, ranging from 0.5 to 20 °C min^{-1}. The temperature was programmed from 20 °C to nearly 250 °C. After cooling the cured sample to room temperature, the cured sample was heated again under the same conditions in order to obtain a reference baseline.

The cylindrical holder had an external radius of 3.4 mm and an internal radius of 2.5 mm.

Heat flux–temperatures curves were obtained using a Hewlett-Packard 86 microcomputer, after storing the data and determining the kinetic parameters with the help of the Freeman–Carroll technique. The validity of these kinetic data was controlled by superimposing the heat flux–temperature curves obtained by experiment and calculation.

Results for the Determination of the Kinetic Parameters

When the sample is so small and the heating rate so low that the temperature throughout the sample is always uniform and about the same as that of the calorimeter, then the reaction rate equation used is

$$\frac{dy}{dT} = \frac{k_0}{b}(1-y)^n \exp\left(-\frac{E}{RT}\right) \tag{13.3}$$

where $y = Q_T/Q_\infty$ represents the extent of heat evolved up to temperature T from the overall reaction, k_0 is the pre-exponential factor, n the order of the reaction, E the activation energy and b the heating rate of calorimeter.

Experimental heat flux–temperature curves obtained with a heating rate of 2 °C min^{-1} were studied using the Freeman–Carroll technique. This method generally gives an accurate value for the activation energy from the slope of the curve (Sect. 13.2.1) and a poor value for the order (ordinate axis intersection). The pre-exponential factor k_0 is determined by adjusting the maximum value of the heat flux calculated with the above values for the order and activation energy and the corresponding heat flux value obtained by experiment. Quite a good agreement can be seen in Fig. 13.2, where experimental and calculated heat flux–time histories are plotted. The kinetic parameters are shown in Table 13.1.

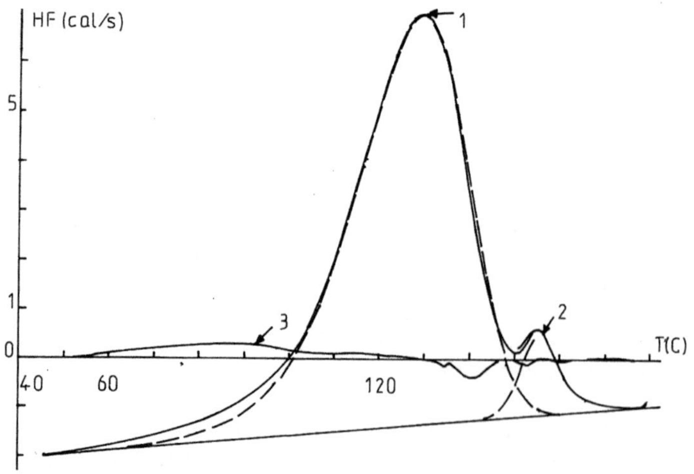

Fig. 13.2. Heat flux–temperature curves obtained at 2 °C min^{-1}. —, experiment; - - -, calculation. Peaks 1 and 2 are calculated. Peak 3 is the difference between the calculated and - - - measured peaks.

Of course the values of the kinetic parameters are surprising, with such high values for the activation energy and the pre-exponential term. It is not our purpose to discuss them.

Results for Peak Separation

The fact of interest in this study is that two phenomena are represented by two peaks in Fig. 13.2, when the heating rate is 2 °C min^{-1}.

By using the kinetic parameters shown in Table 13.1, heat flux–temperature curves were calculated for various values of the heating rate: 10 °C min^{-1} (Fig. 13.3), 20 °C min^{-1} (Fig. 13.4), 1 °C min^{-1} (Fig. 13.5) and 0.5 °C min^{-1} (Fig. 13.6). As clearly shown in these figures, the lower the heating rate, the better the separation obtained for these peaks. For the highest heating rate (20 °C min^{-1}), the small peak is located under the large one as shown with the calculated curves, and the total peak resulting from the addition, which can only be observed experimentally, exhibits a particular shape, and of course no separation of the peak is observed (Fig. 13.4). The shape of the experimental peak, which is also the total peak obtained by calculation, is slightly affected when the heating rate is 10 °C min^{-1} (Fig. 13.3). A visible separation can be appreciated at a heating rate of 2 °C min^{-1} (Fig. 13.2), and a complete separation is attained with lower heating rates: 1 °C min^{-1} (Fig. 13.5) and 0.5 °C min^{-1} (Fig. 13.6).

Table 13.1. Kinetic parameters

Peak	E	n	k_0
Large peak	28 000	1.25	3.6×10^{12}
Small peak	155 000	2.75	4.7×10^{76}

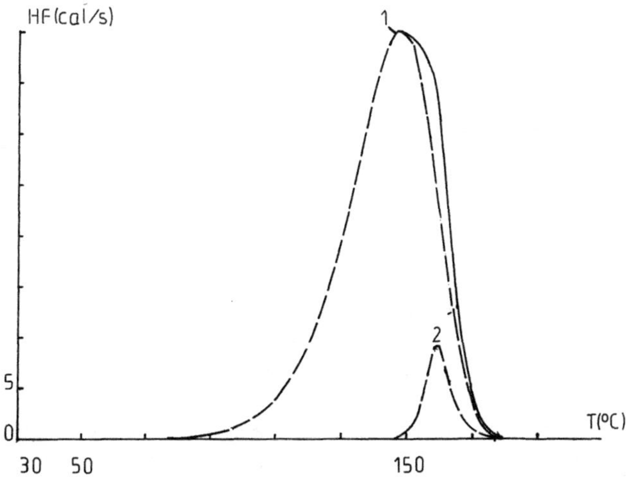

Fig. 13.3. Heat flux–temperature curves obtained at 10 °C min^{-1}. Peak 1 plus Peak 2 = total heat flux curve.

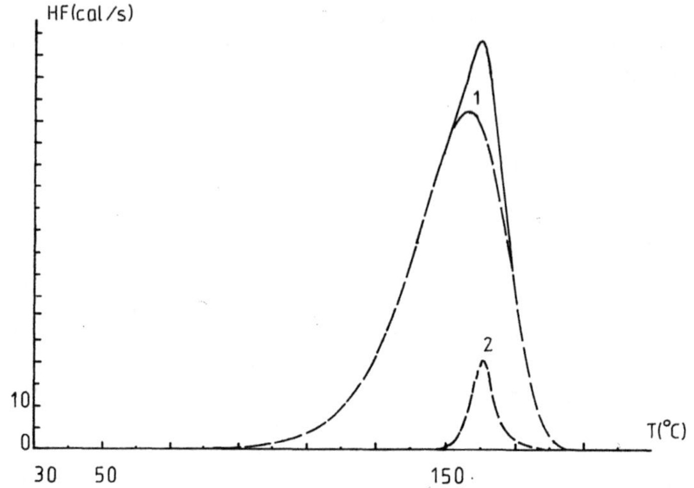

Fig. 13.4. Heat flux–temperature curves obtained at 20 °C min^{-1}. Peak 1 plus Peak 2 = total heat flux curve.

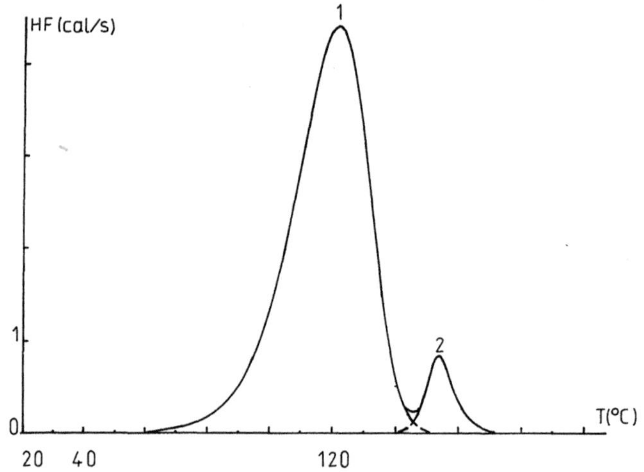

Fig. 13.5. Heat flux–temperature curves obtained at 1 °C min^{-1}. Peak 1 plus Peak 2 = total heat flux curve.

In order to study more precisely this problem of separation of phenomena, a separation factor F is introduced in the same way as in GC:

$$F = \frac{T_{m1} - T_{m2}}{\Delta T_1 + \Delta T_2} \tag{13.17}$$

where T_m is the temperature at which the heat flux reaches its maximum value and ΔT is the width of the peak, at midheight. A value higher than 1 for this separation factor corresponds to a complete separation of the peaks, and a value of about 0.7 allows the operator to distinguish the presence of two distinct

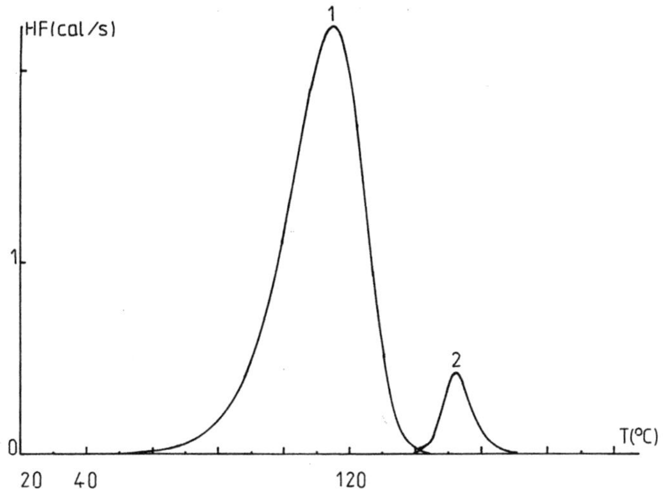

Fig. 13.6. Heat flux–temperature curves obtained at 0.5 °C min^{-1}. Peak 1 plus Peak 2 = total heat flux curve.

phenomena. The variation of the separation factor with the value given to the heating rate is of interest when these phenomena must be distinguished (Fig. 13.7). Of course, the lower the heating rate, the higher is the value of the separation factor.

Another advantage of the DSC technique is that the value obtained for the total cure enthalpy is not dependent on the heating rate, as shown in Fig. 13.7.

Although a lower value selected for the heating rate in DSC improves the separation factor, it is also responsible for the following drawbacks:

1. A simultaneous decrease in the sensitivity of the calorimeter measured by the value of the heat flux, as shown in Fig. 13.7
2. A subsequently large increase in the time needed for experiments

13.3.2 Heating Rate and Sample Size Without Reaction [3]

Heat transfer plays an important role in various cases for industrial reactions, because it appears as the first limitation of the process. Of course, it is widely taken into account in industrial processes and it stands for a large part of the chemical engineering science. But it must also be considered in various other cases on a smaller scale, for instance, in laboratory applications.

The problem of heat transfer limitation also appears in calorimetry, and its role is decisive for this technique. In both DSC and DC techniques, the enthalpy changes accompanying the physical change or chemical reaction in the sample are monitored, the result being the profile of the rate of enthalpy change as a function of temperature in DSC, or time in DC. Some sources of error arise from heat transfer from the specimen to the calorimeter, which apply particularly for reactions of high enthalpy [4, 5]. The recorded and known temperature is the set calorimeter temperature and not the true temperature during the cure, because of local heating arising from exothermic reactions and a low rate of heat transfer

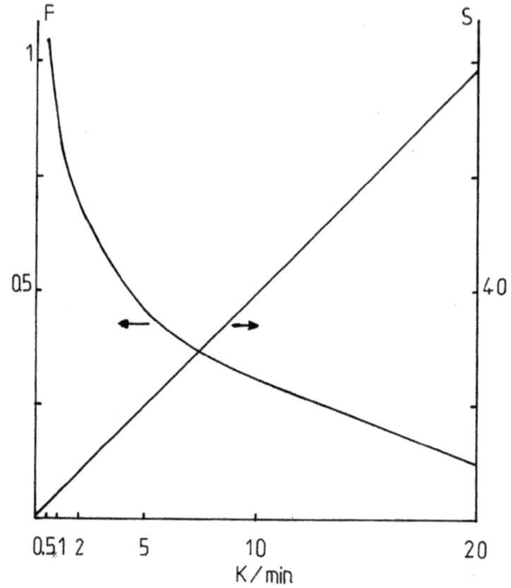

Fig. 13.7. Variation of the separation factor F and maximum heat flux with the value of the heating rate in DSC.

through the sample [6–8]. This phenomenon is of great importance in an isothermal calorimeter (DC) as shown in Chap. 12. The low rate of heat transfer through the sample is responsible for the development of steep gradients of temperature through the sample, and the temperature at the centre of the sample or the mean temperature throughout the sample are far from being equal to that of the calorimeter.

Our purpose is to study the simultaneous effect of the heating rate and of the sample size on the heat transfer through cylindrical samples, when their external surface is heated at a constant rate. In order to simplify the rather complex problem, no reaction takes place in the sample and heat is transferred radially by conduction through the circular cross-section of the cylinder. The choice of a cylinder is of interest in this study, because some experimental information can be obtained from calorimetry, especially the heat flux emitted through the holder–calorimeter interface. The gradients of temperature developed through some samples of various sizes are calculated by using a numerical model [3]. This model, based on an explicit method with finite differences, is able to take into account every kind of heat transfer, e.g., that by conduction in the problem at hand, and especially the quality of contact between the calorimeter and the external surface of the holder.

Numerical Model

Assumptions. The following assumptions are made:

Only radial heat conduction through the circular cross-section of the cylindrical sample is considered.

Thermal properties of the resin, thermal conductivity and heat diffusivity are constant during heating. (Another model is capable of taking into account the variation of these thermal parameters with temperature if the relationships are known).

The contact between the sample and calorimeter is not perfect. The flux across the interface is proportional to the temperature difference between the two sides of the interface.

No sample holder is considered in this case.

Mathematical Treatment. The equation for radial heat conduction is

$$\frac{\partial T}{\partial t} = \alpha \left[\frac{\partial^2 T}{\partial r^2} + \frac{1}{r}\frac{\partial T}{\partial r}\right] \tag{13.18}$$

with the constant thermal diffusivity α.

The rate of heating is constant in the calorimeter:

$$\frac{dT}{dt} = b \quad \text{or} \quad T_m(t) = bt + c \tag{13.19}$$

The initial and boundary conditions are

$$t = 0, \quad 0 \leq r < R \quad (T \text{ sample}) \tag{13.20}$$

$$t > 0, \quad r = R, \quad -\lambda \frac{\partial T}{\partial r}(t, R) = k[T(t, R) - T_m(t)] \tag{13.21}$$

k being the coefficient of heat transfer at the surface between the calorimeter and sample, and R the radius of the sample.

Numerical Analysis. The problem is solved by a numerical method with finite differences. The following circles or radii r, $r + \Delta r$, $r + 2\Delta r$, etc., are considered through the circular cross-section (Fig. 13.8).

From the heat balance determined within the annulus located between the radial abscissae $r - \frac{1}{2}\Delta r$ and $r + \frac{1}{2}\Delta r$, the new temperature TN after time Δt is expressed in terms of the temperature T obtained at the previous time $i\Delta t$.

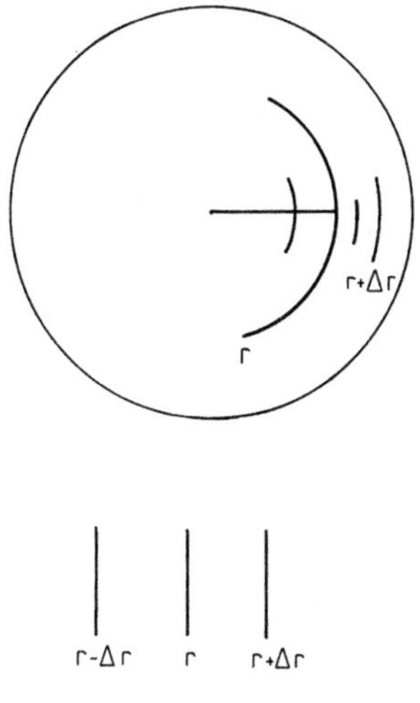

Fig. 13.8. Circular cross-section of the sample, with annuli of thickness Δr.

$$\mathrm{TN}_r = \frac{1}{M}[T_{r-\Delta r} + (M-2)T_r + T_{r+\Delta r}] + \frac{\Delta r}{2Mr}[T_{r+\Delta r} - T_{r-\Delta r}] \quad (13.22)$$

where TN_r and T_r are the new temperature after elapse of time Δt, i.e. at time $(i+1)$. Δt, at position r, and the temperature at time $i\Delta t$ at the same position, respectively.

For the centre of the cylinder ($r = 0$), the following equation is suitable for calculating the new temperature:

$$\mathrm{TN}_0 = \frac{4}{M}[T_{\Delta r} - T_0] + T_0 \quad (13.23)$$

The dimensionless number M is given by

$$M = \frac{(\Delta r)^2}{\Delta t}\frac{1}{\alpha} \quad (13.24)$$

At the interface between the sample and oven ($r = R$), there is the equation obtained in Sect. 8.2.2:

$$\mathrm{TN}_r = T_R + \frac{2}{M}\frac{R - (\Delta r/2)}{R - (\Delta r/4)}(T_{R-\Delta r} - T_R) + \frac{2P}{M}(T_\mathrm{m} - T_R) \quad (13.25)$$

where

$$P = \frac{k\Delta r}{\lambda}\frac{R}{R - (\Delta r/4)}$$

The heat flux entering the sample through the sample–calorimeter interface is obtained by the following equation:

$$\mathrm{HF} = \lambda\frac{3T_R - 4T_{R-\Delta r} + T_{R-2\Delta r}}{2\Delta r} \quad (13.26)$$

where the temperature gradient is evaluated by parabolic approximation.

Experiment

Calorimeter. The apparatus DSC 111 (Setaram) was used in scanning mode, with various values for the heating rate ranging from 2 °C min^{-1} to 20 °C min^{-1}. The sample had a radius of 3.4 mm. Heat flux was continuously recorded as a function of time and temperature.

Materials. A commercial epoxy resin was used. Of course, this resin was cured previously. The thermal properties of the resin are as follows:

Thermal conductivity: $\lambda = 10^{-3}$ cal s^{-1} cm^{-1} deg^{-1}
Thermal diffusivity: $\alpha = 2.5 \times 10^{-3}$ cm^2 s^{-1}

Results for the Temperature–Time Histories in the Centre of the Sample, for Various Sizes

When a material, cylindrical in shape for instance, is heated on its external surface, heat is transferred progressively from the surface to the centre. If the external surface is heated at a constant heating rate, the temperature at the centre follows in a delayed fashion, and it takes some time for the rate of heating at the centre to réach the constant value of heating on the surface. As a result, in DSC, the temperature at the centre of the sample is always lower than that on the surface.

This problem can be solved by mathematical treatment, when the heating rate is constant not only in the calorimeter but also on the surface of the sample [3]. The heating rate on a circle of radius r, within the long cylinder of radius R, is given by

$$\frac{\partial T}{\partial t} = \frac{\alpha}{r}\frac{\partial}{\partial r}\left(r\,\frac{\partial T}{\partial r}\right) \tag{13.27}$$

while the heating rate b on the surface is constant:

$$\frac{\partial T}{\partial t} = b \qquad \text{at position } R \tag{13.19}$$

A simple solution for these two equations is of the form

$$T = ar^2 + ct + d \tag{13.28}$$

with $\alpha = b/4\alpha$ and $c = b$, and it gives

$$T(R, t) - T(0, t) = \frac{bR^2}{4\alpha} = \Delta T \tag{13.29}$$

This last relationship is of practical interest. The difference ΔT between the temperature on the surface and at the centre of the cylinder is proportional to the heating rate and the square of the radius, when the heating rate at the centre attains the constant value b on the surface. Moreover, the numerical model is able to predict the time necessary for the difference in the temperatures on the surface and at the centre to reach this constant value ΔT, for various values of the heating rate and of the radius of the cylinder.

The temperature–time histories on the surface and at the centres of cylinders of various radii are plotted in Fig. 13.9 with a heating rate of 5 °C min^{-1}, in Fig.

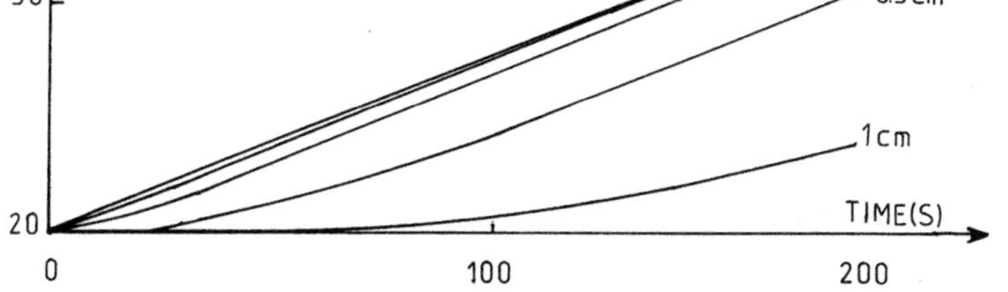

Fig. 13.9. Temperature–time histories at the surface and at the centre of cylinders of various radii, with a heating rate at the surface of 5 °C min^{-1}.

13.10 with 10 °C min^{-1}, and in Fig. 13.11 with 20 °C min^{-1}. The radius of the cylinder is ranged from 0.125 cm to 1 cm in all these cases.

This knowledge explains a feature which appears in DSC at the beginning of the process: an endothermic heat flux increases from zero up to a constant value. This constant value of the endothermic heat flux is obtained when the gradient of temperature in the sample is maximum, and when the centre of the sample is heated at the constant heating rate b.

Results for Profiles of Temperature Through the Sample

The numerical model is also capable of calculating the profiles of temperature expanded through the circular cross-section of the sample at various times. Figs 13.12 and 13.13 illustrate the shape of these profiles obtained for two samples 0.5 and 1 cm in diameter, respectively. As shown in Fig. 13.12, these profiles are parallel to each other for times longer than 30 s. However, because of the larger radius, the profiles of temperature in Fig. 13.13 are only parallel to each other for times higher than 90 s. The curves plotted in these figures are of interest to see how heat progresses through the sample.

Results for the Value of the Endothermic Heat Flux

The great interest of the model described above is that it enables the heat flux emitted through the sample–calorimeter interface to be given as a function of temperature. These curves are of interest in calorimetry because they are usually obtained by experiments.

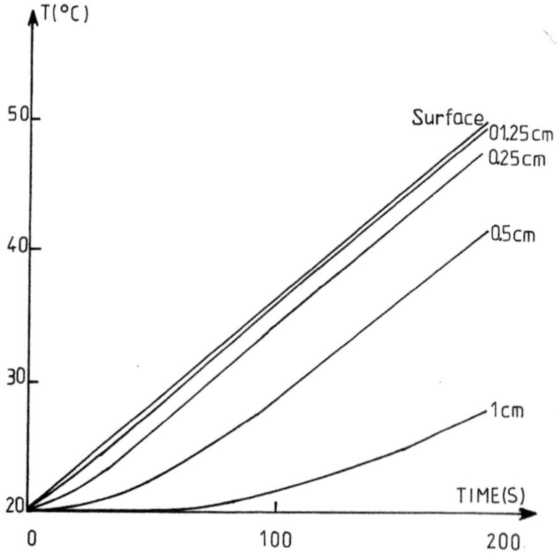

Fig. 13.10. Temperature–time histories at the surface and at the centre of cylinders of various radii, with a heating rate at the surface of 10 °C min^{-1}.

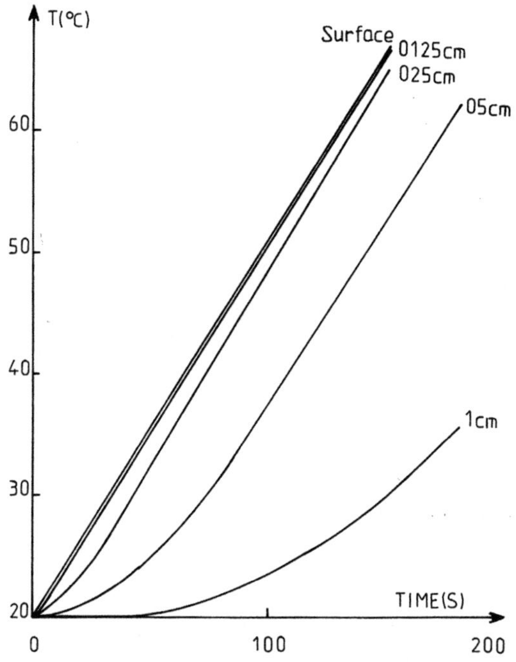

Fig. 13.11. Temperature–time histories at the surface and at the centre of cylinders of various radii, with a heating rate at the surface of 20 °C min^{-1}.

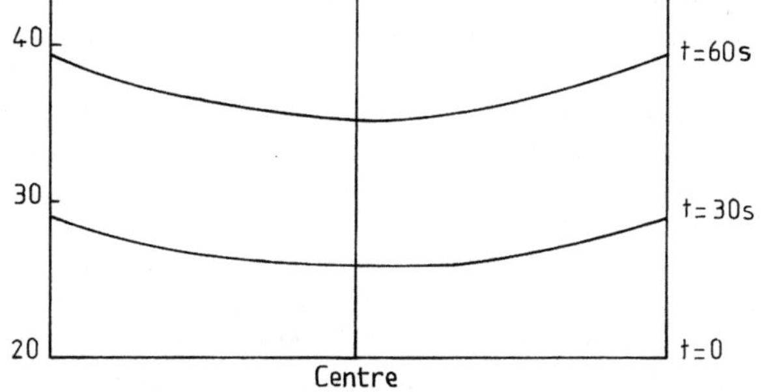

Fig. 13.12. Profiles of temperature developed through the sample with a diameter of 5 mm, at various times. Heating rate at the surface = 20 °C min^{-1}.

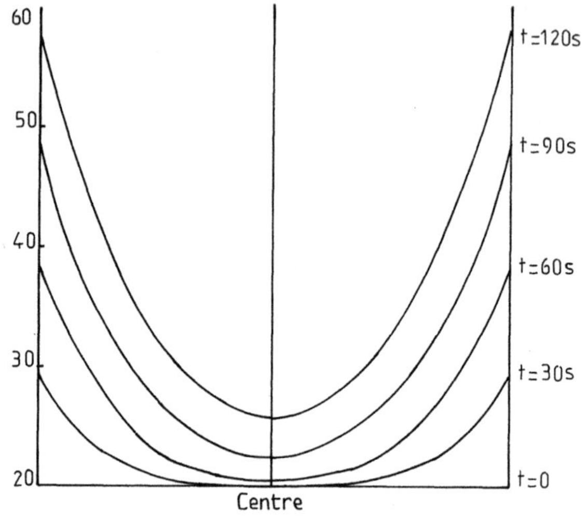

Fig. 13.13. Profiles of temperature developed through the sample with a diameter of 1 cm, at various times. Heating rate at the surface = 20 °C min^{-1}.

At the beginning of the DSC scan, the heat flux is zero, corresponding with the fact that the calorimeter and sample are at the same temperature. After this, the heat flux becomes endothermic, meaning that temperature is higher in the calorimeter. When the difference in the temperatures at the surface and at the centre of the sample reaches its maximum value ΔT, the heat flux then reaches its maximum endothermic value. Of course, this maximum endothermic value is proportional to the maximum difference in temperature ΔT. Thus, the maximum endothermic value is proportional to the heating rate, and to the square of the radius.

13.3.3 Heating Rate and Sample Size with a Reaction of Low Enthalpy [9]

As an extension of the previous study (Sect. 13.3.2), concerned with the effect of the heating rate and sample size on the heat flux emitted through the sample–calorimeter interface when no reaction occurs, the purpose of this present study is to afford fuller insight into the process which takes place in the sample located in the calorimeter when the heating rate and sample size are varied. Thus the aim is to describe the effect of the heating rate and sample size on the production of the heat flux, and on the development of profiles of temperature and SOC through rubber samples heated in a calorimeter working in the scanning mode, when the reaction evolves a low cure enthalpy. Some values are obtained from experiments for one dimension, with a cylindrical sample of external radius 3.4 mm, while other values are determined by calculation. These calculations are made by using an explicit numerical method with finite differences, by taking into account heat transferred by conduction and the exothermic cure reaction [2, 10, 11].

In the same way as for an industrial process, two steps can be considered for the sample heated in the calorimeter:

1. The first step is nearly always the heating stage of the material, by conduction for materials pressed into the mould or the holder of the calorimeter, or also by convection and conduction for materials submitted to pyrolysis. During this step, some temperature profiles are often developed through the sample, because of the low heat conductivity of the material.
2. The second step is the reaction stage with its own kinetics and enthalpy. Following the initial gradients of temperature obtained during the first stage, the rate of the reaction varies through the sample instead of being constant as in a reaction assumed to take place under isothermal conditions. The exothermicity produced by the reaction, which varies with the rate of the reaction, provokes in turn gradients of temperature which are then responsible for the development of other gradients of SOC.

Some emphasis is placed on the phenomena which take place within the sample which is quite large (cylinder of radius of 3.4 mm) heated in the calorimeter working in scanning mode, when the cure enthalpy is rather low. The case of the cure of rubber is chosen for this study. The effect of the heating rate and of the sample size on the process is determined, by considering specifically the heat flux, as well as the profiles of temperature and SOC developed within the sample.

Numerical Model

Assumptions. The following assumptions are made:

1. Heat conduction is only radial through the circular cross-section of the cylindrical sample.
2. The thermal properties of the material, such as thermal conductivity and heat diffusivity, are constant during the process, in spite of the fact that the model can accommodate temperature- or SOC-dependent parameters. It is known that there is a slight change in these thermal parameters which depends on various factors [12–15].
3. The kinetics of the cure of rubber is expressed by the heat evolved from the overall reaction.

Mathematical Treatment. The problem is about the same as that studied in Sect. 13.3.2 with a difference due to the additional heat of reaction. The equation for radial heat conduction through the circular cross-section is

$$\frac{\partial T}{\partial t} = \alpha \left[\frac{\partial^2 T}{\partial r^2} + \frac{1}{r}\frac{\partial T}{\partial r}\right] + \frac{1}{C}\frac{\partial Q}{\partial t} \tag{13.30}$$

where the contribution of conduction heat and reaction heat are shown on the right-hand side of this equation.

The increase in temperature of the calorimeter is constant:

$$\frac{dT_m}{dt} = b \qquad \text{or} \qquad T_m(t) = bt + c \tag{13.19}$$

The initial and boundary conditions are

$$t=0, \quad 0 \leqslant r \leqslant R \quad (T \text{ sample}) \tag{13.20}$$

$$t>0, \quad r = R, \quad -\lambda \frac{\partial T}{\partial r}(t, R) = k[T(t, R) - T_{\mathrm{m}}(t)] \quad \text{(interface)} \tag{13.21}$$

Numerical Analysis. There is no analytical solution for the problem, and it is solved with an explicit numerical method with finite differences.

Circles of radius r, $r+\Delta r$, $r+2\Delta r$, etc., are taken through the circular cross-section of the sample (Fig. 13.8). The heat balance during the increment of time Δt within the annulus of mean radius r and thickness Δr is evaluated by considering heat conduction and heat evolved from the cure reaction. The new temperature after time Δt is thus expressed by

$$\mathrm{TN}_r = \frac{1}{M}[T_{r-\Delta r} + (M-2)T_r + T_{r+\Delta r}] + \frac{\Delta r}{2Mr}[T_{r+\Delta r} - T_{r-\Delta r}] + \frac{\Delta Q_r}{C} \tag{13.12}$$

For the centre of the cylinder ($r = 0$), the following equation must be used:

$$\mathrm{TN}_0 = \frac{4}{M}[T_{\Delta r} - T_0] + T_0 + \frac{\Delta Q_0}{C} \tag{13.13}$$

The dimensionless number M is given by

$$M = \frac{(\Delta r)^2}{\Delta t}\frac{1}{\alpha} \tag{13.24}$$

where α is the thermal diffusivity of the sample.

The following equation is valid for the sample–calorimeter interface:

$$\mathrm{TN}_R = T_R + Z_2(T_{R-\Delta r} - T_R) + \frac{2P}{M}(T_{\mathrm{m}} - T_R) + \frac{\Delta Q_{R-(\Delta r/4)}}{C} \tag{13.31}$$

where

$$Z_2 = \frac{2}{M}\frac{R - (\Delta r/2)}{R - (\Delta r/4)} \quad \text{and} \quad P = \frac{k\Delta r}{\lambda}\frac{R}{R - (\Delta r/4)}$$

The heat flux transferred through the calorimeter–sample interface is given by

$$\mathrm{HF} = \lambda \frac{3T_R - 4T_{R-\Delta r} + T_{R-2\Delta r}}{2\Delta r} \tag{13.26}$$

The rate of heat evolved from the cure reaction is expressed in terms of temperature by a first-order Arrhenius equation:

$$\frac{dy}{dt} = k_0(1-y)\exp\left(-\frac{E}{RT_{\mathrm{t}}}\right) \tag{13.32}$$

where y is the amount of heat evolved up to time t: $y = Q_t/Q_\infty$. By integrating this equation as in Sect. 7.4.2 and defining the function S,

$$S_{\mathrm{t}} = \int_0^{\mathrm{t}} k_0 \exp\left(-\frac{E}{RT_\theta}\right) d\theta \qquad 0 \leqslant \theta \leqslant t \tag{13.33}$$

the following is obtained:

$$y = 1 - \exp(-S_t) \tag{13.34}$$

At any place, the function S is calculated step by step as follows:

$$S_{t+\Delta t} = S_t + k_0 \Delta t \exp\left(-\frac{E}{RT}\right) \tag{13.35}$$

with $S_0 = 0$.
The value of the extent of cure y is thus obtained at any place and any time, and the amount of heat ΔQ evolved during the time Δt is thus given by

$$\Delta Q = Q_\infty (y_{t+\Delta t} - y_t) \tag{13.36}$$

Of course, in the holder where no reaction takes place, Eq. (13.22) without the heat evolved from the cure reaction is used.

Experiment

Calorimeter. A differential calorimeter DSC 111 (Setaram) was used in the scanning mode at various heating rates ranging from 2 to 20 °C min^{-1}. About 100–150 mg of rubber is encapsulated in a hermetically sealed stainless steel holder.

The apparatus is coupled with a microcomputer (HP 86, Hewlett-Packard).

Materials. An industrial rubber compound with a low enthalpy of cure, around 5 cal/g, was used. The kinetics of cure were determined by DSC in scanning mode with a heating rate of 2 °C min^{-1}. The parameters for the kinetics of cure and the thermal properties were as follows:

$$\text{Enthalpy} = 5\ \text{cal g}^{-1}, \quad \text{order } n = 1, \quad E = 25\,400\ \text{cal mol}^{-1}$$

$$k_0 = 3 \times 10^9\ \text{s}^{-1}, \quad C = 0.49\ \text{cal g}^{-1}\,\text{deg}^{-1}, \quad \alpha = 1.83 \times 10^{-3}\ \text{cm}^2\,\text{s}^{-1}$$

$$\lambda = 9.9 \times 10^{-4}\ \text{cal cm}^{-1}\,\text{s}^{-1}\text{deg}^{-1}$$

Results for the Heat Flux Emitted Through the Sample Calorimeter Interface Due to the Exothermic Reaction

These heat flux–temperature curves are of interest in calorimetry because they are usually obtained by experiment. The curves are plotted in Fig. 13.14 for various heating rates, obtained from experiments with the sample of diameter 5 mm and the holder of diameter 6.8 mm, and as calculated with the model. Various scales are used for the heat flux in order to facilitate the comparison. Of course, the baselines corresponding to the heat flux emitted when no reaction occurs are not the same for the same sample heated at different rates [3] (Sect. 13.3.2).

The heat flux–temperature curves were also calculated with the model for larger samples of diameter of 1 cm (Fig. 13.15) and 2 cm (Fig. 13.16). Some results of interest are obtained from these curves, especially regarding the sensitivity of the technique and the positions of the heat flux–temperature curves.

1. The sensitivity of the calorimeter can be expressed by the height of the peak corresponding to the heat flux at the maximum value. As shown in Figs 13.17

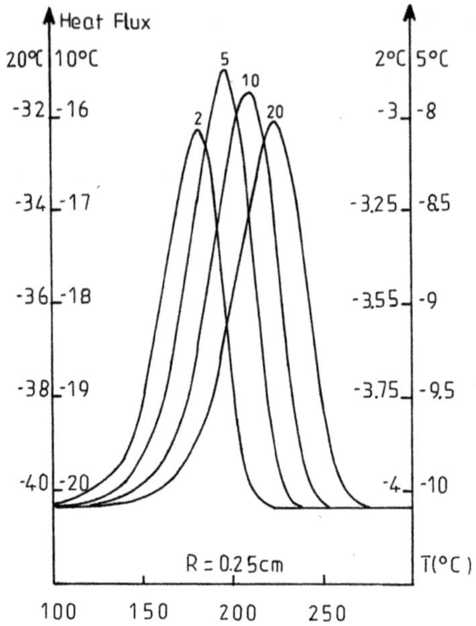

Fig. 13.14. Heat flux–temperature curves for various heating rates ranging from 2 to 20 °C min^{-1}, with various scales for the heat flux. Diameter = 5 mm for the sample and 6.8 mm for the holder.

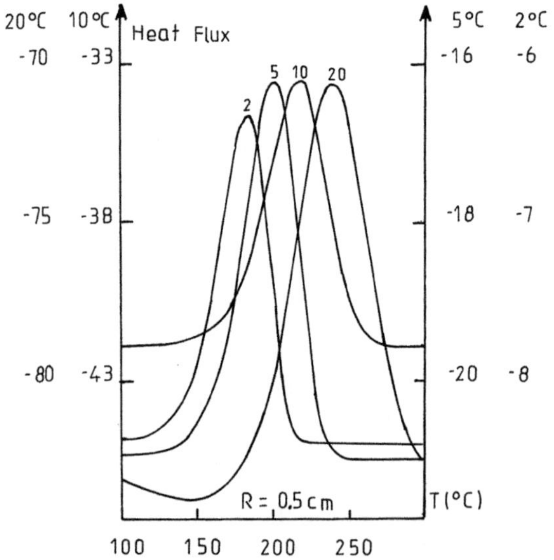

Fig. 13.15. Heat flux–temperature curves for various heating rates ranging from 2 to 20 °C min^{-1}, with various scales for the heat flux. Diameter = 10 mm for the sample and 11.8 mm for the holder.

and 13.18, the sensitivity is found to be directly proportional to the value of the radius of the samples, and also proportional to the heating rate.

2. Another result of interest is related to the position of the heat flux–temperature peaks, and especially the position of the maximum of these peaks

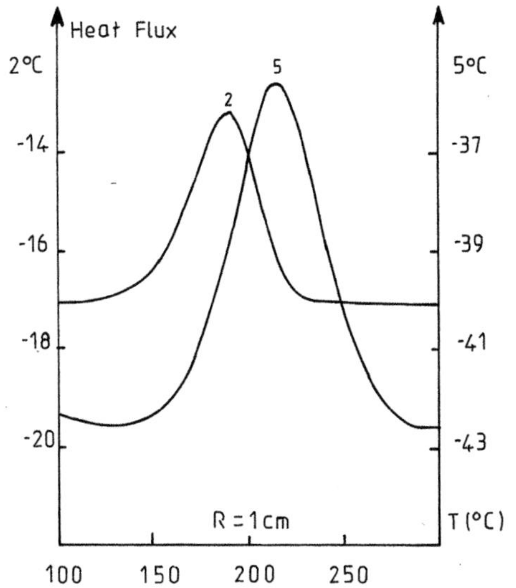

Fig. 13.16. Heat flux–temperature curves for various heating rates ranging from 2 to 5 °C min^{-1}, with various scales for the heat flux. Diameter = 20 mm for the sample and 21.8 mm for the holder.

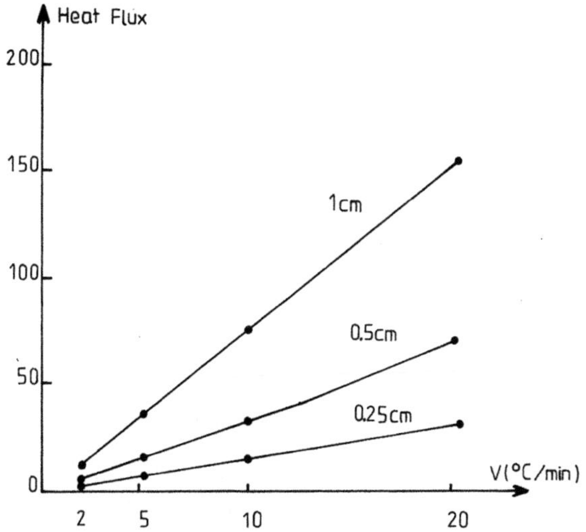

Fig. 13.17. Maximum heat flux as a function of the heating rate for various sample sizes in the cure of rubber.

expressed by the temperature at which they are obtained. The variation of this maximum as a function of the radius of the samples can be appreciated in Fig. 13.19 for various heating rates ranging from 2 to 20 °C min^{-1}. There is no simple relation expressing this variation, and it depends on the kinetic parameters. However, it is clear that the larger the radius of the sample, the

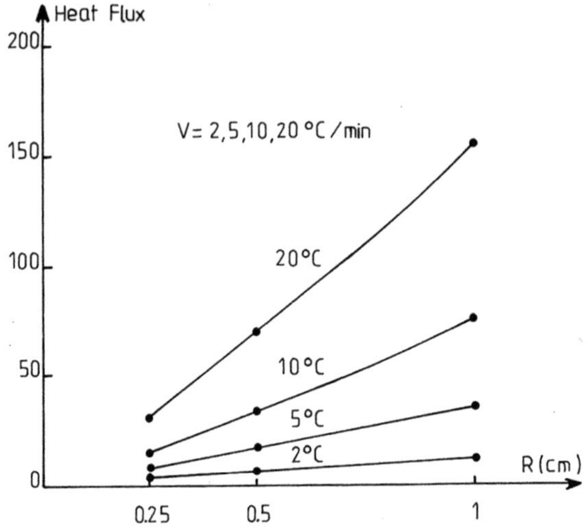

Fig. 13.18. Maximum heat flux as a function of the sample radius, for various heating rates in the cure of rubber.

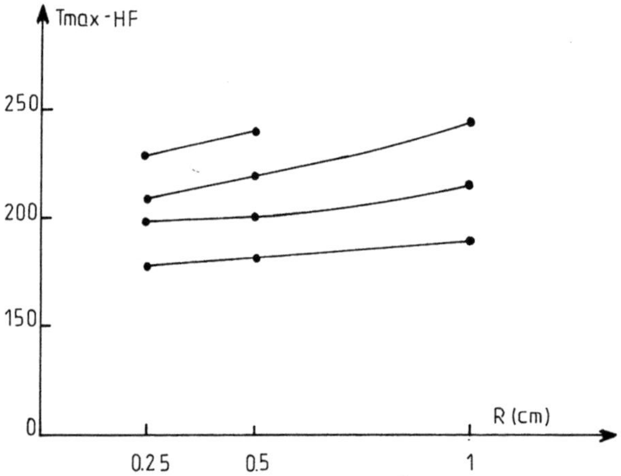

Fig. 13.19. Position of the maximum heat flux (expressed by the temperature at which it is obtained) as a function of the sample radius in the cure of rubber.

higher is the temperature at which the maximum heat flux is attained. In the same way, a significant increase in this temperature is shown when higher heating rates are used.

Results for the Temperature Profiles Within the Sample

As shown in previous papers [3, 16–19] gradients of temperature are developed within the rubber mass during cure because of local heating arising from exothermic reactions and low heat transfer through the rubber. In DSC as well as in DC, the applied temperature is the set calorimeter temperature and not the rubber temperature during cure. As the rate of reaction is largely dependent on the temperature, a precise knowledge of the exact temperature within the sample during calorimetry must be of advantage. In fact the problem is complicated because the heating rate and the rubber temperature are connected with each other, the exothermic reaction provoking an increase in temperature and a following increase in the reaction rate and enthalpy change.

Two facts must be considered in turn:

1. When the material is heated with a constant heating rate on its cylindrical surface, heat is progressively transferred to the centre, so that the temperature at the centre follows in a delayed fashion. As a result, the temperature at the centre is always lower than that on the surface when no reaction occurs, this difference being proportional to the square of the radius and to the heating rate [3].
2. Following the above stage of heating, the exothermic reaction is responsible for the development of other gradients of temperature within the sample, with the highest temperature at the centre of the sample when the enthalpy of reaction is high enough to compensate the first gradient of temperature due to the heating rate.

Because the local temperature results from the superposition of the above two facts, the general problem is complex, the consequence depending largely on the value of the cure enthalpy.

The temperature histories at the centre of the sample ΔT, as a function of the calorimeter temperature are shown in Figs 13.20–13.22 for the various sizes chosen for the samples. Resulting from the cure enthalpy, an increase in the temperature at the centre of the sample is observed (ΔT) and some peaks are obtained, for which ΔT is maximum. The heights of ΔT_{max} are approximately proportional to the heating rate and sample size. The point concerned with the position of these peaks is worth attention, and the values in Table 13.2 allow one

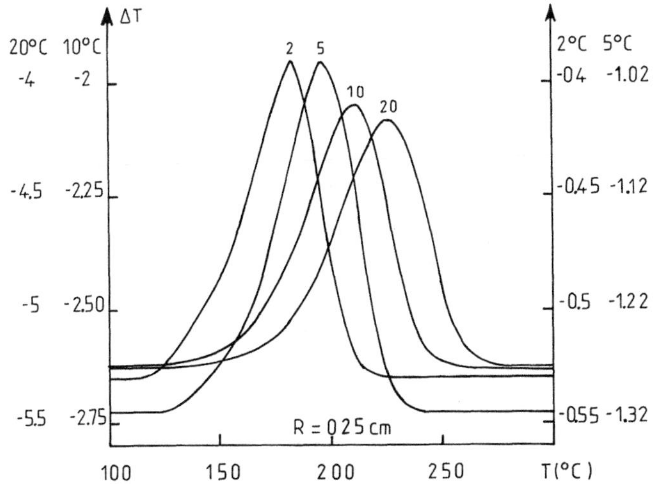

Fig. 13.20. Temperature at the centre of the sample as a function of the calorimeter temperature, in DSC, for various heating rates. Sample diameter 5 mm. (ΔT is the difference of the temperatures obtained at the centre with and without reaction.)

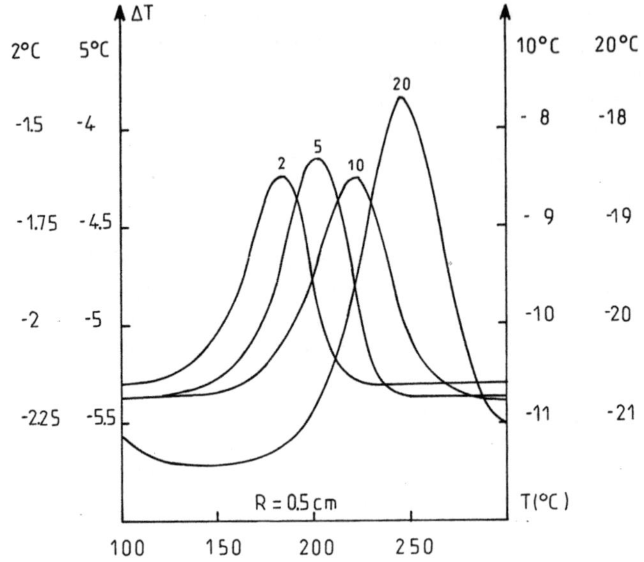

Fig. 13.21. Temperature at the centre of the sample as a function of the calorimeter temperature, in DSC, for various heating rates. Sample diameter 1 cm. (ΔT is the difference of the temperature obtained at the centre with and without reaction.)

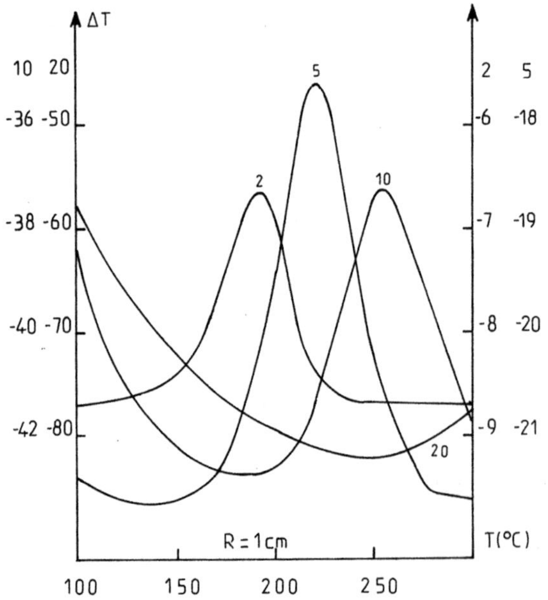

Fig. 13.22. Temperature at the centre of the sample as a function of the calorimeter temperature, in DSC, for various heating rates. Sample diameter 2 cm. (ΔT is the difference of the temperature obtained at the centre with and without reaction.)

to give consideration to this value, compared with that obtained with the heat flux.

The heat flux and the gradients of temperature exhibit a maximum value at about the same temperature of the calorimeter, especially for the lower sample size, at any heating rate.

Results for the SOC Profiles Within the Sample

As the temperature is not the same in the cross-section of the cylindrical sample in DSC, the rate of cure varies through the sample, and subsequently some profiles of SOC must be obtained. These are shown in Figs 13.23–13.25 for the various sizes of the sample, as they are calculated with the help of the model. Of course, the value of the heating rate largely depends on the position of the SOC–temperature curves (which are calculated at the centre of the sample). So the higher the heating rate, the lower the value of the SOC obtained.

By considering the SOC value, and in view of the gradients of SOC developed within the sample, the rubber becomes heterogeneous during the cure. There is no doubt that some significant differences in properties of the final product may

Table 13.2. Temperatures (°C) at which ΔT_{max} and HF_{max} are attained for various heating rates

Radius (cm)	Parameter	Heating rate (°C) min^{-1})			
		2	5	10	20
0.25	HF_{max}	182.3	197.3	211.3	225.9
	ΔT_{max}	183	197	211	226
0.5	HF_{max}	183	201	217	238
	ΔT_{max}	185	201	221	246

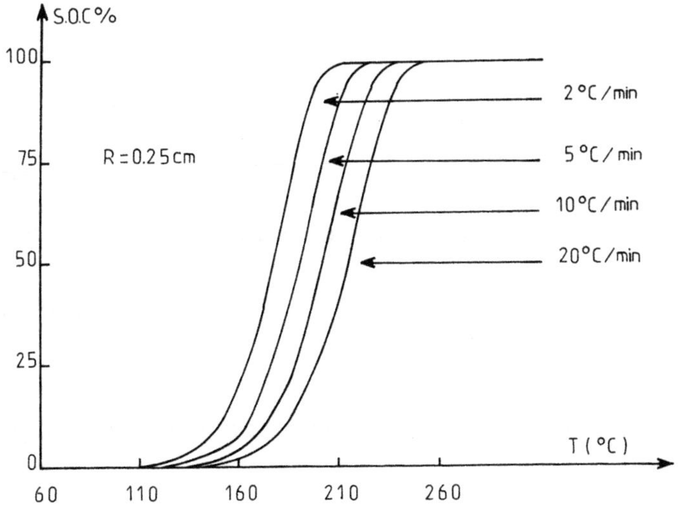

Fig. 13.23. SOC at the centre vs temperature in DSC, for various heating rates. Sample diameter 5 mm.

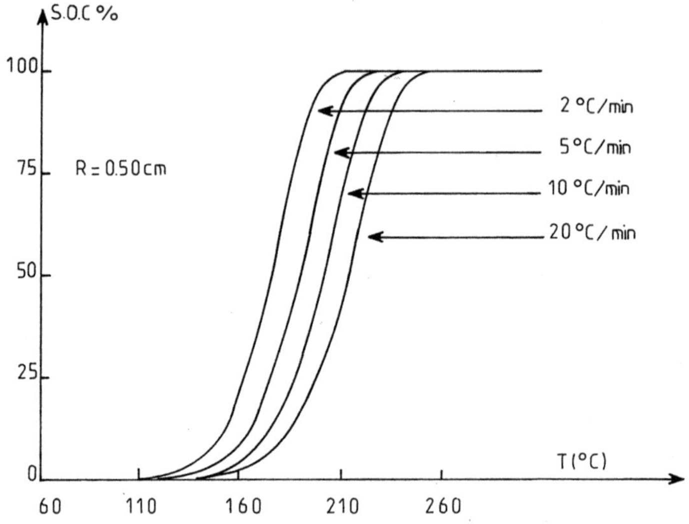

Fig. 13.24. SOC at the centre vs temperature in DSC, for various heating rates. Sample diameter 1 cm.

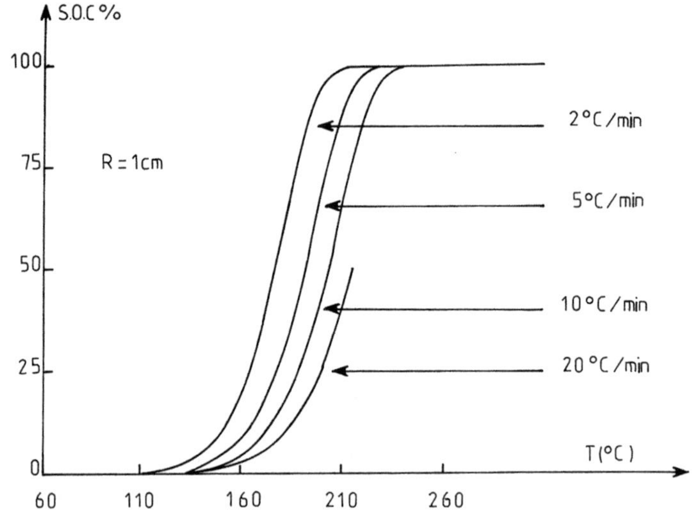

Fig. 13.25. SOC at the centre vs temperature in DSC, for various heating rates. Sample diameter 2 cm.

be obtained if the temperature–time history of the rubber exhibits such differences in the various parts of the sample.

13.3.4 Heating Rate and Sample Size with a Reaction of High Enthalpy [20]

The effect of the heating rate on the process in the sample located in the calorimeter with the DSC mode is studied, by considering a fairly high enthalpy of cure. The effect of the sample size on the process is also examined. The problem is thus about the same as that studied with the cure of rubber and a low enthalpy (Sect. 13.3.3), but in the present case the material is an epoxy resin with a high cure enthalpy.

The theoretical part is the same as that described in the case of the rubber (Sect. 13.3.3). The same numerical model takes into account heat transferred by conduction through the sample and heat evolved from the cure reaction is also used.

The purpose of this section is to study the process of cure of an epoxy resin with a high enthalpy ($76 \, cal \, g^{-1}$), and to determine the effect of the heating rate and of the sample size on the heat flux–temperature histories obtained in DSC. Other additional information such as the profiles of temperature and of SOC developed through the sample are examined.

Two samples, both cylindrical in shape, are considered, one with a diameter of 5 mm and the other with a diameter of 2 mm. The holder diameter is 6.8 mm. As no measurement is possible for the gradients of temperature, the validity of the model is tested by comparing the experimental and calculated heat flux obtained for the larger sample.

Numerical Model

The numerical model is described in Sect. 13.3.3, concerned with the rubber. It takes into account the radial heat transfer by conduction and heat evolved from the cure reaction.

As an epoxy resin with a high enthalpy of cure is used, with a reaction order different from one, the treatment of kinetics of heat evolved from the reaction is different from that shown for the rubber with an order of one. The rate of heat evolved from the cure reaction is assumed to verify the relationship

$$\frac{dy}{dt} = k_0(1 - y)^n \exp\left(-\frac{E}{RT}\right) \tag{13.1}$$

with y, the extent of cure at time t, defined by $y = Q_t/Q_\infty$. The integration of this equation gives

$$y = 1 - [1 + (n - 1)S_t]^{1/(1-n)} \tag{13.37}$$

where the function S_t is defined by

$$S_t = \int_0^t k_0 \exp\left(-\frac{E}{RT_\theta}\right) d\theta \qquad 0 \leqslant \theta \leqslant t \tag{13.33}$$

The values of the function S are calculated step by step as follows:

$$S_{t+\Delta t} = S_t + k_0 \Delta t \exp\left(-\frac{E}{RT}\right) \qquad \text{with} \qquad S_0 = 0 \tag{13.35}$$

The value of the extent of cure y is thus obtained at any place and any time, and the amount of heat evolved during the interval of time $[t, \ t + \Delta t]$ is thus given by

$$\Delta Q = Q_\infty(y_{t+\Delta t} - y_t) \tag{13.36}$$

Of course, the radial heat transfer through the sample and holder, as well as the heat flux are determined in the same way as for the rubber (Sect. 13.3.3).

Experiment

Apparatus. Experiments were carried out using a calorimeter DSC 111 (Setaram) and the 5 mm diameter sample. The sample, which weighed about 150 mg, was located in a cylindrical stainless steel holder with external diameter 6.8 mm. The calorimeter detector was in good contact with the holder, with the help of an aluminium sheet.

Materials. An epoxy resin with two components, epoxy DGEBA and cycloaliphatic hardener, was used. The formulation contained 80 parts of curing agent to 100 parts of resin (w/w). The mixture was previously prepared at room temperature by adding the curing agent with strong stirring until a clear liquid was obtained

The kinetic and thermal parameters were as follows:

$$Q_\infty = 76\ \mathrm{cal\,g^{-1}}, \qquad k_0 = 4 \times 10^{11}\ \mathrm{s^{-1}}, \qquad E = 27\,140\ \mathrm{cal\,mol^{-1}},$$

$$n = 1.5, \qquad \lambda = 10^{-3}\ \mathrm{cal\,cm^{-1}\,s^{-1}\,deg^{-1}},$$

$$C = 0.2 + 0.003 \times T\ \mathrm{cal\,g^{-1}\,deg^{-1}}\ (T \text{ in } ^\circ\mathrm{C})$$

Results for the Heat Flux–Temperature Histories

Heat flux–temperature histories were determined by calculation and experiment for the larger sample, and by calculation only for the smaller sample, both these samples being cylindrical in shape. Other calculations are made with the help of the model. Heat flux–temperature histories are of interest, because they are directly recorded from the calorimeter. They are shown in Fig. 13.26 for the large sample and in Fig. 13.27 for the small sample, for various values of the heating rate ranging from 1° to 5 °C min^{-1}.

Some results are worth noting:

1. The position of the heat flux–temperature peak varies with the value of the heating rate. As shown in a previous paper [21], these peaks are shifted towards a higher temperature when the heating rate is increased.
2. The value of the heat flux depends on the value of the heating rate and the dimensions of the sample. The heat flux is found to be proportional to the heating rate as shown previously [21] in both cases: when no reaction occurs, with an endothermic heat flux below zero; when the reaction takes place in the sample, the heat flux increasing and passing through a maximum.
3. The effect of the sample size is also important. Thus, either when no reaction occurs or when the reaction takes place, the heat flux is proportional to the main dimension of the sample (which is the diameter in the case of a long cylinder).
4. The curves expressing the heat flux as a function of temperature for the large and the small sample are homothetic, when they are obtained at the same

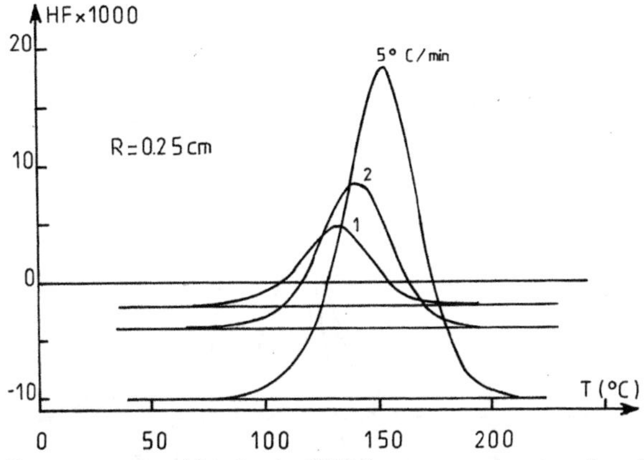

Fig. 13.26. Heat flux–temperature histories in DSC for epoxy resin, at various heating rates (1, 2, 5 °C min^{-1}). Sample diameter 5 mm.

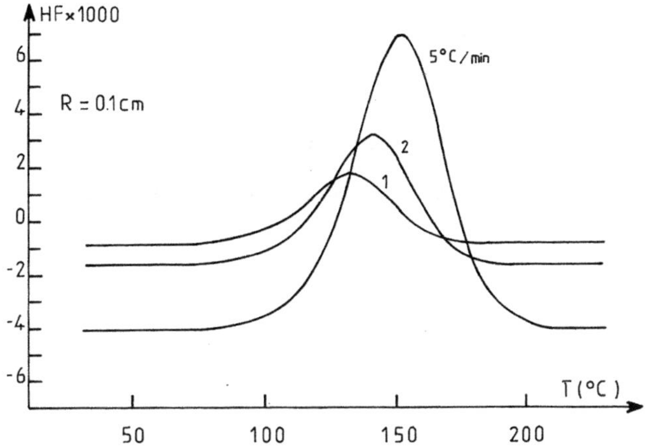

Fig. 13.27. Heat flux–temperature histories in DSC for epoxy resin, at various heating rates (1, 2, 5 °C min^{-1}). Sample diameter 2 mm.

heating rate. In other words, the curves obtained for the two different samples and calorimeters exhibit a maximum at the same temperature.

Results for the Temperature Profiles and Temperature Histories

The temperature histories at the centre of the sample as a function of the temperature of the calorimeter, for various values of the heating rate, are shown in Fig. 13.28 for the larger sample and in Fig. 13.29 for the smaller one. The following conclusions can be drawn from these figures:

1. A good agreement is observed between these curves and the heat flux–temperature histories. In fact the heat flux results from the presence of the gradients of temperature through the sample next to the surface.

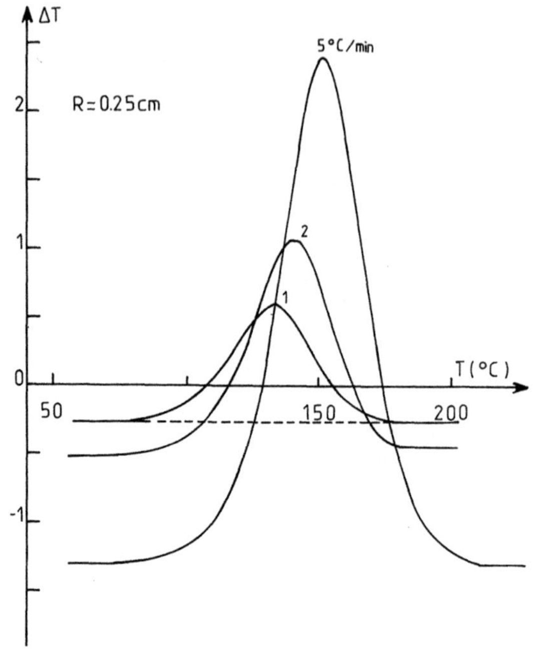

Fig. 13.28. Temperature histories at the centre of the sample (diameter 5 mm) as a function of the calorimeter temperature, for various heating rates (1, 2, 5 °C min^{-1}) (ΔT is the difference of the temperature obtained at the centre with and without reaction).

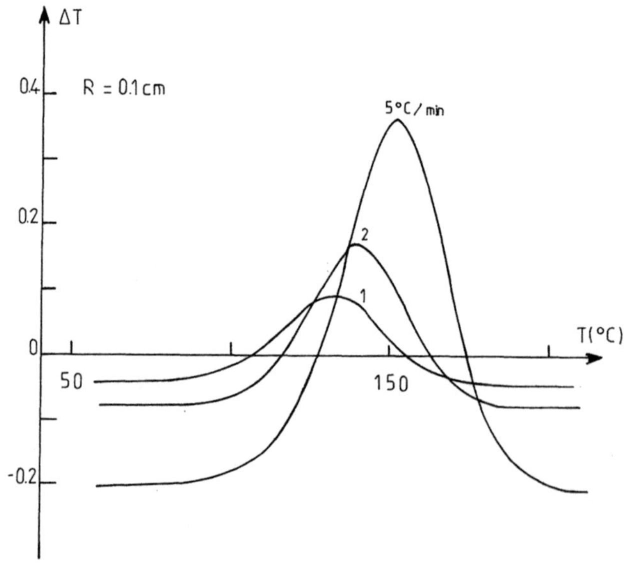

Fig. 13.29. Temperature histories at the centre of the sample (diameter 2 mm) as a function of the calorimeter temperature, for various heating rates (1, 2, 5 °C min^{-1}) (ΔT is the difference of the temperature obtained at the centre with and without reaction).

2. The temperature is higher on the surface than at the centre of the sample when no reaction occurs. This difference is due to the time necessary for heat to be transferred through the sample [21]. This difference in temperature is proportional either to the value of the heating rate of the calorimeter and to the square of the main dimension of the sample (e.g. the radius or diameter for a long cylinder).
3. When the reaction takes place, the temperature at the centre of the sample increases because of the exothermic cure reaction. This increase in the temperature at the centre is denoted ΔT. As a result, the curves drawn in Figs 13.28 and 13.29 exhibit a maximum ΔT_m. This maximum is at the same position along the temperature axis as the maximum of the heat flux. Moreover, the position of the maximum of ΔT depends only on the value of the heating rate and not on the sample size.
4. In fact the temperature profiles at the centre of the sample and the heat flux have not only the same shape, but they are also homothetic, ΔT being proportional to the heat flux. This is true when no reaction occurs with a negative value for ΔT and for the heat flux, and also when the reaction takes place.
5. ΔT, in the same way as the heat flux, is proportional to the value of the heating rate.

Results for the Profiles of Temperature and SOC Through the Sample

The profiles of temperature and SOC along the radius of the sample are shown at various times, for various values of the heating rate either for the larger sample (Fig. 13.30) or for the smaller sample (Fig. 13.31). The effect of the size of the samples can be clearly appreciated by comparing the curves in these two figures. The following conclusions can be drawn:

1. For both samples, the temperature difference ΔT is proportional to the square of the main dimension of the sample at any time, for the same heating rate.

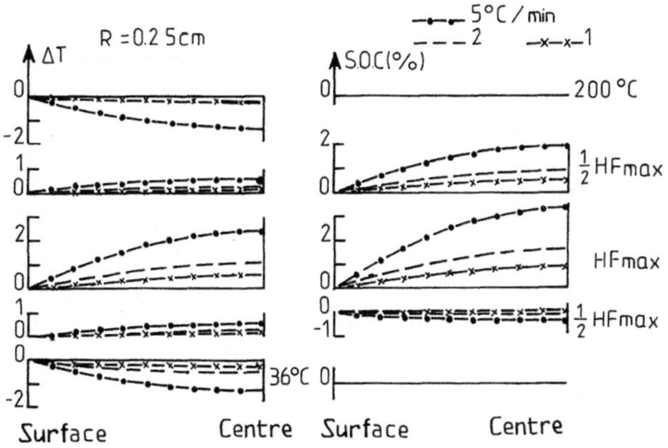

Fig. 13.30. Profiles of temperature (left) and SOC (right) developed through the large sample (diameter 5 mm) for heating rates of 1, 2, 5 °C min^{-1}.

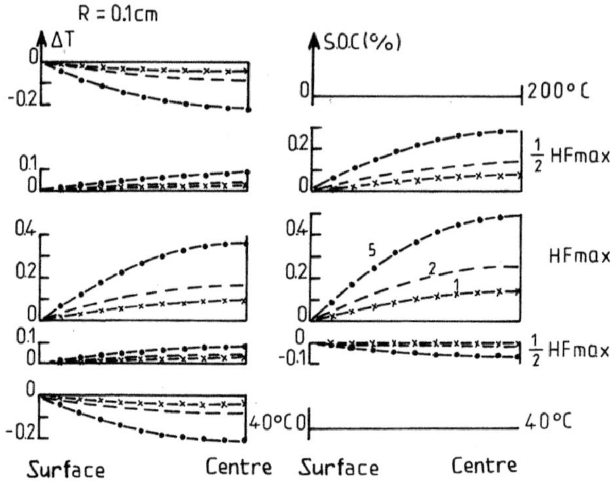

Fig. 13.31. Profiles of temperature (left) and SOC (right) developed through the large sample (diameter 2 mm) for heating rates of 1, 2, 5 °C min^{-1}.

2. The difference between the SOC obtained at the centre and on the surface of the sample is also proportional to the square of the main dimension of the sample (the radius for a long cylinder).

13.3.5 Heating Rate and Cure Enthalpy [21, 22]

In the DSC experiment, the sample is scanned at a constant heating rate up to a temperature at which the reaction exotherm is complete. The first drawback of the technique is obvious, as shown previously: the process is always conducted under transient conditions with a first step necessary for the temperature gradients to be established, followed by a second heating step with a constant temperature gradient [3, 11]. The difference in temperature between the surface and centre of the sample is proportional to the value of the heating rate and to the square of the main dimension of the sample (radius for a cylinder) [3, 9, 22]. Another inconvenience of this technique is, that in the DSC mode, a higher temperature is necessary to achieve the cure than in the DC experiment. Subsequently, this high temperature may be responsible for a degradation of the resin. This fact is especially obvious in the case of a special rubber with low cure enthalpy: a low heat flux is obtained with a low heating rate, and thermal degradation is observed sometimes on the surface of the sample before the completion of the cure.

In spite of these drawbacks, taken for granted, the DSC technique is widely used. However, the problem concerning the temperature gradients developed through the sample remains. This problem is complex because two kinds of temperature gradients are expanded in the sample, one being due to the heating rate and the retardation in the heat propagation, and the other due to the exothermicity of the reaction, these gradients being opposite. The effect of parameters such as the heating rate and cure enthalpy on these gradients are of great importance, for a given sample size.

The purpose of this section is to gain a fuller insight into the problem of these temperature gradients developed through a cylindrical sample. The cure of epoxy resin is chosen for the exothermic reaction, because of the high value of the cure enthalpy when the resin is pure, and because lower values for the cure enthalpy can sometimes be obtained by dispersing a non-reactive material into the resin. This study necessitates good knowledge of the temperature profiles, and is thus conducted with the help of a model built using a numerical method with finite differences. This model takes into account not only the heat evolved from the cure reaction, but also the heat transferred by conduction through the sample. As no measurement is possible for determining these temperature gradients, the validity of the model is tested by comparing the experimental and calculated heat flux [3, 9–11, 20–22].

Special emphasis is placed on the values of the cure enthalpy with values ranging from 15 to 75 $\mathrm{cal\,g^{-1}}$. The heat flux–temperature curves are determined, as well as the difference of temperatures between the surface and the centre of the sample, for various values of the heating rate. The gradients of temperature and SOC developed within the sample during the process are also calculated.

Numerical Model

Assumptions. The following assumptions are made:

1. Heat transfer is radial through the sample by conduction.
2. The kinetics of the heat evolved from the cure reaction is expressed by an Arrhenius equation.
3. The enthalpy of cure is proportional to the weight per cent of resin.
4. The thermal properties of the diluting material are about the same as that of the resin.

Mathematical Treatment

The increase in temperature is described by the classical equation of radial transient heat transfer, where the contributions due to heat conduction and internal heat generated by the reaction are taken into account:

$$\rho C \frac{\partial T}{\partial t} = \frac{1}{r}\frac{\partial}{\partial r}\left(r\lambda \frac{\partial T}{\partial r}\right) + \rho \frac{\partial Q}{\partial t} \tag{13.7}$$

with the parameters shown above (Sect. 13.2).

In contrast to some workers who used equations with two constant pre-exponential factors and activation energy [23–26], the heat evolved from the overall cure reaction is expressed by the following equation with one energy of activation [1, 27].

$$\frac{dy}{dt} = k_0(1 - y)^n \exp\left(-\frac{E}{RT}\right) \tag{13.1}$$

where $y = Q_t/Q_\infty$ is the extent of cure at time t.

Numerical Analysis

The problem is the same as that shown in the case of the cure of rubber in DSC with a first-order reaction (Sect. 13.3.3) and in the case of the cure of epoxy resin in DSC with a *n*th order reaction (Sect. 13.3.4). In fact, the numerical model for the heat transfer and the heat flux is exactly the same as that described for the rubber, while the treatment of the heat generated by the reaction is described in the case of the epoxy resin. As a result, it is possible to determine the following information:

1. The heat flux emitted through the surface of the calorimeter, which can also be obtained by experiments
2. The profiles of temperature and SOC developed within the sample

Experiment

Apparatus. The 5 mm diameter sample (150 mg) was placed in a cylindrical aluminium holder, external diameter 6.8 mm. The calorimeter DSC 111 (Setaram) was used for this work, but the results obtained are also of interest for any other apparatus if the dimensions and size of the holder are considered.

Materials. An epoxy resin with two components: epoxy DGEBA (Lopox 200 ST 01) and hardener (cycloaliphatic D 2605 from CDF Chimie) was used. The formulation contains 80 parts (w/w) curing agent to 100 parts resin. The epoxy hardener mixture was prepared at room temperature by adding the curing agent with continuous stirring until a clear mixture was obtained. Grains of sand were continuously added to the epoxy mixture with appropriate stirring, in order to vary the cure enthalpy of the sample [28]. The compositions were experimentally determined in order to obtain the desired values of cure enthalpy ranging from 15 to 75 cal g^{-1}.

The values of the kinetic and thermal parameters were as follows:

$$\text{Cure enthalpy} = 15, 25, 50 \text{ and } 75 \text{ cal g}^{-1},$$

$$n = 1.5, \quad k_0 = 4 \times 10^{11}\ \text{s}^{-1}, \quad E = 27\,140\ \text{cal mol}^{-1},$$

$$\lambda = 10^{-3}\ \text{cal cm}^{-1}\,\text{s}^{-1}\,\text{deg}^{-1}, \quad C = 0.5\ \text{cal g}^{-1}, \quad \rho = 1.4\text{–}1.8\ \text{cal g}^{-1}$$

Results for the Effect of Cure Enthalpy on the Heat Flux–Temperature Curves

The problem is rather complex because various parameters such as the enthalpy of cure, the kinetic and thermal characteristics of the sample and the heating rate are of importance to the process. For this reason the problem was simplified in a previous paper [3] (Sect. 13.3.2), in which heat transfer in the sample was considered only as a function of the heating rate and sample size. As no reaction occurs in this case, the heat flux and temperature histories are defined as a function of the size of the sample and of the value selected for the heating rate. The main result obtained in this study is that the process in this case is conducted in transient conditions, with the following two steps: one when the temperature gradient along the radius of the cylindrical sample is increasing from zero at the beginning up to a constant value; the other with an established constant temperature gradient through the sample while the temperatures on the sample surface and at any place in the sample are increased at the same rate as the calorimeter. Another result of interest is that during the second stage, when the gradient of temperature is established, the value of the endothermic heat flux is proportional to the heating rate of the calorimeter and to the square of the main dimension of the sample (diameter for a long cylinder, thickness for a sheet).

As the enthalpy of cure is the principal parameter studied in this section, the effect of the value of this enthalpy on the process in particular is examined, by considering successively the heat flux–temperature histories, the profiles of temperature and SOC developed within the sample.

The heat flux–temperature history was determined for various values of the heating rate ranging from 0.5 to 5 °C min^{-1}, and for the following values of the enthalpy of cure: 15 cal g^{-1} (Fig. 13.32); 25 cal g^{-1} (Fig. 13.33); 50 cal g^{-1} (Fig. 13.34); and 75 cal g^{-1} (Fig. 13.35). As shown in these figures, these two parameters have a marked effect on the heat flux–temperature curves.

The heating rate of the calorimeter has the following effects on these curves:

1. The height of heat flux–temperature peaks depends largely on the value of the heating rate. As shown in a previous paper [9] and also in Sect. 13.3.3, the sensitivity of the calorimeter to the reaction expressed by the value of the heat flux is directly proportional to the heating rate.
2. Another result of interest is related to the position of the heat flux–temperature peak, and especially the position of the maximum of this peak. This position, defined by the temperature of the calorimeter at which the

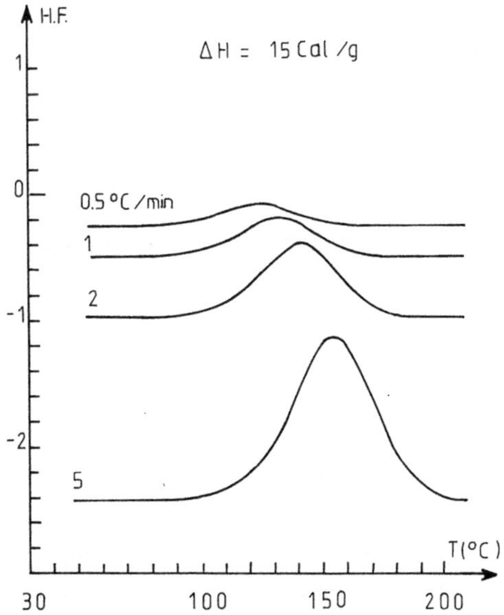

Fig. 13.32. Heat flux–temperature history in DSC for various heating rates ranging from 0.5 to 5 °C min^{-1}, with a cure enthalpy of 15 cal g^{-1}.

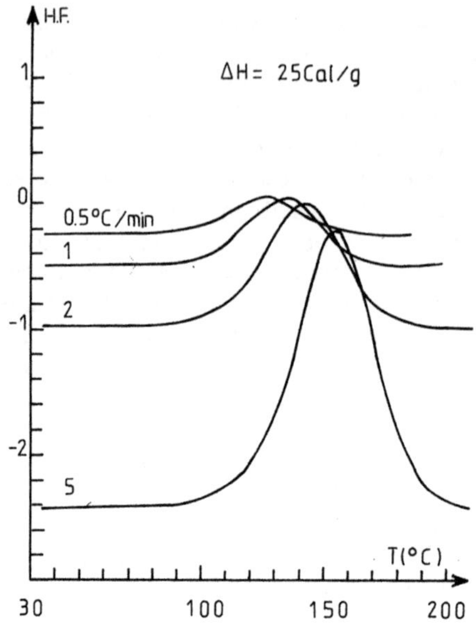

Fig. 13.33. Heat flux–temperature history in DSC for various heating rates ranging from 0.5 to 5 °C min^{-1}, with a cure enthalpy of 25 cal g^{-1}.

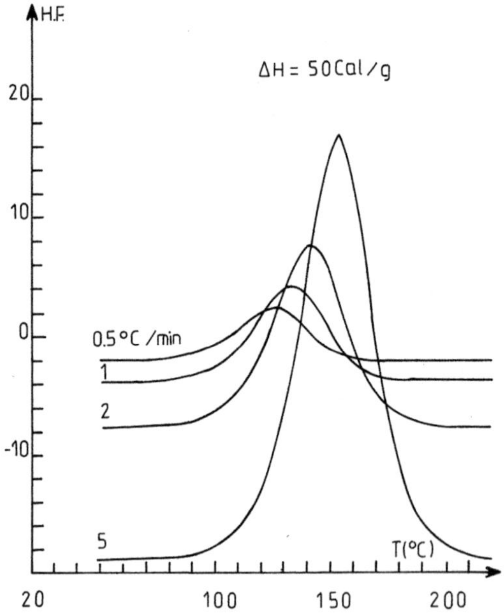

Fig. 13.34. Heat flux–temperature history in DSC for various heating rates ranging from 0.5 to 5 °C min^{-1}, with a cure enthalpy of 50 cal g^{-1}.

maximum of heat flux is attained, is stated as a function of the value of the heating rate, as shown in Fig. 13.36. The heat flux–temperature peaks are shifted towards higher temperature when the heating rate is increased, essentially because the true temperature of the sample is not the calorimeter

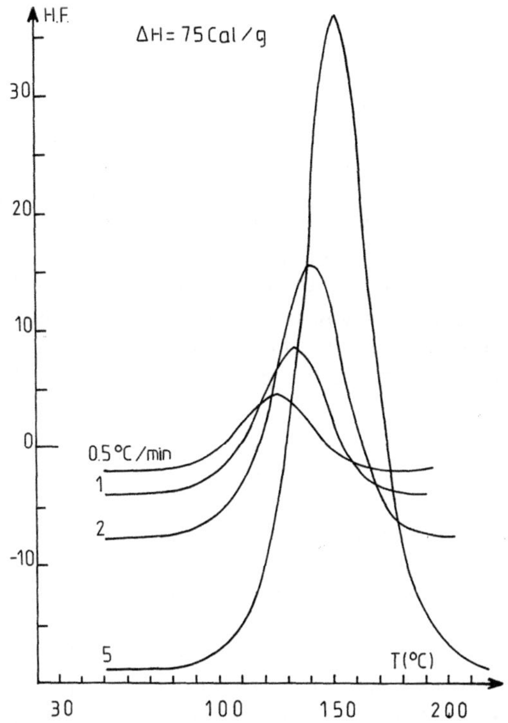

Fig. 13.35. Heat flux–temperature history in DSC for various heating rates ranging from 0.5 to 5 °C min^{-1}, with a cure enthalpy of 75 cal g^{-1}.

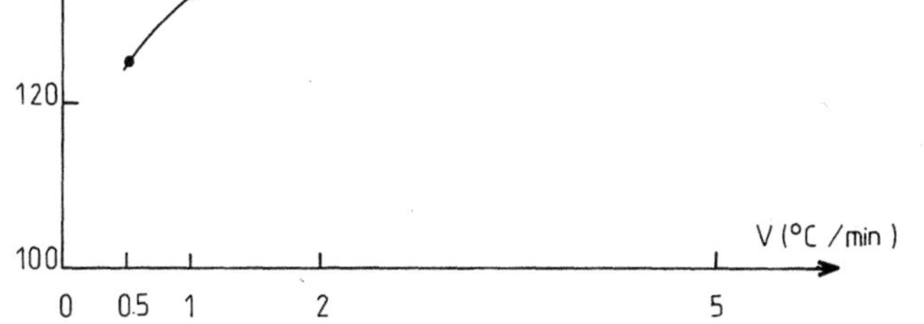

Fig. 13.36. Variation of the position of the maximum of heat flux (expressed by the temperature at which it is attained) as a function of the heating rate, for various values of the cure enthalpy.

temperature, the rate of heat transfer through the sample playing the role of a limiting stage, being low compared to the rate of increase in temperature in the calorimeter and on the surface of the holder.

3. The position of the heat flux–temperature curve along the heat flux axis also depends on the value of the heating rate of the calorimeter. This fact is due to the temperature difference between the surface and the centre of the sample, resulting from the time taken for heat to progress through the sample. This difference is found to be proportional to the value of the heating rate [3].

The role played by the cure enthalpy on the heat flux–temperature histories is clearly illustrated in Figs 13.32–13.35, obtained for different values of the enthalpy, as well as in Fig. 13.37 where the maximum of the heat flux is plotted as a function of the enthalpy for various values of the heating rate. On the whole, the effect of the enthalpy of cure on the process can be defined by the following points:

1. Of course, the heat flux is increased with increasing enthalpy, with a constant factor of proportionality as proved in Fig. 13.37, concerned with the maximum value of the heat flux.
2. A linear relationship obtained between the maximum heat flux value and the enthalpy is found for various values of the heating rate, the slope of these straight lines being proportional to the value of the heating rate (Fig. 13.37).
3. As a result, the position of the heat flux–temperature curves along the heat flux axis largely depends on the value of the enthalpy. As shown in Figs 13.32–13.35, and especially in Fig. 13.37, the heating process taking place may be observed either as endothermic or exothermic depending on the values of the enthalpy and of the heating rate.

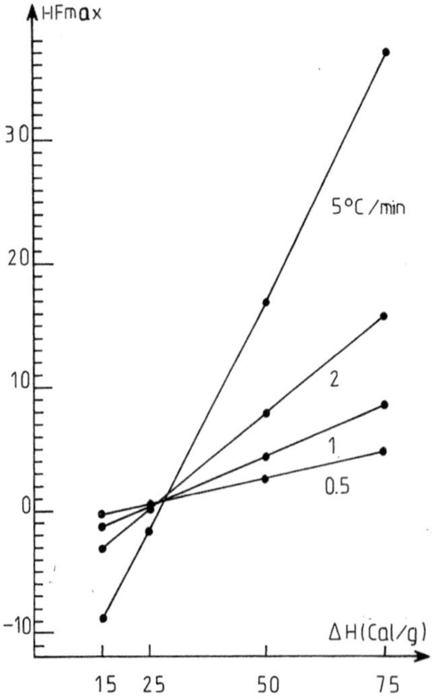

Fig. 13.37. Maximum heat flux as a function of the cure enthalpy for various heating rates.

Results for the Effect of Cure Enthalpy on the Temperature History at the Centre of the Sample

The temperature histories at the centre of the sample were calculated, and plotted as a function of the calorimeter temperature for various values of the cure enthalpy (Figs 13.38–13.41). Various conclusions can be drawn from these curves:

1. A good correspondence between these curves and the heat flux–temperature curves is observed.
2. When no reaction occurs, the temperature is higher on the surface than at the centre of the sample, because of the time necessary for heat to be transferred through the sample. This difference in temperature ΔT is negative and proportional to the heating rate of the calorimeter.
3. When the reaction takes place, the temperature in the sample is increased, because of the exothermicity of the reaction. The highest increase in temperature is attained at the centre of the sample as shown in Figs 13.42–13.44.
4. The increase in temperature at the centre of the sample ΔT is proportional to the heat evolved from the reaction, and so to the cure enthalpy. For a value of 25 cal g^{-1}, the value of the temperature at the centre reaches the value of the calorimeter temperature. Of course, when the cure enthalpy is higher than 25 cal g^{-1}, the value of the temperature at the centre of the sample is higher than that of the calorimeter, when the temperature history at the centre ΔT rises to its maximum.

Results for the Effect of Cure Enthalpy on the Profiles of Temperature and SOC Developed Through the Sample

The profiles of temperature and SOC developed through the sample are plotted at various times, for the various values of the cure enthalpy, in Fig. 13.42 with a heating rate of 1 °C min^{-1}, in Fig. 13.43 with 2 °C min^{-1} and in Fig. 13.44 with 5 °C min^{-1}. These times are chosen in the following way: the first before the reaction takes place, the second when the heat flux is half the maximum value of the heat flux, the third at the maximum value of the heat flux, the fourth when the flux is half the maximum heat flux again, and last when the reaction is complete.

The following conclusions are drawn:

1. When no reaction occurs, the temperature profiles are the same, whatever the value of the cure enthalpy. In fact, they only depend on the heating rate of the calorimeter, with the statement: the higher the heating rate of the calorimeter, the steeper the temperature gradients.
2. During the cure period, the steepest temperature gradients are attained when the heat flux rises to its maximum, when the cure enthalpy is high (higher than 25 cal g^{-1}).
3. In the same way, the steepest SOC gradients are developed through the sample when the heat flux rises to its maximum, and of course, the higher the cure enthalpy, the steeper the gradient of the SOC.

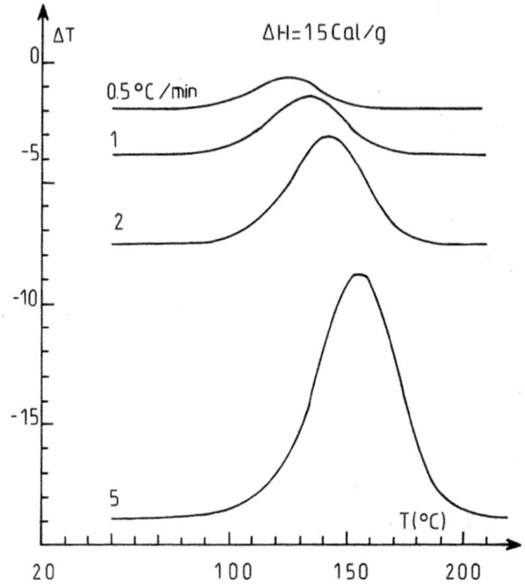

Fig. 13.38. Temperature histories at the centre of the sample, ΔT, as a function of the calorimeter temperature, for various values of the heating rate, with a cure enthalpy of 15 cal g^{-1}. (ΔT is the difference of the temperatures obtained with and without reaction.)

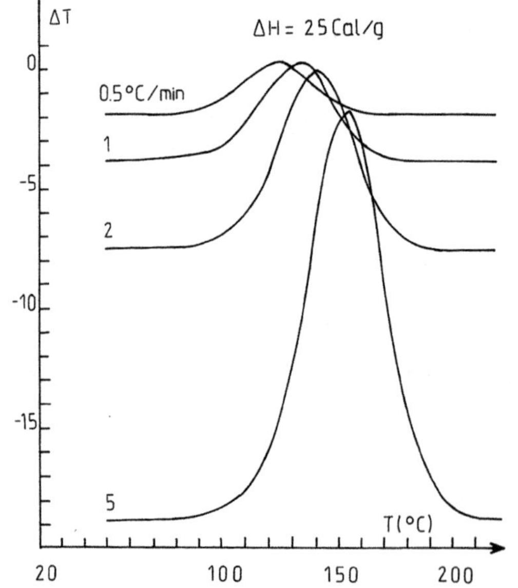

Fig. 13.39. Temperature histories at the centre of the sample, ΔT, as a function of the calorimeter temperature, for various values of the heating rate, with a cure enthalpy of 25 cal g^{-1}. (ΔT is the difference of the temperatures obtained with and without reaction.)

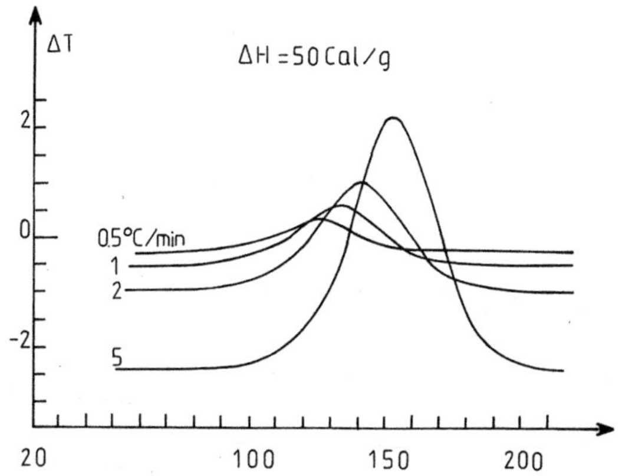

Fig. 13.40. Temperature histories at the centre of the sample, ΔT, as a function of the calorimeter temperature, for various values of the heating rate, with a cure enthalpy of 50 cal g^{-1}. (ΔT is the difference of the temperatures obtained with and without reaction.)

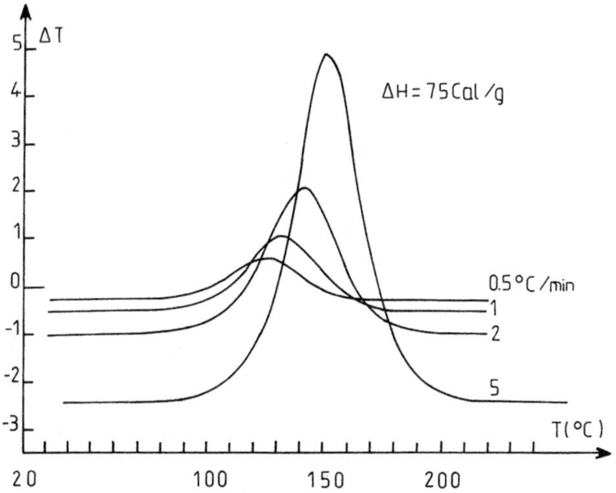

Fig. 13.41. Temperature histories at the centre of the sample, ΔT, as a function of the calorimeter temperature, for various values of the heating rate, with a cure enthalpy of 75 cal g^{-1}. (ΔT is the difference of the temperatures obtained with and without reaction.)

13.3.6 Heating Rate and Kinetic Parameters [29]

In calorimetry techniques, the enthalpy changes accompanying chemical events in the sample are monitored. The result is the curve of the rate of enthalpy changes as a function of temperature when the calorimeter is heated at a constant rate in DSC. These enthalpy profiles allow certain critical parameters relevant to the actual manufacturing process to be rapidly determined, and the profiles can sometimes be treated as a fingerprint of the system for purposes of quality control. Moreover, a detailed kinetic analysis of the enthalpy profiles is possible especially in the case of simple reactions, and the determination of meaningful

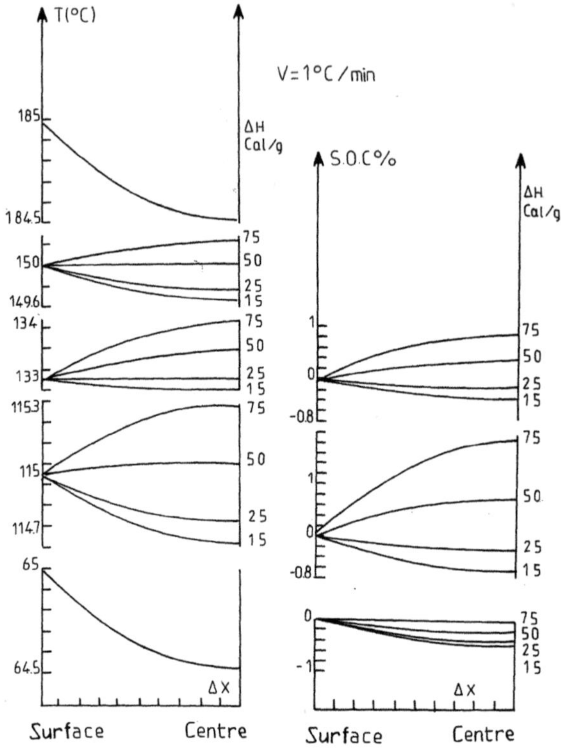

Fig. 13.42. Profiles of temperature (left) and SOC (right) through the sample at various times and various cure enthalpies, with a heating rate of 1 °C min^{-1}.

kinetic parameters is more difficult for complex reactions. For the latter however, it is possible to obtain kinetic equations able to describe the whole process, as has been shown in the case of sulphur vulcanisation processes [5, 17] or in the case of epoxy resin [1, 23–27].

In the DSC experiment, the fresh unreacted sample is scanned at a selected heating rate up to a temperature at which the reaction exotherm is complete. The following drawbacks for this technique may occur:

1. After cure in the calorimeter, the sample is cooled and rescanned to give the temperature dependence of the thermal parameters of the reacted product formed after the initial DSC experiment. The rescan curve is superimposed on the initial curve to give a baseline which would be observed in the absence of reaction. A shift in baseline appears when the thermal parameters of the initial and final materials differ significantly.
2. Because of the heating rate imposed by the calorimeter and low heat conductivity of the material, gradients of temperature are developed throughout the sample. These gradients are proportional to the heating rate when there is no reaction, and are also proportional to the enthalpy of reaction and to the heating rate when the reaction takes place [3, 10, 11, 21].

The purpose of this study is to show that the heating rate in DSC not only plays an important role, but also must be controlled by using a sufficiently low value. The combination of the heat of reaction and low heat conductivity is responsible

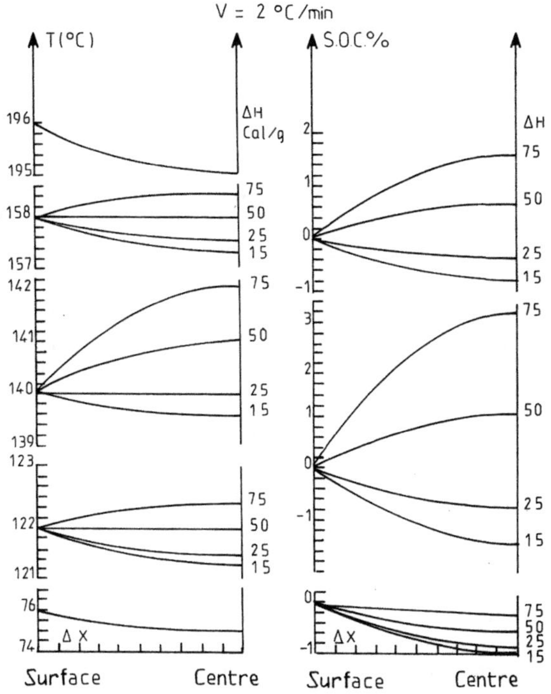

Fig. 13.43. Profiles of temperature (left) and SOC (right) through the sample at various times and various cure enthalpies, with a heating rate of 2 °C min^{-1}.

for an increase in temperature within the sample, and this higher temperature produces in turn a higher rate of reaction. The result is a thinner and higher peak for the heat flux–temperature history, this peak being thus described by kinetic parameters which may be different from the values obtained under slower conditions of heating. In order to attain this objective, the same material was scanned with various heating rates ranging from 2 to 20 °C min^{-1}, and the kinetic parameters determined by using the Freeman–Carroll technique in each case. Moreover, these heat flux–temperature curves are calculated in two ways: One uses the Arrhenius equation and assumes that the temperature throughout the sample is uniform and always the same of that of the calorimeter. For each value of the heating rate, the values used for the kinetic parameters are those determined from the experimental curves at the same heating rate. The other uses the numerical model with finite differences which takes into account not only heat evolved from the cure reaction expressed by the Arrhenius equation but also heat transferred by conduction. With this model, the actual temperature within the sample is calculated and is accounted for in evaluating the heat flux and the SOC. The values chosen for the kinetic parameters in all cases are those obtained with a low value of the heating rate (1 °C min^{-1}), as shown in Table 13.3.

Experiment

Apparatus. A DSC 111 calorimeter (Setaram) was used, with various heating rates ranging from 2 to 20 °C min^{-1}. The 5 mm diameter sample of about 150 mg

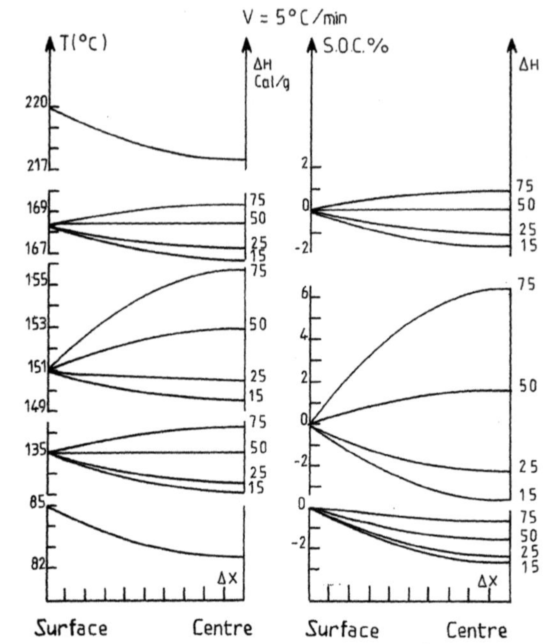

Fig. 13.44. Profiles of temperature (left) and SOC (right) through the sample at various times and various cure enthalpies, with a heating rate of 5 °C min^{-1}.

was placed in a cylindrical stainless steel holder, external diameter 6.8 mm. The programming of temperature ranged from room temperature up to a temperature at which full exotherm is achieved.

Material. An unsaturated polyester of high enthalpy was chosen for this study. The thermal properties of this material were previously determined. The kinetic parameters were calculated from the DSC curve obtained at a low rate of heating (1 °C min^{-1}):

$$Q_\infty = 152 \text{ cal g}^{-1},\ k_0 = 4 \times 10^{11} \text{ s}^{-1},\ E = 27\,140 \text{ cal mol}^{-1},\ n = 1.5,$$

$$\lambda = 10^{-3} \text{ cal cm}^{-1} \text{ s}^{-1} \text{ deg}^{-1},\ C = 0.2 + 3 \times 10^{-3} T \text{ cal g}^{-1} \text{ deg}^{-1},\ T(°\text{C})$$

Numerical Model

Assumptions. The following assumptions are made:

1. Heat transfer is radial through the sample by conduction.
2. The kinetics of the heat evolved from the overall cure reaction is expressed by an Arrhenius equation, with the kinetic parameters above.

Mathematical and Numerical Treatment. The problem is the same as that described in the case of epoxy resin in DSC with an nth-order reaction (Sect. 13.3.4). As a result, the following information is provided by the model:

1. The heat flux emitted through the calorimeter sample interface. As this heat flux is also determined by experiment, a comparison between these curves is of interest not only for testing the validity of the model, but also the accuracy of the kinetic parameters.
2. The profiles of temperature and SOC developed through the sample.

Results for Heat Flux–Temperature Curves at Various Heating Rates

Two types of heat flux–temperature histories are calculated and compared to the experimental heat flux–temperature curves, for various heating rates ranging from 1 to 20 °C min^{-1}.

1. The heat flux–temperature history calculated by using the Arrhenius equation and making the following two assumptions: the temperature throughout the sample is uniform, the temperature of the sample is always the same as that of the calorimeter. For each value of the heating rate, the kinetic parameters considered are those determined with the Freeman–Carroll technique from the experimental curve obtained with the same heating rate.
2. The heat flux temperature history calculated with the help of the numerical model with finite differences, taking into account the heat transferred by conduction and heat generated by the reaction. The kinetic parameters used for all values of the heating rate are those determined with the Freeman–Carroll technique from the experimental curve obtained with a low heating rate (1 °C min^{-1}).

The results are shown in Figs 13.45–13.48 for heating rates of 2, 5, 10 and 20 °C min^{-1}. There are three types of curves:

Experimental curve

Curve calculated with the first method

Curve exhibiting the difference between the experimental and calculated curve

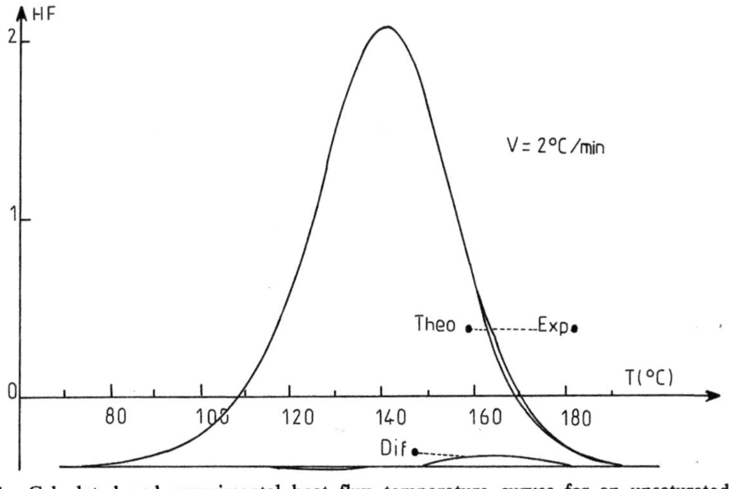

Fig. 13.45. Calculated and experimental heat flux–temperature curves for an unsaturated polyester with a heating rate of 2 °C min^{-1}.

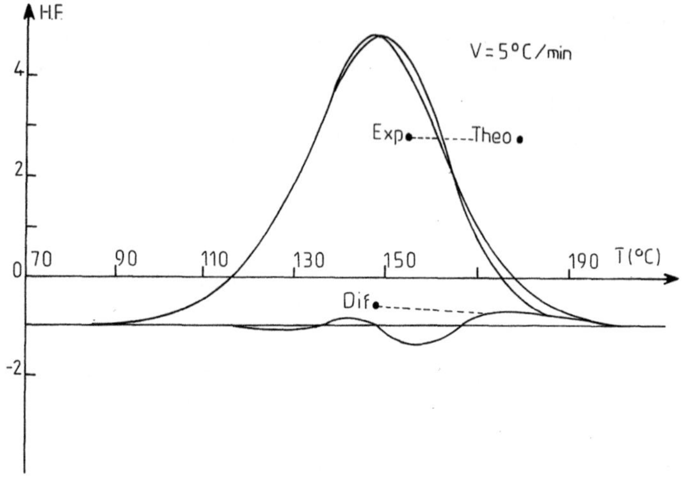

Fig. 13.46. Calculated and experimental heat flux–temperature curves for an unsaturated polyester with a heating rate of 5 °C min^{-1}.

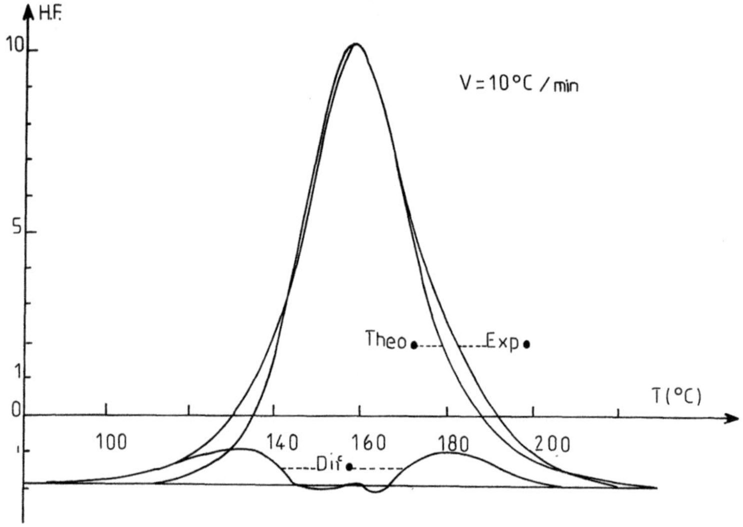

Fig. 13.47. Calculated and experimental heat flux–temperature curves for an unsaturated polyester with a heating rate of 10 °C min^{-1}.

The following conclusions can be drawn from these curves, and from the kinetic parameters obtained from these curves by the Freeman–Carroll technique (Table 13.3).

1. Perfect agreement is obtained between these curves for the low heating rate of 1 and 2 °C min^{-1}. The kinetic parameters determined by the Freeman–Carroll technique for these two heating rates are about the same.
2. When the heating rate is 5 °C min^{-1}, the agreement between the two curves is still acceptable. The position of the maximum of the heat flux is well placed, as

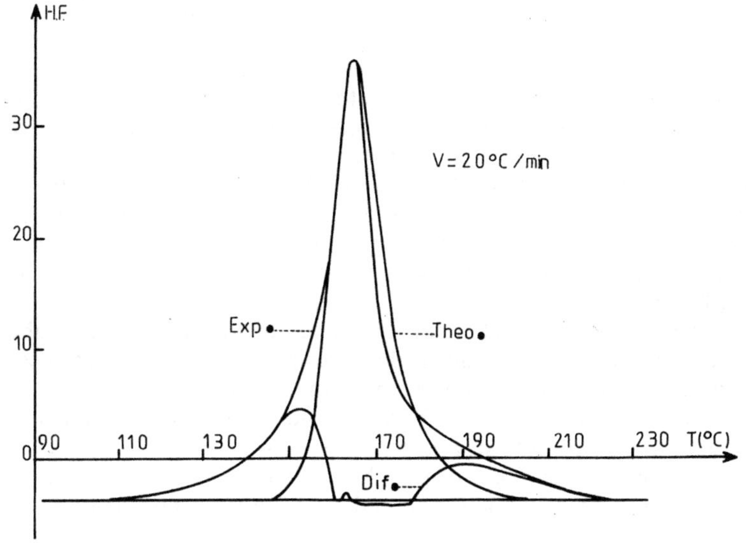

Fig. 13.48. Calculated and experimental heat flux–temperature curves for an unsaturated polyester with a heating rate of 20 °C min^{-1}.

Table 13.3. Kinetic parameters from the Freeman–Carroll technique

Heating rate (°C min^{-1})	E	k_0	n
1	27 140	4×10^{11}	1.5
2	27 600	1×10^{12}	1.5
5	28 300	1.8×10^{12}	1.54
10	47 000	2×10^{22}	2.5
20	100 000	5×10^{58}	3.3

well as the positions of the points corresponding to half this maximum value. However, some slight differences appear, and other kinetic parameters may give a better result as shown in Table 13.3.

3. The difference between the two curves appears to be significant in the case where heating rates are higher than 5 °C min^{-1}. For instance, with a heating rate of 10 °C min^{-1}, the maximum of the heat flux is well placed as well as the points corresponding to half this maximum value. Moreover, and this following fact is very relevant, higher values for the activation energy and the pre-exponential factor are obtained with the Freeman–Carroll technique as shown in Table 13.3, in order to describe the heat flux peak which becomes very thin.
4. For the highest value of the heating rate (20 °C min^{-1}) it is clear that it is difficult to obtain kinetic parameters able to describe the heat flux–temperature history.

As a conclusion, the value chosen for the heating rate must be sufficiently low, or the kinetic parameters obtained from the experimental curves are incorrect.

Results for Temperature Profiles at the Centre of the Sample

In order to obtain an explanation for these results, especially for the change in the kinetic parameters with the heating rate, the temperature at the centre of the sample over the scanning process was calculated by using the numerical model taking into account the conduction heat transfer and kinetics of heat of reaction.

As shown in Fig. 13.49 for heating rates of 2 and 5 °C min^{-1}, and Fig. 13.50 for the higher heating rates of 10 and 20 °C min^{-1}, these curves have about the same shape and position as the heat flux–temperature histories drawn in Figs 13.45–13.48. This fact has already been reported in the case of lower enthalpy of reaction [30, 31].

ΔT expressed in Table 13.4 is the difference of the temperatures obtained at the centre of the sample, with and without reaction. Of course, ΔT is zero when no reaction occurs, and becomes positive during the stage of reaction.

ΔHF represents the difference between the heat fluxes obtained with and without reaction. Of course ΔHF is zero when no reaction occurs, and becomes positive when the reaction takes place. The value of the heat flux HF is constantly below zero when no reaction appears, and its value is proportional to the heating rate.

The following conclusions can be drawn from Figs 13.49 and 13.50 and from Table 13.4.

1. The shape of the heat flux and ΔT–temperature histories are about the same, with the same position of the maximum, for various heating rates.

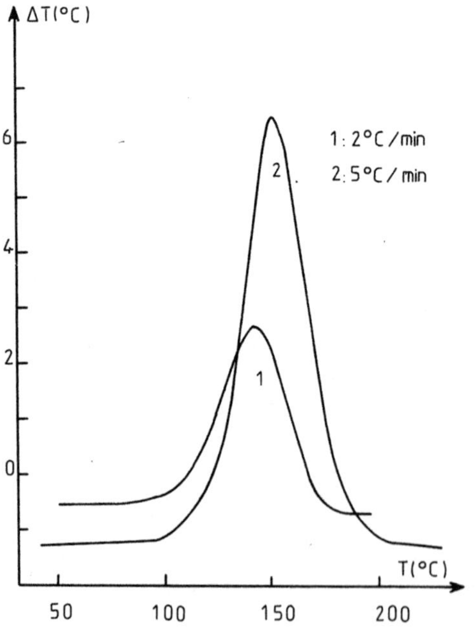

Fig. 13.49. Temperature history at the centre of the sample, ΔT, as a function of the temperature of the calorimeter, for a heating rate of 2 and 5 °C min^{-1}. (ΔT is the difference of the temperatures obtained with and without reaction.)

2. The value of ΔT is proportional to the value of ΔHF obtained during the reaction, and especially at its maximum.
3. The maximum values for ΔHF and ΔT are proportional to the value of the heating rate.

Results for the Profiles of Temperature and SOC Through the Sample

The numerical model is able to provide one with the profiles of temperature and SOC developed through the sample either during the heating period or during the period of reaction. These profiles are plotted for the heating rate of 2 °C min^{-1} in Fig. 13.51 and 20 °C min^{-1} in Fig. 13.52. From these curves it may be seen that:

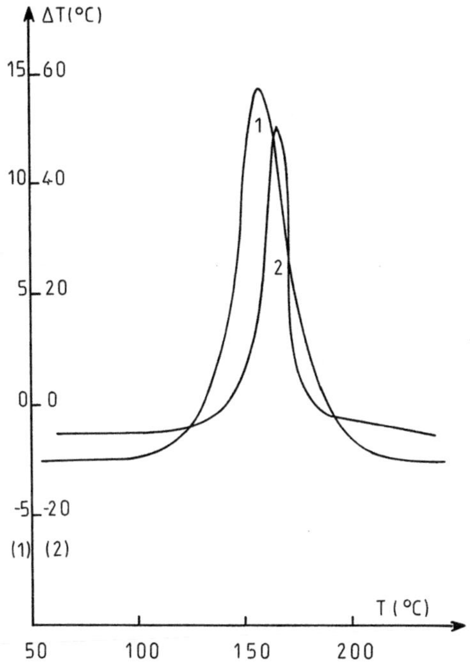

Fig. 13.50. Temperature history at the centre of the sample, ΔT, as a function of the temperature of the calorimeter, for a heating rate of 10 and 20 °C min^{-1}. (ΔT is the difference of the temperatures obtained with and without reaction.)

Table 13.4. Characteristics of heat flux histories

	Heating rate (°C min^{-1})			
	2	5	10	20
Temp. for max ΔHF (°C)	142.5	151.7	158	164.8
Temp. for max ΔT (°C)	142.5	151.7	157.3	163.5
Maximum ΔHF	2.1	4.8	10.2	36
Maximum ΔT (°C)	2.7	6.5	14.6	53.7

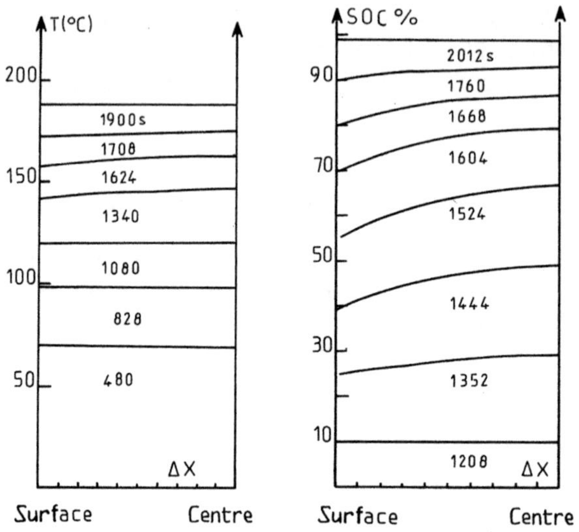

Fig. 13.51. Profiles of temperature (left) and SOC (right) in the sample, with a heating rate of 2 °C min^{-1}.

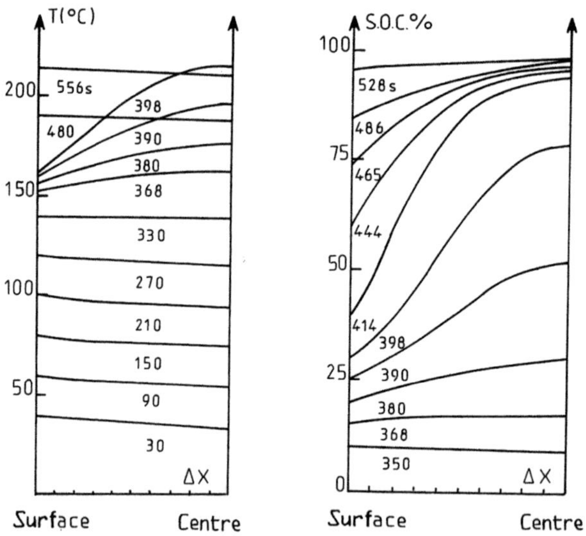

Fig. 13.52. Profiles of temperature (left) and SOC (right) in the sample, with a heating rate of 20 °C min^{-1}.

1. For the low heating rate, rather low gradients of temperature are developed within the sample. The highest value for these gradients is attained when the heat flux is maximum as shown in Figs 13.45 and 13.51.
2. For the lower heating rate, the value of the SOC, indicating the actual extent of the reaction, is about the same throughout the sample.
3. For higher heating rates, high gradients of temperature are grown within the sample. When no reaction occurs, the temperature is higher on the surface

than at the centre of the sample. As shown previously this retardation is proportional to the heating rate and to the square of the main dimension of the sample (diameter for a cylinder, thickness for a sheet) [3]. Other gradients of temperature are expanded during the reaction, the temperature increasing at the centre of the sample.

4. For higher heating rates, the values of the SOC are quite different from one place to another at each time. High gradients for the SOC are developed during the reaction period, with the highest value attained at the centre of the sample.

13.4 Conclusions on DSC and DC

Calorimetry is the best technique for determining the kinetic parameters for the cure reaction, by measuring the heat evolved, which is of high interest in the heating process of cure. With this technique, the rate of enthalpy change due to the reaction is recorded either as a function of time when the calorimeter is held at constant temperature (DC), or as a function of temperature when the calorimeter is scanned at a constant heating rate (DSC).

These two ways of working are quite different, and it is of interest to sum up the advantages and drawbacks of the DC and DSC modes. Much work has been done in order to compare these two techniques and the accuracy of the results obtained (e.g. [30–33]). As a result, the DSC mode is found to be better than the DC mode, if the value of the heating rate is selected carefully.

13.4.1 Drawbacks of DC

These important drawbacks appear in the DC mode.

Narrow Temperature Window

A narrow temperature window over which meaningful data can be obtained exists in the DC mode, whatever the value of the cure enthalpy.

At low temperature, a low rate of enthalpy change is obtained, and the exothermicity of the reaction is expanded over a long period of time. A lack of accuracy also results from the low rate of reaction.

At high temperature, high rate of enthalpy is developed, and steep gradients of temperature are expanded.

Problems from Temperature Gradients

Steep gradients of temperature are developed within the sample during the heating stage, followed by other gradients during cure.

The temperature in the sample always differs from that of the calorimeter, except at two times: when the heat flux reaches the zero value, and at the end of the process.

Quality of Contact between the Sample and Calorimeter

Some emphasis is placed in Chap. 12 on the quality of contact, not only between the sample and the holder, but also between the holder and calorimeter. It is difficult to determine the quality of the contact between the sample and holder in the case of a small sample, but is surely poor. In the case of a fairly large sample, it is already difficult to fill the holder completely. In this latter case, the holder and sampler can be sawn through at the end of the process, allowing some investigation of the cross-section.

The presence of a thin air layer has been considered for a fairly large sample, but it is clear that it is easier to adjust the dimensions and shape of the sample to those of the calorimeter in a better way when the sample is large than when it is very small. Worse still, it is impossible to determine the quality of the contact in the case of a small sample when it is heated on one surface only.

13.4.2 Drawbacks of DSC and Best Working Method

The DSC mode in calorimetry has to face two difficulties, one with the apprehension of some workers to work with a changeable temperature, the other more serious with the choice of a right value of the heating rate.

Heat Flux–Temperature and Heating Rate

The heating rate is responsible for the establishment of a gradient of temperature with higher temperature at the sample surface, and subsequently for the presence of an endothermic heat flux.

The value of this endothermic heat flux is proportional to the heating rate.

The heating rate is in connection with the value of the cure enthalpy. For a low cure enthalpy, the heat flux increases during the reaction period, and it may remain so low that the heat flux is still endothermic (below the zero value). For a high cure enthalpy, the heat flux is increased to such an extent that it is above zero.

Profiles of Temperature and Heating Rate

Another parameter of interest is introduced, with the difference of the temperatures at the centre of the sample obtained with and without reaction, ΔT. This means that not only is the temperature of the sample different from that of the calorimeter, but also that it may vary strongly within the sample.

This difference ΔT is positive during the heating period, and is increased by increasing the heating rate.

Kinetic Parameters and Heating Rate

Because of the changes in temperature within the sample due to the heat transferred by conduction through the sample and to the heat generated by the reaction, the shape and position of the heat flux–temperature curve may depend on the value selected for the heating rate. A high value for the heating rate is thus responsible for the formation of thin peaks shifted to a higher calorimeter temperature. As a result of this fact, too high a value is obtained for the activation energy, as well as for the pre-exponential factor.

The best way of working consists of using two samples scanned at two different low values for the heating rate, and giving the same kinetic parameters.

References

1. Kretzschmar K, Hoffmann KW. Kinetic aspects arising during the differential scanning calorimetric investigation of curing epoxy resins with anhydrides. Thermochim Acta 1985; 83: 139
2. Chataing G, Vergnaud JM. Effect of heating rate on the separation of various phenomena in the DSC technique. Thermochim Acta 1985; 94: 379
3. Armand JY, Gonnet R, Jones R, Bouzon J, Touchard M, Vergnaud JM. Effect of heating rate and sample size on heat transfer through the sample in DSC. Thermochim Acta 1986; 103: 341
4. Manley TR. In: Chiu J (ed.) Polymer characterization by thermal methods of analysis, M. Dekker, New York, 1974, p 53
5. Brazier DW. In: Grassie N (ed.) Developments in polymer degradation, Applied Science Publishers, London, 1981, p 27
6. Rochette JB, Sadr A, Abdul M, Vergnaud, JM. Effect of variation in the values of the thermal properties of rubber on the vulcanization of thin and thick sheets. Thermochim Acta 1984; 80: 287
7. Abdul M, Vergnaud JM. Vulcanization progress in rubber sheets during cooling in motionless air after extraction from the mould. Thermochim Acta 1984; 76: 161
8. El Brouzi A, Bouzon J, Vergnaud JM. Modelling of the cure of fibre glass-epoxy resin composite heated by liquid oil at constant temperature. Effect of parameters on the process. Plast Rubber Process Applic 1989; 12: 71
9. Armand JY, Bourgois J, Touchard M, Vergnaud JM. Effect of sample size and heating rate on cure reaction of rubber in DSC. Thermochim Acta 1986; 108: 345
10. Chater M, Chataing G, Vergnaud JM. Enhanced study of differential scanning calorimetry with determination of heat flux-time curves, profiles of temperature and state of cure. Thermochim Acta 1985; 90: 135
11. Khouider A, Bouzon J, Vergnaud JM. Modelization of the DSC technique by considering heat transfer and kinetics of reaction in the case of rubber cure. Thermochim Acta 1986; 98: 285
12. Gehman SD. Heat transfer in processing and use of rubber. Rubber Chem Technol 1967; 40: 36
13. Hands D. The thermal transport properties of polymers. Rubber Chem Technol 1977; 50: 480
14. Prentice GA, Williams MC. Numerical evaluation of the state of cure in a vulcanizing article. Rubber Chem Technol 1980; 53: 1023
15. Sircar AK, Wells JL. Thermal conductivity of elastomer vulcanizates by differential scanning calorimetry. Rubber Chem Technol 1982; 55: 191
16. Accetta A, Vergnaud JM. Upgrading of scrap rubber powder by vulcanization without new rubber. Rubber Chem Technol 1981; 54: 302
17. Accetta A, Le Parlouer P, Vergnaud JM. Kinetic parameters of the overall reaction of scrap rubber vulcanization by 2% sulfur. Thermochim Acta 1982; 59: 149
18. Sadr A, Granger R, Vergnaud JM. Calculation of profiles of temperature and state of cure developed within the rubber mass in injection molding process. J Appl Polym Sci 1984; 29: 955
19. Accetta A, Vergnaud JM. Calculation of the temperature and extent of reaction during the vulcanization of powdered rubber. Rubber Chem Technol 1983; 56: 689
20. Liu H, Armand JY, Bouzon J, Vergnaud JM. Effect of sample size and heating rate on the DSC process for reactions of high enthalpy. Thermochim Acta 1988; 126: 81

21. Armand JY, Bourgois J, Vergnaud JM. Fundamental study of heating process in DSC. Effect of the cure enthalpy and heating rate. Thermochim Acta 1987; 116: 301
22. Armand JY, Vergnaud JM. Effect of the enthalpy sample size and heating rate on the cure reaction of rubber in DSC. Thermochim Acta 1987; 112: 313
23. Kamal MR. Thermoset characterization for moldability analysis. Polym Engng Sci 1974; 14: 231
24. Ryan ME, Dutta A. Kinetics of epoxy cure: a rapid technique for kinetic parameter estimation. Polymer 1979; 20: 203
25. Sickfield J, Mielke W. Applications of thermal analysis for the investigation of epoxy resins. Progr Organ Coat 1984; 12: 27
26. Nixon JA, Hutchinson JM. Analysis of the cure of sheet moulding compound. I. Development of the model. Plast Rubber Applic 1985; 5: 349
27. Chater M, Lalart D, Michel-Dansac F, Vergnaud JM. Variation of enthalpy and kinetic parameters of the cure of epoxide resin with the composition of the binary system. Eur Polym J 1986; 22: 805
28. Robert G, Rochette JB, Bouzon J, Vergnaud JM. Determination of the state of cure of epoxy resin used in composite with sand. Thermochim Acta 1986; 102: 93
29. Armand JY, Vergnaud JM. Effect of the value of heating rate in DSC on the kinetic parameters, when there is high enthalpy of reaction. Thermochim Acta 1988; 131: 15
30. Vergnaud JM. Modelling in calorimetry working under isothermal conditions or in scanning mode. Thermochim Acta 1987; 114: 15
31. Armand JY, Vergnaud JM. Modelling of processes in calorimetry by taking into account heat transfer and reaction. Thermochim Acta 1987; 121: 381
32. Armand JY, Vergnaud JM. Comparative study of a reaction of low enthalpy (rubber cure) by DC and DSC. Thermochim Acta 1987; 116: 111
33. El Haij E, Armand JY, Bouzon J, Vergnaud JM. Values of kinetic parameters of cure reaction of epoxy in calorimetry by DC and DSC. Thermochim Acta 1989; 146: 329

Chapter 14

Variation of Enthalpy and Kinetic Parameters of the Cure of Epoxy Resin with the Composition of the Binary System

14.1 Introduction

Obtaining perfect mixtures is nearly always difficult, particularly for solid components, but also for liquids and especially when they are highly viscous. Thus an uneven distribution of components may occur in some places of the mixture, which could be responsible for disturbances in the development of the process, not only when it is conducted on an industrial scale with the troublesome variation in mechanical properties of the final products, but also when it is carried out on a laboratory scale, e.g., in samples used in calorimetry, these samples being sometimes of the order of a few milligrams.

It follows that the distribution of the two components for an epoxy resin, the resin and hardener, is of great interest. It may have some important effects on the mechanical properties of the final product, and also on the value of the enthalpy of cure as well as on the kinetic parameters of the overall reaction. The main purpose of this chapter is to determine the variation in the kinetic parameters of the overall reaction of cure and the value of the cure enthalpy, as a function of the composition of the binary mixture.

In the very small samples sometimes used in calorimetry, it is necessary that all ingredients are very well dispersed, otherwise there is a lack of reproducibility. Following this, a question always arises on how representative a 5 mg sample is, when taken from a large batch, of the quality of distribution of the components in the batch. As in calorimetry, studies on the cure reaction present specific problems, both in sample handling and in obtaining reproducible results and meaningful data, when an apparatus with fairly large samples of about 100–200 mg is used.

On the other hand, another source of error results from the possibility of resistance to heat transfer between the holder and sample [1, 2], which plays an important role for reactions of high enthalpy as in the case of epoxy resin. It has been shown (Chaps 12, 13) that the calorimeter temperature is sometimes far from the sample temperature, essentially because of the internal heat generated by the cure reaction and of the low heat conductivity of the sample [3–6]. Moreover, the large size of the sample may provoke the development of steep gradients of temperature within the sample, depending on the value of the cure enthalpy and of the value of heating rate in DSC.

In order to solve this problem due to the high exothermicity of the reaction and low heat conductivity, the DSC technique with a low heating rate is used [7].

14.2 Theory

The kinetics of the cure reaction of thermosets has been studied with the help of various techniques as shown in Chap. 11, but generally the kinetic parameters are determined by using calorimetry either under isothermal conditions with DC or in scanning mode with DSC [8, 9]. The determination of the kinetic reaction of epoxy cure by means of the DSC technique has been widely discussed [10, 11].

In a first attempt [9], the curing reaction of epoxy systems was described by way of the following kinetic equation:

$$\frac{d\alpha}{dt} = (k + k_1\alpha^m)(1 - \alpha)^n \tag{14.1}$$

expressing the rate of a reaction as a function of the degree of cure α, k and k_1 being the kinetic rate constants, and m and n kinetic exponents; k and k_1 vary with temperature by following an Arrhenius equation.

Following this study, this type of kinetic law was thus widely used [10–12].

The particular case where k_1 is equal to zero, has been tested more recently [3–7, 13] and it gives results in good agreement with experiments. The rate of heat generated by the overall cure reaction is expressed as a function of the heat of reaction as follows:

$$\frac{dY}{dt} = k_0 \exp\left(-\frac{E}{RT}\right)(1 - Y)^n \tag{14.2}$$

where

$Y_t = Q_t/Q_\infty$ is the extent of cure at time t, Q_t being the heat which has evolved from the cure reaction up to time t, and Q_∞ the total enthalpy which has evolved after infinite time.

n is the order of the reaction.

k_0 is the pre-exponential factor.

E is the activation energy.

R is the ideal gas constant.

T is the temperature expressed in kelvin.

Several methods can be used to determine the values of the kinetic parameters from experiments carried out with calorimetry in scanning mode [13]. The Freeman–Carroll technique is used, and to ensure that these kinetic parameters are obtained with sufficient efficiency, the heat flux–temperature curves, calculated with the help of a simple model are superimposed with the experimental curves. The model used in this case with a low heating rate is based on the fact that the temperature is uniform throughout the sample and constantly equal to the calorimeter temperature (Chap. 13).

14.3 Experiment [14]

Materials

The components of the mixture are:

1. Epoxy resin DGEBA (Lopox 200, CDF Chimie).
2. Hardener (D 2605, CDF Chimie) containing a mixture of an anhydride of MeTHP and an accelerator containing 0.75% of an aromatic amine (A 105, CDF Chimie).

Various compositions for the binary mixture were used, ranging from 20 to 160 parts (by weight) of the resin and 160 to 20 parts of the hardener, the total amount being 180 parts by weight in every case. The mixture was prepared at room temperature by adding the hardener with continuous stirring until a clear mixture was obtained.

Apparatus

The cylindrical sample, diameter 5 mm, was placed in a hermetically sealed cylindrical iron–steel holder, external diameter 6.8 mm. The weight of the sample was about 150 mg in each case. Heat flux was measured as a function of temperature with the DSC 111 (Setaram), from 30 to 250 °C, using a constant heating rate of 2 °C min^{-1}.

After cooling the calorimeter and sample to 30 °C, the sample was scanned again under conditions identical to that of the first heating, in order to obtain the baseline for the heat flux–temperature curves.

A microcomputer (HP86, Hewlett–Packard) coupled with the DSC stored the data (heat flux, temperature, time) at given times. After the end of the first run, allowing calculation of the total enthalpy of cure Q_∞, the following runs allowed the determination of the kinetic parameters through various methods. The last run allowed the validity of these parameters to be tested by superimposing the heat flux–temperature obtained by experiment and calculation.

14.4 Results

After testing the validity of the kinetic equation with one pre-exponential factor and one activation energy, the kinetic parameters are determined for the various formulations of the epoxy resin.

14.4.1 Validity of the Kinetic Equation

The heat flux–temperature curve was initially determined experimentally, with a formulation of the mixture giving a final material with good properties. The kinetic parameters thus obtained by the Freeman–Carroll technique were then used for calculating the heat flux as a function of the temperature with the help of Eq. (14.2), by considering the temperature to be uniform throughout the sample each time.

As shown in Fig. 14.1, the heat flux–temperature curves obtained either by experiment or by calculation are very well superimposed, proving the accuracy of the kinetic parameters and the validity of Eq. (14.2), as well as the validity of the method used for calculating these heat flux–temperature profiles.

In fact, it is necessary to draw a third curve representing the difference between the experimental and calculated heat flux as a function of temperature, in order to appreciate the slight differences between these data.

Each of the other formulations of the epoxy resin is experimented in this way, and the values of the kinetic parameters obtained are also tested by this method.

14.4.2 Kinetic Parameters as a Function of the Resin Formulation

Different formulations of the epoxy resin are tested by DSC. The various heat flux–temperature profiles obtained with these formulations are drawn in Figs 14.1–14.4. More information is also given by the computer, and especially the SOC value, expressed by the extent of heat evolved from the reaction at various temperatures [14].

For these various compositions of the binary mixture, the following data are given in Table 14.1: the cure enthalpy Q_∞ ($\text{cal}\,\text{g}^{-1}$), the activation energy E ($\text{cal}\,\text{mol}^{-1}$), the pre-exponential factor k_0, the reaction order, the temperature at

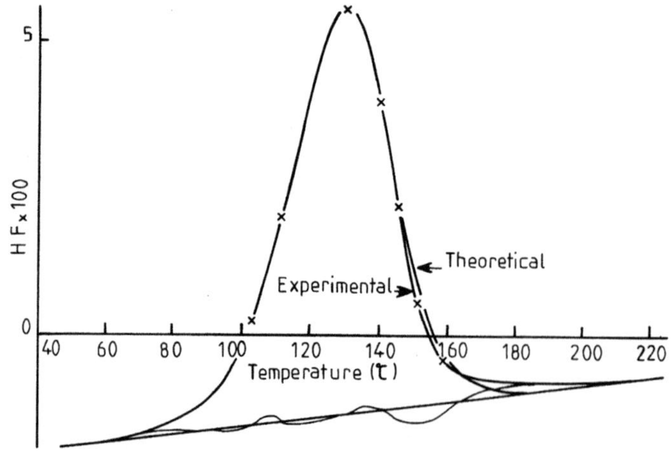

Fig. 14.1. Experimental and calculated heat flux–temperature curves for an epoxy resin with the composition 100 parts resin plus 80 parts hardener by weight.

which the heat flux is maximum, and a correlation factor. This correlation factor is the ratio of the calculated and the experimental heat flux at the maximum.

As indicated by the value of the correlation factor, good agreement is obtained for the formulations ranging from 80 to 110 parts of the resin. For compositions with resin below 60 and above 110, poor agreement is found.

A parameter of interest is the cure enthalpy. Its variation with the composition of the binary mixture is illustrated in Fig. 14.5. It is clear that the cure enthalpy is almost constant at about 79 cal g^{-1} when the composition ranges from 80 to 110 parts for the resin. Outside this range, the cure enthalpy decreases considerably with the composition.

The order of the overall reaction, appearing in Eq. (14.2) initially varies little from around 1.5, and decreases to less than 1 when the amount of resin is increased from 20 to 160 (Fig. 14.6).

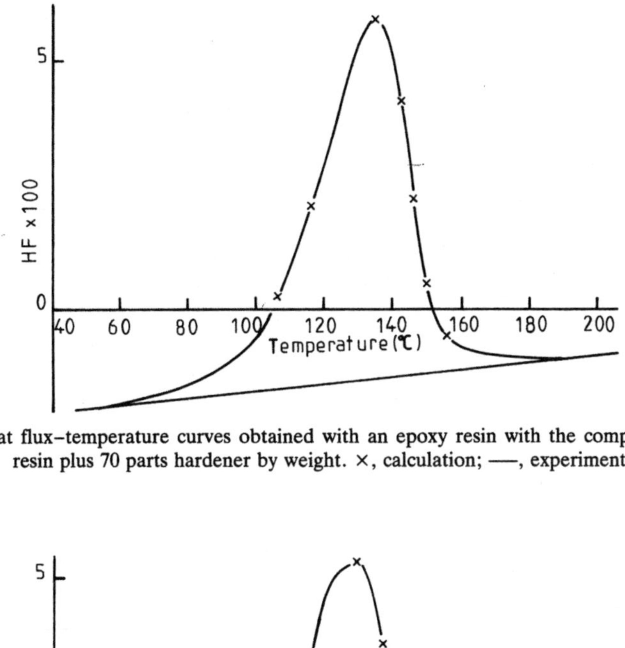

Fig. 14.2. Heat flux–temperature curves obtained with an epoxy resin with the composition 110 parts resin plus 70 parts hardener by weight. ×, calculation; —, experiment.

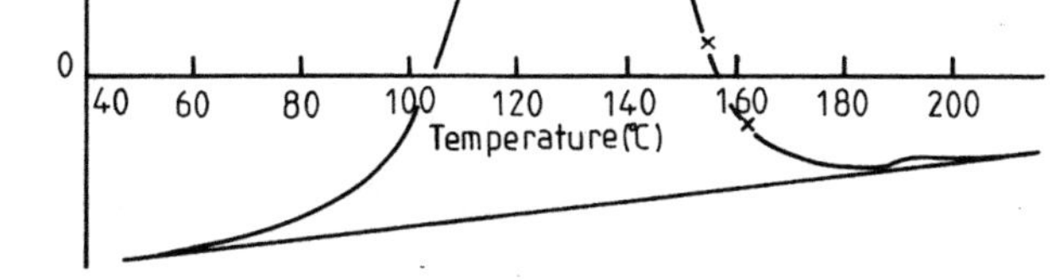

Fig. 14.3. Heat flux–temperature curves obtained with an epoxy resin with the composition 90 parts resin plus 90 parts hardener by weight. ×, calculation; —, experiment.

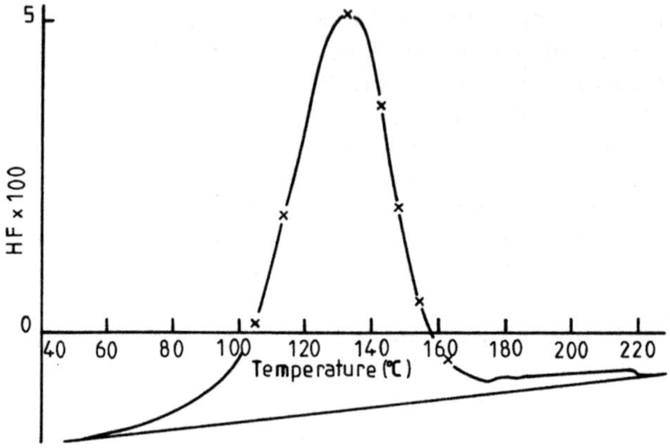

Fig. 14.4. Heat flux–temperature curves obtained with an epoxy resin with the composition 80 parts resin plus 100 parts hardener by weight. ×, calculation; —, experiment.

Table 14.1. Variation of the kinetic parameters with the formulation

Resin–hardener ratio (w/w)	Q_∞ (cal g^{-1})	E (kcal m^{-1})	k_0 (10^{11} s^{-1})	n	HF_{max} T(°C)	SOC (%)	Correlation factor
20/160	21	32.5	5×10^3	1.5	137	58	0.91
60/120	61	32	6×10^3	1.6	132	56	0.91
80/100	77.4	27.7	17.5	1.5	132	53	1.00
90/90	79.8	27.3	12.3	1.5	134	61	0.99
100/80	80	26.9	11.5	1.5	130	54	0.99
110/70	75	25.4	0.8	1	134	62	0.98
160/20	24	25	0.5	0.75	136	70	0.94

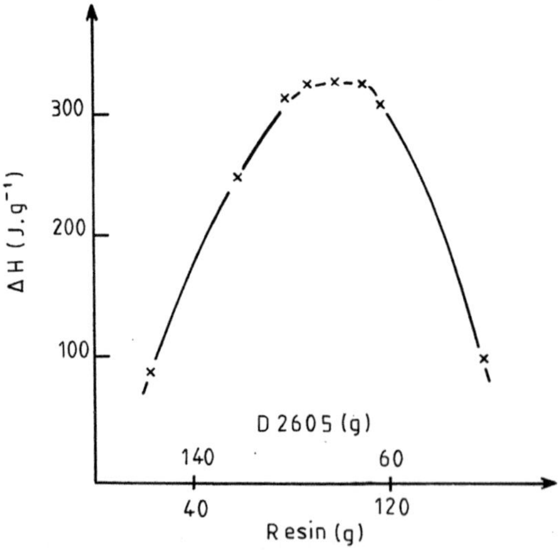

Fig. 14.5. Variation of the cure enthalpy (J g^{-1} resin) as a function of the composition.

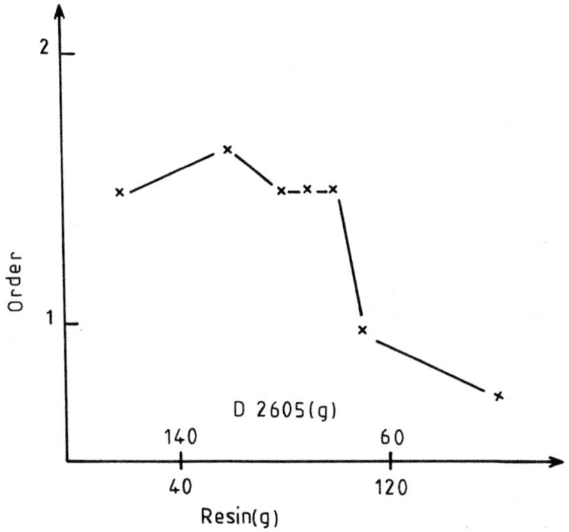

Fig. 14.6. Variation of the order in the simple kinetic equation as a function of the composition.

The value of the activation energy increases regularly as the percentage of hardener is increased.

By also considering some results obtained in other studies [15–17], the following conclusions can be drawn:

1. The value given to the order of the overall reaction has no effect on the position of the heat flux–temperature curves, because it has little effect on the value of the reaction rate when the extent of cure is below 50%. Moreover, this parameter has a strong effect on the shape of the tail of the curve when the extent of cure exceeds 80% [16].
2. The activation energy as well as the pre-exponential factor has a strong effect on the position of the heat flux–temperature curves [15–17]. Of course, too high a value obtained for the activation energy is followed by an increase in the value of the pre-exponential factor.

14.5 Conclusions

The effect of the composition of the binary mixture is proved to be of great importance to the value of the cure enthalpy and kinetic parameters. When the composition is varied slightly from 40 to 55 wt % of hardener, the cure enthalpy remains almost constant at 79 cal mol^{-1}. Within the same range for the composition, the other kinetic parameters are about constant.

The problem of mixing components is difficult, especially when these components are highly viscous, and when the amount of each component is small.

References

1. Brazier DW. In: Grassie N (ed.) Developments in polymer degradation, vol. 3, Applied Science Publishers, London, 1981, p 27
2. Manley TR. In: Chiu J (ed.) Polymer characterization by thermal methods of analysis, M. Dekker, New York, 1974, p 53
3. Accetta A, Vergnaud JM. Calculation of the temperature and extent of reaction during the vulcanization of powdered rubber. Rubber Chem Technol 1983; 56: 689
4. Chataing G, Chater M, Vergnaud JM. Determination of heat flux and profiles of temperature and state of cure in isothermal calorimetry with reaction of high enthalpy. Thermochim Acta 1985; 89: 107
5. Chater M, Chataing G, Vergnaud JM. Model for studying calorimetry with programmed temperature followed by isothermal heating. Calculation of heat flux and profiles of temperature with cure reaction of high enthalpy. Thermochim Acta 1985; 89: 119
6. Chater M, Chataing G, Vergnaud JM. Enhanced study of differential scanning calorimetry with determination of heat flux-time curves, profiles of temperature and state of cure. Thermochim Acta 1985; 90: 135
7. Chataing G, Vergnaud JM. Effect of heating rate on the separation of various phenomena in the DSC technique. Thermochim Acta 1985; 94: 379
8. Prime RB. In: Turi EA (ed.) Thermal characterization of polymeric materials, Academic Press, New York, 1981, p 435
9. Kamal MR. Thermoset characterization for moldability analysis. Polym Engng Sci 1974; 14: 231
10. Ryan ME, Dutta A. Kinetics of epoxy cure: a rapid technique for kinetic parameter estimation. Polymer 1979, 20: 203
11. Sickfeld, J Mielke W. Application of thermal analysis for the investigation of epoxy resins. Progr Organ Coat 1984, 12: 27
12. Nixon JA, Hutchinson JM. Analysis of the cure of sheet moulding compound. I. Development of the model. Plast Rubber Process Applic 1985; 5: 349
13. Kretzschmar K, Hoffmann KW. Kinetic aspects arising during the differential scanning calorimetric investigation of curing epoxy resins with anhydrides. Thermochim Acta 1985; 83: 139
14. Chater M, Lalart D, Michel-Dansac F, Vergnaud JM. Variation of enthalpy and kinetic parameters of the cure of epoxide resin with the composition of the binary system. Eur Polym J 1986; 22: 805
15. Ferradou C, Vergnaud JM. Effect of variation in the kinetic parameters on the vulcanization of rubber sheet. Thermochim Acta 1983; 68: 283
16. Rochette JB, Sadr A, Abdul M, Vergnaud JM. Effect of a variation in the order of the overall reaction of cure on the profiles of temperature and state of cure developed at the midplane of rubber sheets. Thermochim Acta 1984; 79: 187
17. Ferradou C, Rochette JB, Vergnaud JM. Effect of a variation in kinetic parameters (rate constant, activation energy) on the vulcanization of rubber sheets in injection molding process. J Appl Polym Sci 1985; 30: 2663

Chapter 15

Cure of Epoxy Resin in a Long Cylindrical Mould Heated by Air

15.1 Introduction

One important purpose of this chapter is to build a model able to describe the curing reaction of epoxy resin during the process of heating and cure in a long cylindrical mould. The aim is to be capable of predicting both the temperature–time profile at any position within this cylinder at any time, and the state of cure (SOC)–time profiles at the same place and time. The model described here can be perfectly general and thus applied to moulding of any geometry. The model, based on a numerical method with finite differences, takes into account all the facts: the heat transfer by conduction through the resin and the mould, the heat convection at the external surface of the mould, and the heat generated by the cure reaction.

In this chapter, a long cylinder is considered for the mould, so that the heat transfer is radial only. This case is of interest because it is easy to determine the temperature at the middle of the resin along the longitudinal axis, and the temperature depends only on the radial position. The mould made of a glass cylinder is very simple, and it is heated by air. This type of mould is chosen because it is particularly able to solve the two problems of importance for epoxy resin:

1. It can supply the resin with heat.
2. It can extract heat from the resin as fast as possible when necessary, by varying the temperature of the fluid.

Other objectives are considered in this chapter, either with the help of experiments or with the model:

1. The determination of the effect of the kinetic and thermal parameters on the process of cure.
2. The effect of the size of the sample, and especially of the diameter of the long cylinder.
3. The effect of the temperature of the heated air.

All these parameters are of great importance for the process of cure, and it is necessary to have good knowledge on their effect for optimising the operational conditions.

15.2 Modelling of the Process

15.2.1 Mathematical Treatment

A few assumptions are made in order to clarify the problem.

1. Only radial heat transfer by conduction through the circular cross-section is considered, as the cylindrical mould is very long. No matter transfer takes place.
2. Heat is transferred by free convection through the air–mould interface.
3. The cure reaction is expressed by a single Arrhenius equation, as obtained from calorimetric determination.
4. The heat capacity of the resin varies with temperature.

The equation for the rate of heat transfer by radial conduction through the circular cross-section is

$$\rho C \frac{\partial T}{\partial t} = \frac{1}{r}\frac{\partial}{\partial r}\left(r\lambda \frac{\partial T}{\partial r}\right) + \rho \frac{\partial Q}{\partial t} \tag{15.1}$$

where the contribution due to the exothermic reaction can be seen in the right-hand term and where C and λ are the thermal parameters, and ρ the density.

The rate of heat evolution from the cure reaction at time t is expressed as a function of the heat evolved up to time t, Q_t, and the total enthalpy after infinite time, Q_∞, by the Arrhenius equation

$$\frac{1}{Q_\infty}\frac{dQ_t}{dt} = k_0\left(1 - \frac{Q_t}{Q_\infty}\right)^n \exp - \frac{E}{RT} \tag{15.2}$$

where

n is the order of the overall reaction,
k_0 is the pre-exponential factor.
E is the activation energy.
R is the ideal gas constant.
T is the absolute temperature.

The boundary condition at the external surface of the mould is

$$-\lambda \frac{\partial T}{\partial t}(R_m, t) = h(T(R_m, t) - T_{air}) \tag{15.3}$$

where R_m is the radius of the mould.

15.2.2 Modelling and Numerical Analysis [1]

Because of the heat generated by the cure reaction, no analytical solution can be found for the problem defined by the above two equations. The problem must be solved by using a numerical model with finite differences.

The circular cross-section is divided by NR concentric circles of radii $i\Delta r$, i being an integer between 0 and NR, and Δr the increment of radius defined by

$$R_s = \mathrm{NR}\Delta r \tag{15.4}$$

where R_s is the radius of the resin.

The space–time diagram for the temperature is shown in Fig. 15.1. The temperature is thus calculated at any place and any time, as well as the extent of heat evolved Q_t, or Q_t/Q_∞.

Within the Resin, $1 \leqslant i \leqslant \mathrm{NR} - 1$

The energy balance is calculated within the annulus of mean radius $i\Delta R$ and thickness Δr during the time increment Δt, by considering the heat conduction and the internal heat generated by the reaction. The new temperature TN_i after time Δt is thus expressed as a function of the temperatures obtained at the previous time $i\Delta t$.

$$\mathrm{TN}_i = \frac{1}{M_i}[T_{i-1} + (M_i - 2)T_i + T_{i+1}] + \frac{1}{2iM_i}[T_{i+1} - T_{i-1}] + \frac{\Delta Q_i}{C_i} \tag{15.5}$$

where the dimensionless number M_i is given by

$$M_i = \frac{(\Delta r)^2}{\Delta t}\frac{\rho C_i}{\lambda} \tag{15.6}$$

where Δr is the thickness of each annulus in the resin and ρ, C_i, λ are the density, heat capacity at time i and position r, and thermal conductivity of the resin, respectively.

Centre of the Resin

The heat balance is calculated in the cylinder of radius $\frac{1}{2}\Delta r$ at the centre of the resin during the time increment Δt. The new temperature at the centre of the resin after time Δt is thus

$$\mathrm{TN}_0 = T_0 + \frac{4}{M_0}(T_1 - T_0) + \frac{\Delta Q_0}{C_0} \tag{15.7}$$

Within the Mould, $\mathrm{NR} + 1 \leqslant i \leqslant \mathrm{NT} - 1$

The equation is in the same form as within the resin for heat conduction, without heat evolution from the reaction:

$$\mathrm{TN}_i = \frac{1}{M_\mathrm{m}}[T_{i-1} + (M_\mathrm{m} - 2)T_i + T_{i+1}] + \frac{(T_{i+1} - T_{i-1})\Delta r_\mathrm{m}}{2M_\mathrm{m}[R + (i - \mathrm{NR})\Delta r_\mathrm{m}]} \tag{15.8}$$

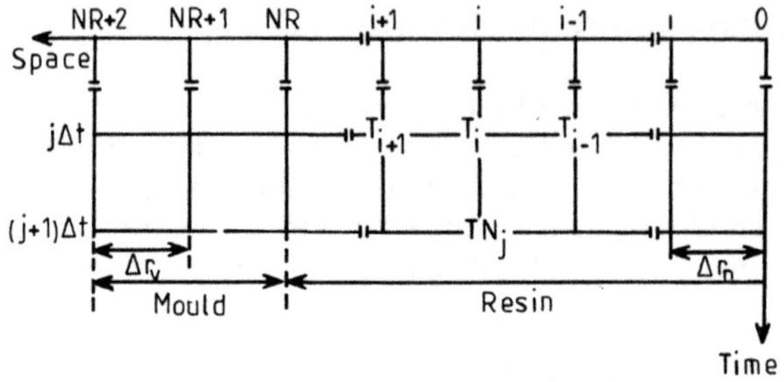

Fig. 15.1. Schematic diagram for numerical analysis.

NR being the number of annuli in the resin and NT the total number of annuli in the resin and mould. The dimensionless number M_m is

$$M_m = \frac{(\Delta r_m)^2}{\Delta t}\left(\frac{\rho C}{\lambda}\right)_m \tag{15.9}$$

$(\rho, C, \lambda)_m$ being the parameters for the mould and Δr_m the thickness of each annulus in the mould.

Mould–Resin Interface

The heat balance is calculated within the two annuli of thicknesses $\frac{1}{2}\Delta r$ and $\frac{1}{2}\Delta r_m$ on both sides of the mould–resin interface, during the increment of time Δt. The new temperature on the interface TN_{NR} after time Δt is thus obtained:

$$\begin{aligned} TN_{NR} = {} & T_{NR} + BM[T_{NR+1} - T_{NR}] - BR[T_{NR} - T_{NR-1}] \\ & + BC[3\Delta Q_{NR} + \Delta Q_{NR-1}] \end{aligned} \tag{15.10}$$

where BM, BR and BC are expressed as follows:

$$BM = \frac{(2R_s - \Delta r)\lambda \Delta t}{A\Delta r} \tag{15.11}$$

$$BR = \frac{(2R_s + \Delta r_m)\lambda_m \Delta t}{A\Delta r_m} \tag{15.12}$$

$$BC = \frac{(4R_s - \Delta r)\Delta r \rho}{16A} \tag{15.13}$$

with

$$A = \left[R_s + \frac{\Delta r_m}{4}\right]\rho_m C_m \Delta r_m + \left[R_s - \frac{\Delta r}{4}\right]\rho C_{NR}\Delta r \tag{15.14}$$

and ΔQ_{NR} is the amount of heat generated by the reaction at the position NR, per unit mass, during the increment of time Δt.

Surface of the Mould

As heat transfer through the gas–mould interface is by free convection, the coefficient of surface heat transfer is calculated by using the Grashof and Prandtl numbers, as well as the Nusselt number:

$$h = h_0|\Delta T|^{N_1} \tag{15.15}$$

where h_0 is a constant depending on the size of the mould and its position (vertical, horizontal) and ΔT is the difference between the temperature of the heated air far away from the mould and temperature of the surface of the mould. N_1 is a coefficient which depends on whether the convection is laminar or turbulent.

The heat balance is evaluated during the time increment Δt within the annulus of thickness $\frac{1}{2}\Delta r_m$ in the mould next to the external surface, by considering the heat convection from the surrounding air and the heat conduction within the mould. The new temperature after time Δt on the surface of the mould is thus given by

$$\mathrm{TN}_{\mathrm{NT}} = T_{\mathrm{NT}} + hZ_1(T_{\mathrm{air}} - T_{\mathrm{NT}}) - Z_2(T_{\mathrm{NT}} - T_{\mathrm{NT}-1}) \tag{15.16}$$

with the coefficients Z_1 and Z_2 defined by

$$Z_1 = 2R_m\Delta t\left[\Delta r_m\rho_m C_m\left(R_m - \frac{\Delta r_m}{4}\right)\right]^{-1} \tag{15.17}$$

$$Z_2 = (2R_m - \Delta r_m)\left[M_m\left(R_m - \frac{\Delta r_m}{4}\right)\right]^{-1} \tag{15.18}$$

where M_m is the dimensionless number defined by Eq. (15.9).

Extent of Cure

Upon defining the function S,

$$S_j = \int_0^{j\Delta t} k_0 \exp\left(-\frac{E}{RT_\theta}\right) d\theta \qquad 0 \leqslant \theta \leqslant t \tag{15.19}$$

j being an integer characterising the time ($t = j\Delta t$) with the kinetic parameters k_0, E. The function S can be obtained after time Δt at each place, by the following equation:

$$S_{j+1} = S_j + k_0 \exp\left(-\frac{E}{RT_j}\right) \tag{15.20}$$

with

$$S_0 = 0 \tag{15.21}$$

The extent of cure Q_j/Q_∞ at time $j\Delta t$ is thus obtained by the following:

$$\frac{Q_j}{Q_\infty} = 1 - [1 + (n-1)S_j]^{1/(1-n)} \tag{15.22}$$

when the order of the reaction n is different from one, and the SOC is usually given by

$$\mathrm{SOC}_j = 100\,\frac{Q_j}{Q_\infty} \tag{15.23}$$

Conditions of Stability for Calculation

In the equation expressing the new temperature TN, the coefficient of the previous temperature T must be positive:

$$M > 2 \qquad \text{(Eq. (15.5))} \tag{15.24}$$

$$M_m > 4 \qquad \text{(Eq. (15.8))} \tag{15.25}$$

$$M > 4 \qquad \text{(Eq. (15.7))} \tag{15.26}$$

$$1 - \text{BM} - \text{BR} > 0 \qquad \text{(Eq. (15.10))} \tag{15.27}$$

or

$$\Delta t < \left[\frac{(2R_s - \Delta r)\lambda}{A\Delta r} + \frac{(2R_s + \Delta r_m)\lambda_m}{A\Delta r_m} \right]^{-1} \tag{15.27'}$$

$$1 - hZ_1 - Z_2 > 0 \qquad \text{(Eq. (15.16))} \tag{15.28}$$

or

$$\Delta t < \rho_m C_m \left(R_m - \frac{\Delta r_m}{4} \right) \left[\frac{2R_m h}{\Delta r_m} + \frac{(2R_m - \Delta r_m)\lambda_m}{\Delta r_m^2} \right]^{-1} \tag{15.28'}$$

Remark: Natural Convection. The factors involved in the natural convection of heat from a vertical plane to a fluid were given in 1881 by Lorenz. It was assumed that heat is transferred by conduction to a layer of fluid upward in laminar motion under the influence of buoyancy. The wall temperature is assumed constant. The Lorenz equation is of the form

$$\text{Nu} = a(\text{GrPr})^p \tag{15.29}$$

where Nu, Gr and Pr are dimensionless numbers. The Grashof number Gr is

$$\text{Gr} = \frac{L^3 \rho^2 g \beta |\Delta T|}{\mu} \tag{15.30}$$

L denotes the height of a vertical surface, or length of a horizontal surface, and the diameter in the case of a horizontal cylinder.

ρ and μ are the density and the viscosity of the fluid.

β is the coefficient of volumetric expansion of the fluid.

ΔT is the temperature potential, the difference in the temperatures at the surface of the solid and in the surrounding fluid.

g is the acceleration owing to gravity.

The Prandtl number Pr is

$$\text{Pr} = \frac{\mu C}{\lambda} \tag{15.31}$$

where μ, C, λ are the viscosity, the heat capacity and the thermal conductivity of the fluid, respectively.

The Nusselt number Nu is

$$\text{Nu} = \frac{hL}{\lambda} \tag{15.32}$$

where h is the surface coefficient of heat transfer.

The coefficients a and p are given in Table 15.1.

The laminar and turbulent range are characterised by the value of the Grashof and Prandtl numbers:

Laminar range $10^4 < \mathrm{GrPr} < 10^9$
Turbulent range $\mathrm{GrPr} > 10^9$

These results are essential for the calculation of the coefficient of surface heat transfer h. More information can be obtained in specialised books [2].

15.3 Experiment

Materials

The resin used in this chapter was obtained by mixing epoxy resin (Lopox 200 from CDF Chimie) and hardener (D 2605 from CDF-Chimie) containing an anhydride of Me THP, and an accelerator with 0.75% of an aromatic amine (type A 105 from CDF Chimie). The binary mixture was prepared at room temperature with continuous stirring, and then poured into the cylindrical glass mould.

The best composition (in weight percent) is [3] epoxy resin 55.6%, hardener 44.4%.

Mould and Temperature Measurements

The cylindrical glass mould with the resin sample was put in an oven heated by air kept at constant temperature. The mould was thus heated by free convection.

The temperature was recorded in various places:

At the centre of the resin sample
In the exterior air
On the external surface of the mould

These temperature measurements are necessary for testing the validity of the model.

The centre of the resin, along the longitudinal axis, was chosen for measuring the temperature for two reasons:

1. Better accuracy is obtained at this place, as the gradient of temperature is

Table 15.1. Coefficients a and p in natural convection

Surface	Dimension	Laminar		Turbulent	
		a	p	a	p
Vertical cylinder	Height	0.56	0.25	0.12	0.33
Horizontal cylinder	Diameter	0.47	0.25	0.10	0.33

equal to zero, because of the symmetry. Thus a slight change in the position for the thermocouple does not alter the temperature measurement too much.

2. The lowest temperature is obtained at the centre of the resin during the stage of preheating, and the highest temperature is attained during the stage of cure reaction. The highest amplitude of temperature is observed at this place.

Kinetic and Thermal Parameters

The kinetic parameters were calculated from the heat flux–temperature curves obtained by differential scanning calorimetry with a DSC 111 (Setaram), scanned at a constant heating rate of 2 °C min^{-1}. The sample of about 150 mg was encapsulated in a sealed metal holder. The heat flux–temperature curves were registered and analysed with the help of a microcomputer (HP 86, Hewlett–Packard) connected to the calorimeter, by using the Freeman–Carroll technique and considering that the temperature of the sample is constantly equal to that of the calorimeter.

These experimental curves were compared with the heat flux–temperature curves calculated by using the kinetic parameters obtained, and an excellent superimposition of these curves is considered as a test for the validity of these parameters. The heat flux curves were calculated by also assuming that the temperature is always uniform throughout the sample and equal to that of the calorimeter, as the value of the heating rate is very low.

The enthalpy of cure was also obtained by DSC, by integrating the exotherm. It was found previously [4–6] that the cure enthalpy obtained by the DSC technique is generally higher than those measured by isothermal calorimetry, because a significant part of the enthalpy is lost in DC at the beginning of the stage of the cure reaction and at the end of the first stage of heating, when the highly negative heat flux cannot be recorded without a computer.

The kinetic and thermal parameters were as follows:

Kinetics of reaction: $k_0 = 4 \times 10^{11}\ \mathrm{s}^{-1}$, $E = 27140\ \mathrm{cal\,mol}^{-1}$,

$$n = 1.5,\ Q_\infty = 76\ \mathrm{cal\,g}^{-1}$$

Thermal parameters:

Resin (sample), $\lambda_s = 10^{-3}\ \mathrm{cal\,cm}^{-1}\,\mathrm{s}^{-1}\,\mathrm{deg}^{-1}$

$$C_s = 0.2 + 0.003T\ (\mathrm{cal\,g}^{-1}\,\mathrm{deg}^{-1})\ (T \text{ in } ^\circ\mathrm{C})$$

Glass, $\lambda_g = 2.66 \times 10^{-3}\ (\mathrm{cal\,cm}^{-1}\,\mathrm{s}^{-1}\,\mathrm{deg}^{-1})$

$$C_g = 0.3\ \mathrm{cal\,g}^{-1}\,\mathrm{deg}^{-1}$$

Coefficient of heat transfer by convection:

$$h = 7.12 \times 10^{-4}|\Delta t|^{0.25}\ \mathrm{cal\,cm}^{-2}\,\mathrm{s}^{-1}\,\mathrm{deg}^{-1}$$

The coefficient of heat transfer by convection h was obtained either by calculation with the dimensionless numbers shown in the remark on natural convection (Sect. 15.2.2), or by selecting the value which enabled the temperature–time history at the centre calculated with the model to fit the experimental results. The experiment was performed with the same mould with the precured resin in it either by heating the mould previously at room temperature in heated air under conditions identical to the cure, or by letting the heated mould cool down in air at room temperature. The temperature at the centre of the resin was recorded during these processes with the precured resin, and the value of h was obtained by fitting the experimental temperature–time curves with the calculated ones.

15.4 Results

After testing the validity of the model in various cases, the following parameters of interest were studied:

1. Kinetic parameters which depend on the resin
2. Thermal parameters which depend on the resin, and especially the inert additives
3. Diameter of the cylindrical mould
4. Temperature of the heated air

The results obtained either from experiments or from calculation with the help of the numerical model, are expressed in three ways:

1. Temperature–time histories at the centre of the resin
2. SOC–time histories at the centre of the resin
3. Profiles of temperature and of SOC developed within the circular cross-section of the resin

As the kinetic parameters determined from DSC experiments (or DC), and the thermal parameters by other experiments, are obtained with various sources of

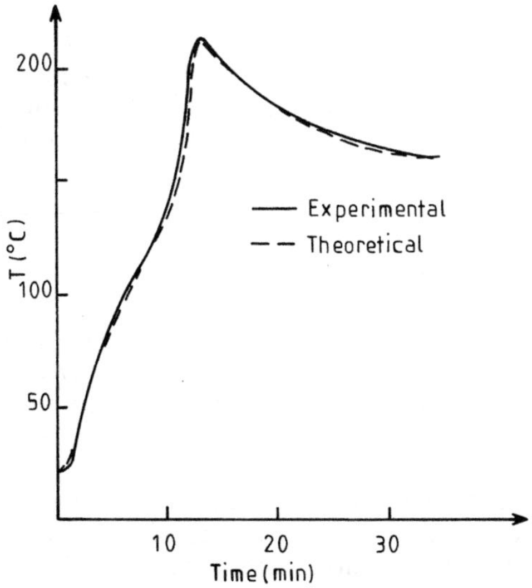

Fig. 15.2. Temperature–time history at the centre of the resin proving the validity of the model, for a sample diameter of 1.5 cm and an air temperature of 150 °C.

error, which can be traced to the instrument, the composition of the resin, the representativeness of the sample taken from a larger amount, it is of interest to determine the effect of these parameters on the process of cure.

15.4.1 Validity of the Model [7]

The validity of the model was tested by comparing the temperature–time history obtained at the centre of the resin either from experiments or from calculation. This temperature–time history is shown in Fig. 15.2. The curves are very well superimposed, proving the validity of the model, as well as the accuracy of the kinetic and thermal parameters shown above (Sect. 15.3). The air temperature in the oven was 150 °C.

In fact, the validity of the model was also tested in many other cases, with different values of the radius of the mould and with different temperatures.

It is of interest to describe the temperature–time history obtained at the centre of the resin. Two, or perhaps three, steps can be considered during the process:

1. The heating period, from room temperature to nearly 120 °C. It takes some time for heat to progress through the resin, and at the very beginning of the process, the temperature at the centre remains constant for around 1 min.
2. The cure reaction which starts at a temperature of around 125 °C. Because of the exothermicity of the reaction, an increase in temperature at the centre of the resin is observed. The temperature at the centre passes thus through a maximum, and then decreases rather slowly down to the air temperature.
3. Between these two stages, an inflection point can easily be observed.

15.4.2 Effect of the Kinetic Parameters of the Resin [7]

Four parameters have been shown to be of interest for the cure of rubber: the order of the overall cure reaction [8], the cure enthalpy [9, 10], the activation energy and the pre-exponential factor k_0 [11, 12]. These four parameters are also worth attention for the cure of epoxy resin, and especially the enthalpy of reaction because of its high value.

Order of the Overall Cure Reaction

In a previous paper [8] devoted to the cure of rubber sheets, the effect of the order of the overall reaction on the rate of cure and on the profiles of temperature was proved to be almost negligible when the extent of cure was below 50%, and became increasingly of interest when the extent of cure increased from 50% to 100%.

The temperature–time histories at the centre of the resin are shown for various values of the order of the overall cure reaction of the reaction of the epoxy resin (Fig. 15.3). Of course, these curves are obtained either from experiment or calculation when the order is 1.5, and they are calculated for the other values of 1.3 and 1.7.

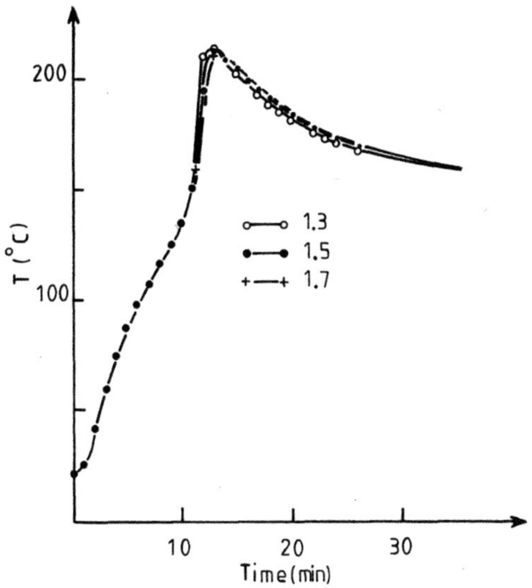

Fig. 15.3. Effect of the order of the overall cure reaction on the temperature–time histories at the centre of the resin.

As shown in these three curves drawn in Fig. 15.3, the effect of the order of the overall reaction of cure on the temperature–time history at the centre of the resin is very small. This conclusion is in line with the results obtained with the rubber.

Variation in the Pre-Exponential Factor

The following three values were selected for the pre-exponential factor: 3.2×10^{11}, 4×10^{11} and $4.8 \times 10^{11}\ s^{-1}$, while the other kinetic and thermal parameters were kept constant. The temperature–time as well as the state of cure–time histories were calculated for each of these three values. The temperature–time history for the value of $4 \times 10^{11}\ s^{-1}$ was also obtained by experiments. As shown by these curves (Fig. 15.4) the effect of the pre-exponential factor k_0 on the process is of great importance, and the following conclusions are pointed out:

1. Of course, the higher the pre-exponential factor, the faster the reaction of cure. This increase in the rate of cure can be appreciated by the steeper slope of the temperature–time history calculated with the highest value of the pre-exponential factor. This change in the slopes of these curves appears clearly around 120 °C, as soon as the reaction starts.
2. The maximum attained by the temperature at the centre is slightly higher for the highest value of the pre-exponential factor. The difference in these extreme values does not exceed 3 °C.
3. Corresponding with the first conclusion, a higher value for the pre-exponential factor is accountable for an increase in the SOC.
4. The SOC–time curves are shifted to the shortest times in the same way as the temperature–time histories when the pre-exponential factor is increased. For

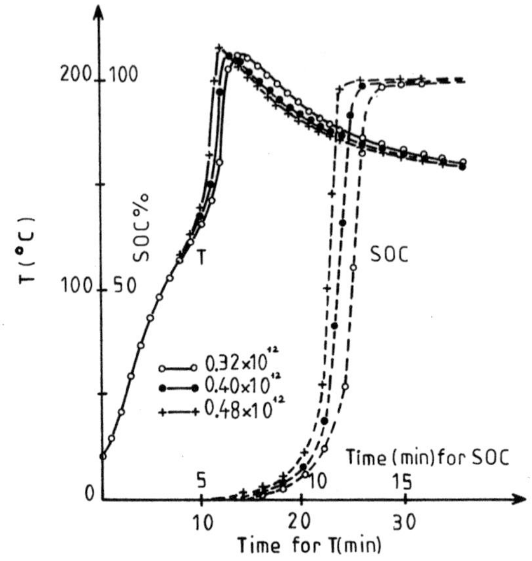

Fig. 15.4. Effect of the pre-exponential factor on the temperature–time history and SOC–time history at the centre of the resin.

instance, an increase in 20% for the pre-exponential factor provokes a reduction of about 45 s in the time necessary for the SOC to reach a given value. The same conclusions were also drawn for the cure of rubber [11, 12].

Variation in the Activation Energy

The following three values were chosen for the activation energy E: 26 986, 27 140 and 27 330 cal mol^{-1}, while the other kinetic and thermal parameters were kept constant. These values were calculated by keeping $k_0 \exp(-E/RT)$ constant at the temperature of the mould (150 °C), while the pre-exponential factor k_0 was varied by 20% around the mean value obtained by experiment.

The effect of a slight variation in the activation energy on the temperature–time and SOC–time histories at the centre of the resin are shown in Fig. 15.5. The following conclusions are drawn:

1. Of course, the lower the activation energy, the faster the rate of cure. In line with this fact, the slopes of the temperature–time histories are steepest for the lowest value of the activation energy. Subsequently, a decrease in the activation energy provokes a reduction of the time necessary for the SOC to attain a given value.
2. A similarity exists between the curves plotted for the effect of the activation energy and the pre-exponential factor. However, the point of concern is that a slight change in the value of the activation energy, between 154 and 190 cal mol^{-1}, causes the same effect as a change of 20% in the pre-exponential factor.

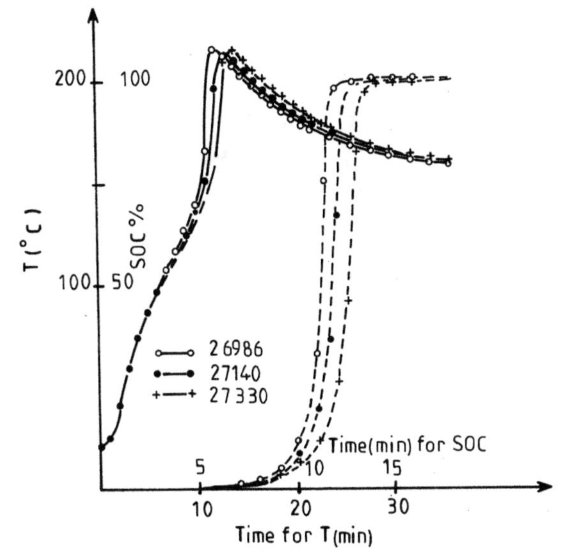

Fig. 15.5. Effect of the activation energy E (cal mol^{-1}) on the temperature and SOC–time histories at the centre of the resin.

Enthalpy of Cure

The effect of the cure enthalpy on the process of cure is determined by giving the following values 60, 76 and 92 cal g^{-1} for this enthalpy, and keeping the other kinetic and thermal parameters constant. The value of 76 cal g^{-1} is obtained from experiments.

The temperature–time histories are shown in Fig. 15.6 for the various values of the cure enthalpy, as well as the SOC–time histories at the centre of the resin (Fig. 15.7).

Some conclusions can be drawn:

1. The effect of the cure enthalpy is different from the effect of the other kinetic parameters k_0 and E, the difference being essentially the change in the value attained by the temperature, especially at the maximum of temperature. An increase in 20% for the cure enthalpy produces an increase in the maximum temperature of nearly 18 °C.
2. This high increase in temperature due to the internal reaction heat, has adverse consequences on the properties of the final material.
3. Due to the heat generated by the cure reaction, the slopes of the temperature–time curves are steeper when the cure enthalpy is increased.
4. The effect of the cure enthalpy on the SOC–time history at the centre of the resin is more pronounced than the effect of the other kinetic parameters.

15.4.3 Effect of the Thermal Parameters of the Resin [7]

Two thermal properties of the resin are of interest: the thermal conductivity and heat capacity, which characterise the resin. There is also a third parameter, which

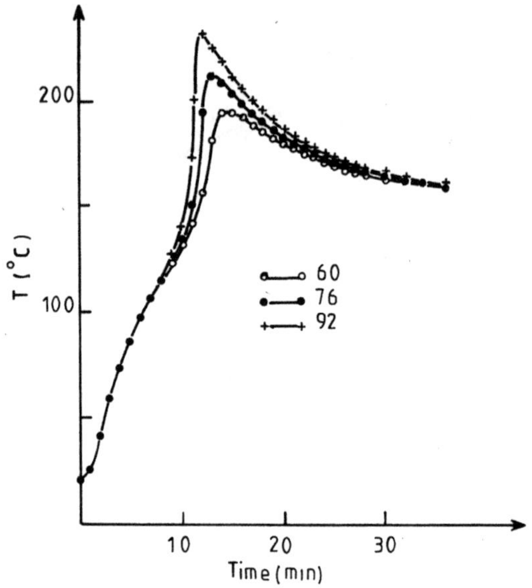

Fig. 15.6. Effect of the cure enthalpy ($cal\,mol^{-1}$) on the temperature–time history at the centre of the resin.

characterises the mould and the system of heating: the coefficient of the surface heat transfer on the external surface of the mould.

A previous paper concerned with the cure of rubber showed the significant effect of the thermal diffusivity on the process [13]. As the heat diffusivity $\alpha = \lambda/\rho C$ is a function of various parameters which vary diversely, it is better to consider the effect of thermal conductivity and the heat capacity separately, in succession.

Thermal Conductivity

The effect of the thermal conductivity of the resin on the process of cure was determined by varying its value by 20% around the mean value obtained by experiment, while the other parameters were kept constant. The temperature and SOC–time histories at the centre of the resin are shown in Fig. 15.8 for the various values of the thermal conductivity of the resin.

Some conclusions can be drawn:

1. The effect of the thermal conductivity on the process is not very important.
2. The effect of this parameter on these curves is apparent not only during the cure stage, but also during the first stage of heating and during the cooling period of the resin beyond the maximum temperature.
3. All these curves can be obtained by translation along the time axis, a higher value of the thermal conductivity allowing a shorter time to reach the same temperature and state of cure.

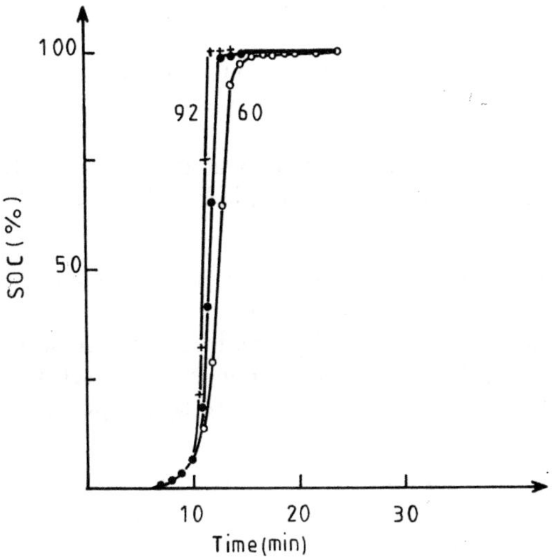

Fig. 15.7. Effect of the cure enthalpy (cal mol^{-1}) on the SOC–time history at the centre of the resin.

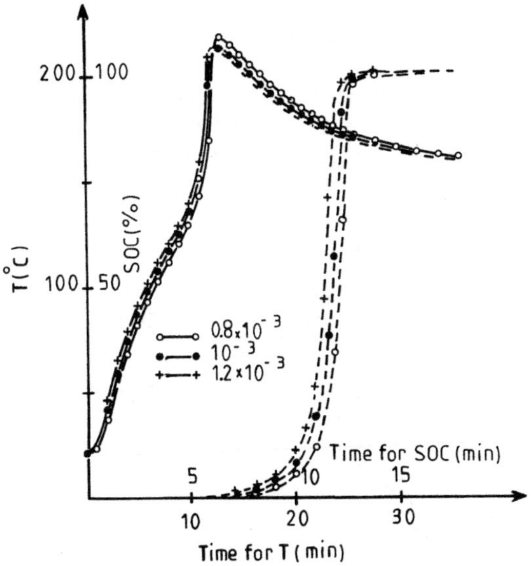

Fig. 15.8. Effect of the thermal conductivity (cal cm^{-1} s^{-1} deg^{-1}) on the temperature and SOC–time histories at the centre of the resin.

Heat Capacity

The heat capacity plays an important role on the process of cure, as it acts not only on the heat transfer by conduction in the same way as the thermal conductivity but also on the heat evolved from the reaction in the same way as the cure enthalpy.

The effect of the heat capacity of the resin on the process of cure is calculated by varying the value of the heat capacity by 20% around the mean value obtained by experiment. As the heat capacity is found to be a function of temperature (Sect. 15.3), both coefficients are varied: $(0.16 + 0.0024T)$, $(0.2 + 0.003T)$ and $(0.24 + 0.0036T)$.

The temperature and SOC–time histories at the centre of the resin are plotted for the various values of the heat capacity, in Fig. 15.9. The following conclusions are worth noting:

1. The effect of this parameter on the process is very important, as it acts on both the heat conduction and the heat of cure.
2. On the one hand, the heat capacity acts in an inverse way to the enthalpy of cure. An increase in the heat capacity reduces the effect of the cure enthalpy.
3. On the other hand, the heat capacity acts in an inverse way to the thermal conductivity.
4. Both these effects are superimposed. As a result, a decrease in the value of the heat capacity is responsible for a faster heating and faster cure, and for a higher temperature in the resin.
5. The effect of the heat capacity on the SOC–time histories is also of high significance.

Coefficient of Surface Heat Transfer

The coefficient of surface heat transfer characterises the rate of the heat transferred by convection to the external surface of the mould.

Three values for the coefficient k_0, obtained by varying by 20% around the

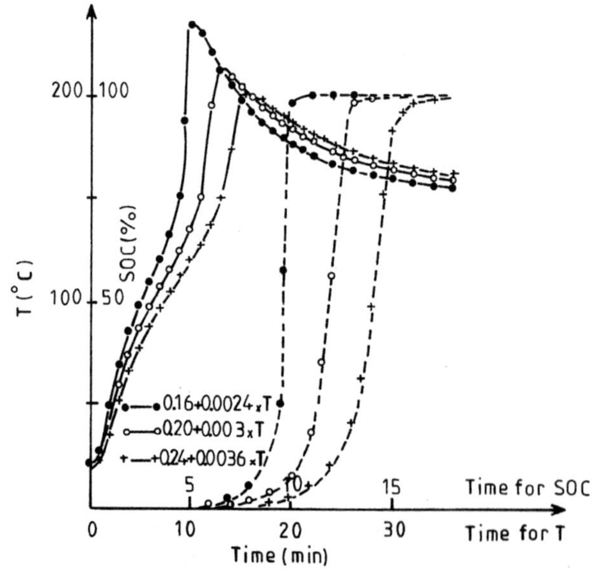

Fig. 15.9. Effect of the heat capacity (cal g^{-1} deg^{-1}) on the temperature and SOC–time histories at the centre of the resin.

mean experimental value (5.7×10^{-4}, 7.12×10^{-4} and 8.54×10^{-4}), were used for calculation, by keeping the other parameters constant.

As shown in Fig. 15.10 for the temperature and SOC–time histories, the effect of the coefficient of surface heat transfer on the process is very important:

1. Acting on the rate of heat transferred at the surface of the mould, this coefficient plays a significant role during the whole process.
2. An increase in this coefficient reduces the time for heat transferred into or out of the mould.
3. The temperature at the maximum is not significantly increased by increasing the coefficient of heat transfer.
4. The effect of this coefficient on the SOC–time histories is very important.

15.4.4 Effect of the Size of the Mould [14]

The mould used for this study consisted of a thin glass cylinder heated by an external fluid, this kind of mould being particularly suitable either for supplying the resin with heat or by extracting heat from the resin as fast as possible when necessary by varying the temperature of the fluid.

The validity of the model was tested by comparing experimental and theoretical values for the temperature at various places in the mould and the resin, especially at the centre of the resin, and by considering cylindrical moulds of various internal diameters ranging from 1 to 2 cm.

The numerical model is described in Sect. 15.2; the epoxy resin and its thermal and kinetic parameters are described in Sect. 15.3.

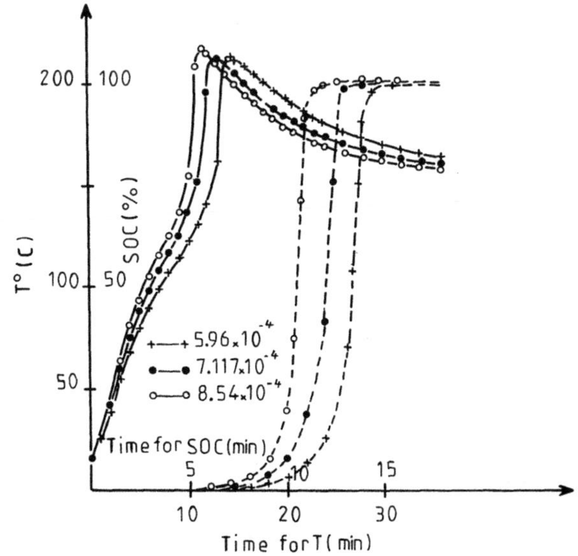

Fig. 15.10. Effect of the coefficient of surface heat transfer by convection (10^4 cal cm^{-2} s^{-1} deg^{-1}) on the temperature and SOC–time histories at the centre of the resin.

The results are given by considering on the one hand the temperature–time and SOC–time histories at the centre of the resin, and on the other hand the profiles of temperature and SOC developed through the resin during the process.

Temperature–Time Histories at the Centre of the Resin

The temperature–time histories obtained either by experiment or by calculation are plotted for various internal diameters of the long cylindrical moulds: 1 cm (Fig. 15.11), 1.4 cm (Fig. 15.12) and 2 cm (Fig. 15.13). The thickness of the glass was always 0.2 cm. As shown in these three figures, good agreement between the experimental and calculated curves can be determined, proving the validity of the model and the accuracy of the values used for the kinetic and thermal parameters.

Three steps can be distinguished during the whole process:

1. The heating stage of the mould and resin when the temperature of the surrounding air is higher than that of the mould. Of course, it takes a longer period of time for the larger mould to be heated. The following statement can be made: the time necessary for heat to be transferred through a medium is proportional to the square of the thickness of the medium (the radius in this case). This statement is also true for the period of time over which the temperature at the centre of the resin remains about constant, at the beginning of the process.

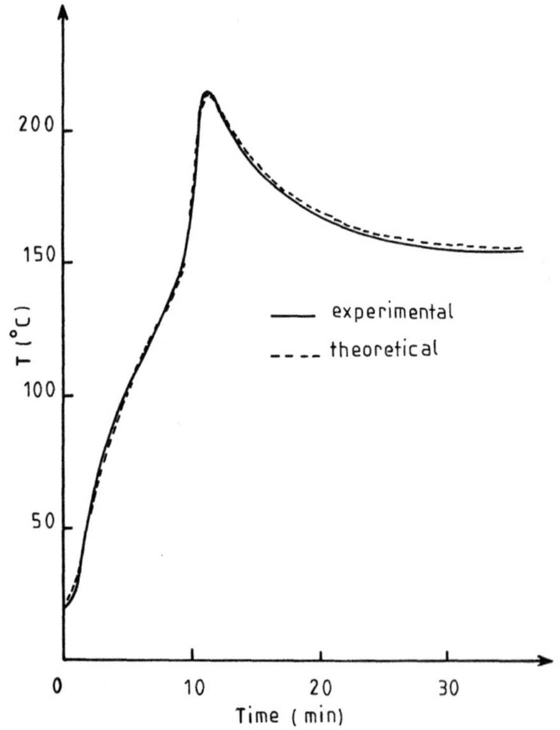

Fig. 15.11. Temperature–time history at the centre of the long cylindrical sample of 1 cm diameter. Validity of the model proved by superimposition of experimental and calculated curves.

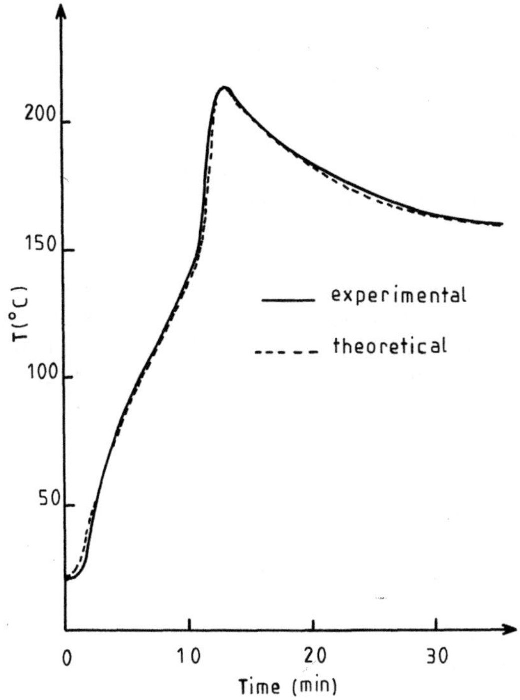

Fig. 15.12. Temperature–time history at the centre of the long cylindrical sample of 1.4 cm diameter. Validity of the model proved by superimposition of experimental and calculated curves.

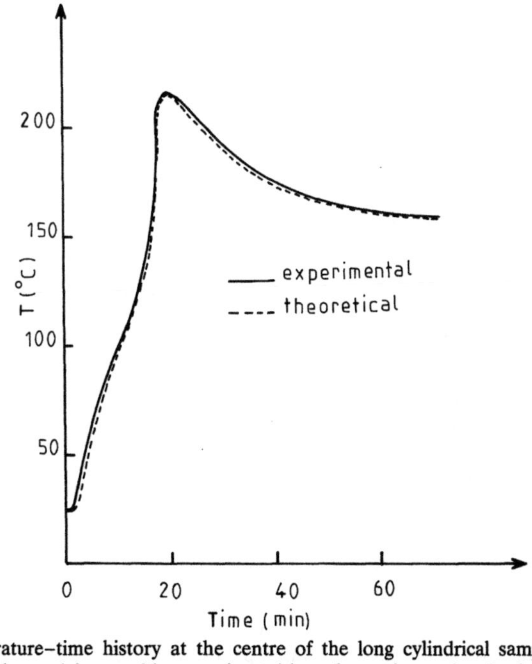

Fig. 15.13. Temperature–time history at the centre of the long cylindrical sample of 2 cm diameter. Validity of the model proved by superimposition of experimental and calculated curves.

2. The stage of cure reaction, which can be seen from the point of inflection on the temperature–time curve to the end of the process. Because of this internal heat generated by the reaction, an increase in temperature is observed. This temperature reaches a maximum. This increase in temperature varies with the size of the sample with the obvious statement: the larger the sample, the higher the increase in temperature. For instance, for the smaller sample with a radius of 1 cm, the temperature at the centre rises to a maximum of 213 °C, while it reaches a maximum of 217 °C in the case of the larger sample with a radius of 2 cm.
3. As shown in Figs 15.14 and 15.15, where the temperature–time histories in various places in the sample are plotted, this maximum temperature is observed at any place in the resin. The highest maximum is attained at the centre of the resin. Moreover, the maximum temperature is obtained initially at the centre of the resin.

SOC–Time Histories at the Centre of the Resin

One of the advantages of this model is its ability to provide information on the SOC as well as the temperature. The SOC–time history is illustrated in Figs 15.16 and 15.17 for the moulds of internal diameter 1 and 2 cm.

The three stages already shown for the temperature–time history can also be considered for the SOC:

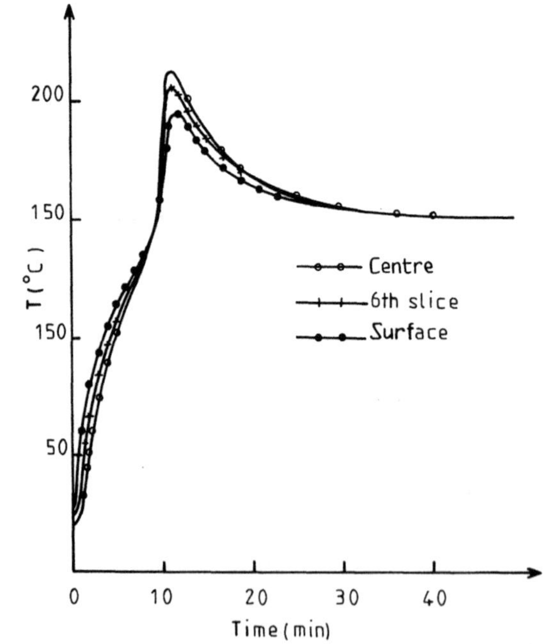

Fig. 15.14. Temperature–time histories at various places within the resin, for the sample of 1 cm diameter: centre ($r = 0$); $r = \frac{1}{2}R$; surface ($r = R$).

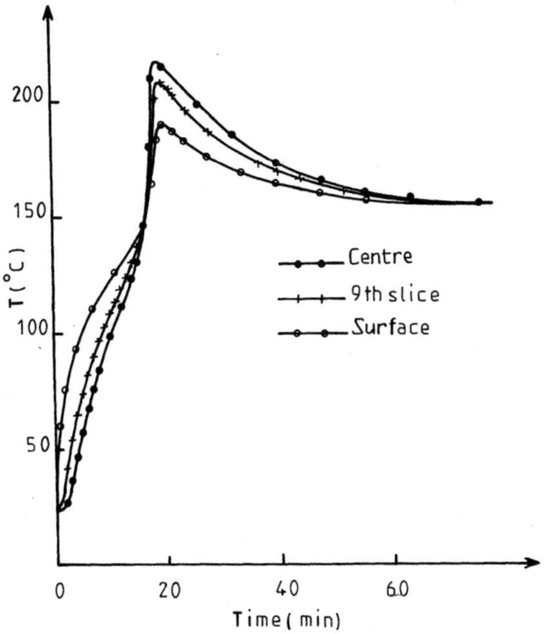

Fig. 15.15. Temperature–time histories at various places within the resin, for the sample of 2 cm diameter: centre ($r = 0$); $r = \frac{1}{2}R$; surface ($r = R$).

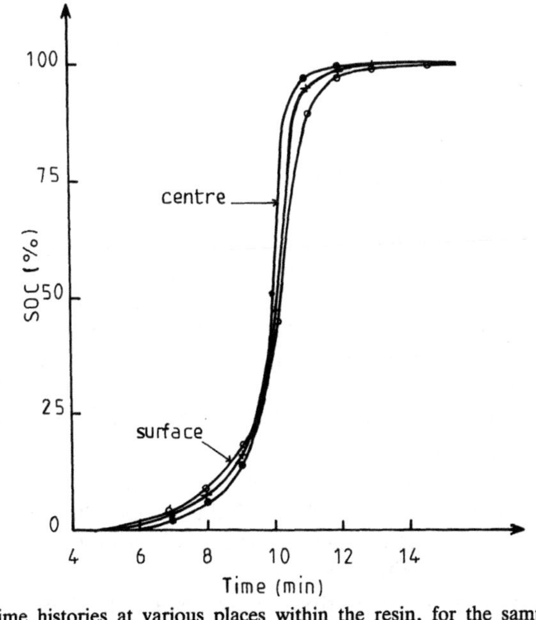

Fig. 15.16. SOC–time histories at various places within the resin, for the sample of 1 cm diameter: centre ($r = 0$); $r = \frac{1}{2}R$; surface ($r = R$).

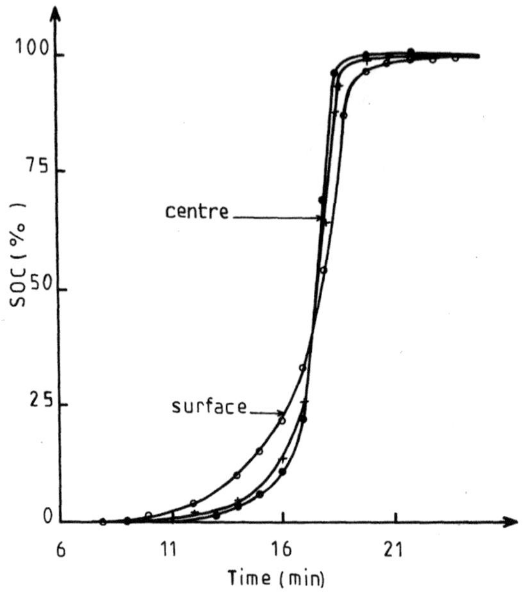

Fig. 15.17. SOC–time histories at various places within the resin, for the sample of 2 cm diameter: centre ($r = 0$); $r = \frac{1}{2}R$; surface ($r = R$).

1. During the first stage of heating, there is no reaction anywhere in the resin. The reaction then starts initially on the surface, and progressively within the resin.
2. During the second stage, the reaction takes place. The rate of the reaction increases very quickly up to a maximum value.
3. After this maximum value of the rate of cure reaction, a third stage can be distinguished with the decrease in the rate of cure. At the end of the process, the SOC is about 100%.

Profiles of Temperature and SOC

A piece of complementary information of great importance is obtained from the calculation of the profiles of temperature and SOC developed through the cross-section of the resin. The calculated profiles of temperature and SOC are shown in Fig. 15.18 for the 1 cm diameter mould and in Fig. 15.19 for the 2 cm diameter mould. Comparison between the profiles of temperature (left) and (right) can afford further insight into the process of cure. The following steps can thus be defined precisely. For instance, in the small mould, they are as follows:

1. Up to 8 min, during the heating period, or rather during the time necessary for the resin to reach a temperature next to that of the surrounding air, the temperature is higher on the surface of the mould than at the centre of the resin. Subsequently, a small value of the SOC is attained, while the SOC is highest on the surface of the resin.
2. During the following 2 min period, between 9 and 11 min, the temperature gradients become inverted with a highest value at the centre of the resin.

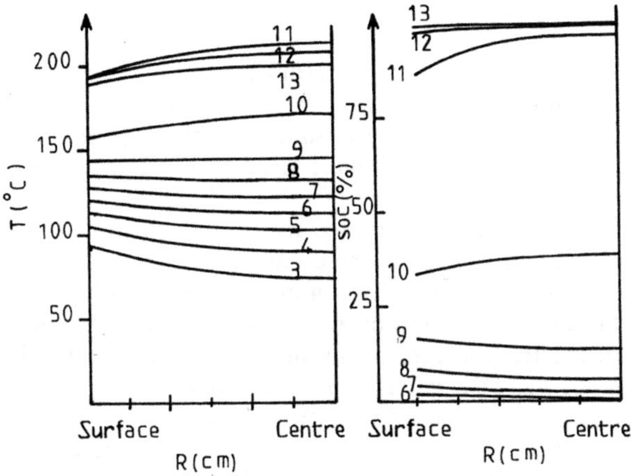

Fig. 15.18. Profiles of temperature (left) and SOC (right) developed through the cross-section of the resin within the sample of 1 cm diameter.

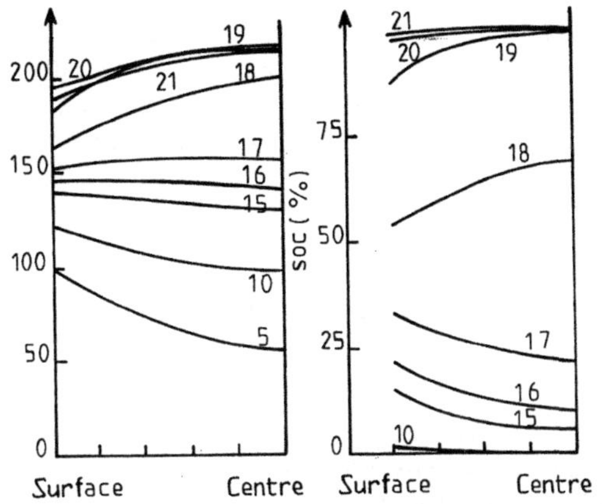

Fig. 15.19. Profiles of temperature (left) and SOC (right) developed through the cross-section of the resin within the sample of 2 cm diameter.

Obviously, this fact is due to the development of the cure reaction producing internal heat. In the same way, the SOC is higher at the centre of the resin than elsewhere.

3. At the maximum temperature, at around 11 min for the small mould and 19 min for the large one, the cure reaction is not complete. At these times, the SOC ranges from 87% to 95% through the cross-section of the resin with the small mould, and from 88% to 97% with the large mould. The cure reaction is found to proceed to completion during the cooling period, over about 2–5 min.
4. The effect of the dimension of the mould on the process appears very clearly with the profiles of temperature and SOC developed through the resin. Steeper gradients of temperature and SOC are obtained with the large mould either during the heating period or during the reaction period.

15.4.5 Effect of Temperature [15]

The main purpose of this study is to determine precisely the effect of a change in the mould temperature on the process, and especially on the profiles of temperature and SOC developed within the sample. This information is necessary not only to enhance the understanding of the process but also to determine the best operational conditions for the process.

One aspect of fundamental importance in the process for epoxy resin or other thermosetting polymers is the exothermic nature of the cure reaction which gives rise to substantial temperature gradients and to excessively high temperatures [16, 17] which may cause discolouration or even degradation of the material [18–21].

So the effect of the temperature of the mould on the process of cure is of great interest, because an increase in temperature reduces the time of cure but also enhances the gradients of temperature in the sample.

The model described above (Sect. 15.2) was also tested for these different temperatures by comparing some experimental and calculated results, essentially the temperature–time histories obtained at the centre of the resin. Of course, another feature of the model is its ability to give more information than experiment on the profiles of temperature and of SOC developed throughout the sample.

As a long cylinder is used in this study, heat transfer is radial only, and the circular cross-section of the sample is considered specifically. A cylindrical glass mould with a diameter of 1.4 cm was used. The mould was heated by convection by the help of heated motionless air. The air temperature was chosen within the range 142°–164 °C.

Mould Temperature and Temperature–Time Histories at the Centre of the Resin [15]

The temperature–time histories at the centre of the resin were determined either by experiments or calculation with the help of the model, for three values of the air temperature: 142, 153 and 164 °C.

As shown in Figs 15.20 (142 °C), 15.21 (153 °C), and 15.22 (164 °C), good agreement is obtained between the experimental and theoretical curves, proving the validity of the model, as well as the accuracy of the values of the kinetic and thermal parameters.

The effect of the value given to the air temperature on the process of cure can be appreciated by comparing the temperature–time histories obtained at these three temperatures. The following conclusions can be drawn:

1. During the period over which the sample is heated up to the air temperature, the rate of heating depends on the temperature of the air, as the coefficient of heat transfer is a function of ΔT, the difference between the temperatures of air and of the mould surface.
2. Following the heating period of the sample, the reaction occurs with the higher slope of the temperature–time curve. The temperature at the centre of the sample rises to a maximum, and the time necessary for the temperature to reach this maximum is found to depend on the air temperature. Of course, the

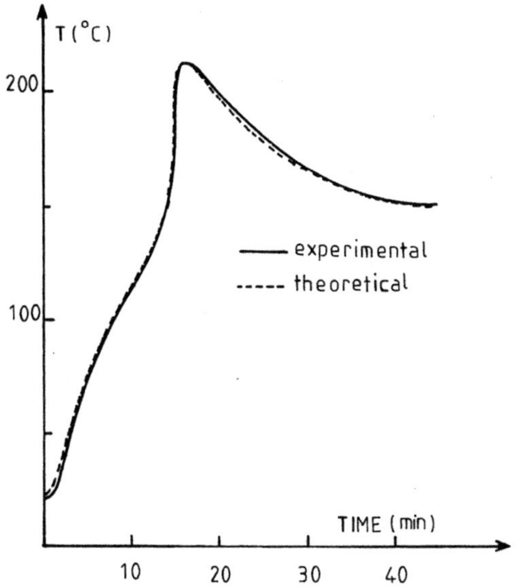

Fig. 15.20. Temperature–time history at the centre of the sample. Temperature of heated air = 142 °C. Diameter = 1.4 cm.

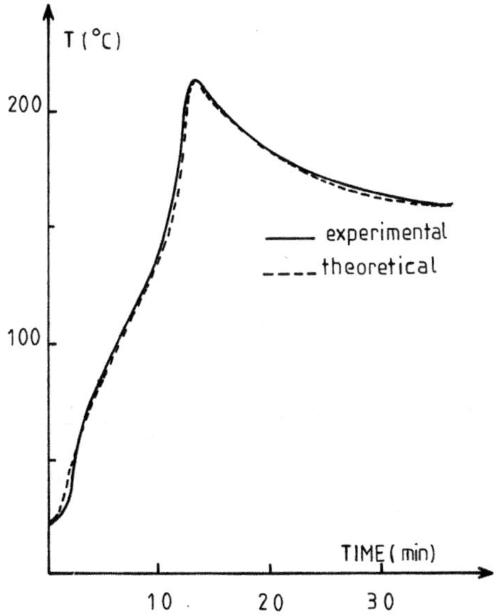

Fig. 15.21. Temperature–time history at the centre of the sample. Temperature of heated air = 153 °C. Diameter = 1.4 cm.

higher the air temperature, the shorter is this time: 16 min for 142 °C, 13.4 min for 153 °C and 12 min for 164 °C.

3. It is difficult to find a relation between the time at which the maximum temperature at the centre of the resin is attained and the mould temperature,

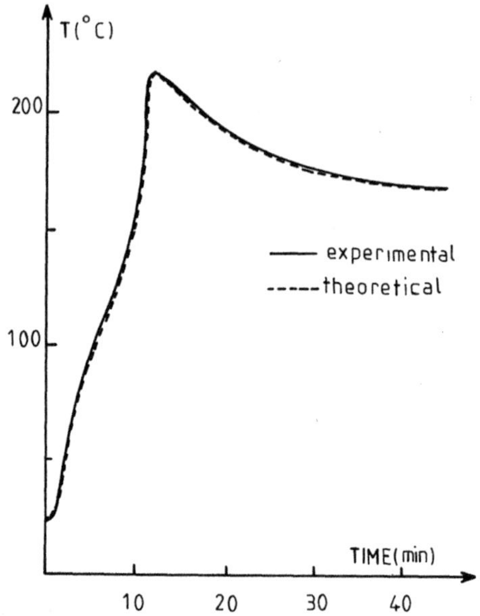

Fig. 15.22. Temperature–time history at the centre of the sample. Temperature of heated air = 164 °C. Diameter = 1.4 cm.

because of the complexity of the process due to the presence of heating by convection and conduction and heat evolved from the cure reaction.

4. The third stage is the cooling period when the temperature at the centre of the resin has passed through the maximum value.

Mould Temperature and the SOC–Time History at the Centre of the Resin

The model used in this chapter has some advantages over various other models [16, 17], because it is capable of giving some information not only about the temperature in the sample but also about the value of the SOC.

The effect of the temperature of heated air on the SOC–time history at the centre of the resin is illustrated in Fig. 15.23. The effect of the temperature on this history is clearly shown, with the obvious statement: the higher the temperature, the higher the rate of cure and the shorter the time of cure.

Profiles of Temperature and SOC

The effect of the temperature of heated air on the profiles of temperature and SOC developed within the cross-section of the sample is shown in Fig 15.24 at 142 °C, in Fig. 15.25 at 153 °C and in Fig. 15.26 at 164 °C. The following conclusions are of interest:

1. During the heating period, the reaction takes place already, with a higher value of the SOC on the surface. A steeper gradient is obtained with the

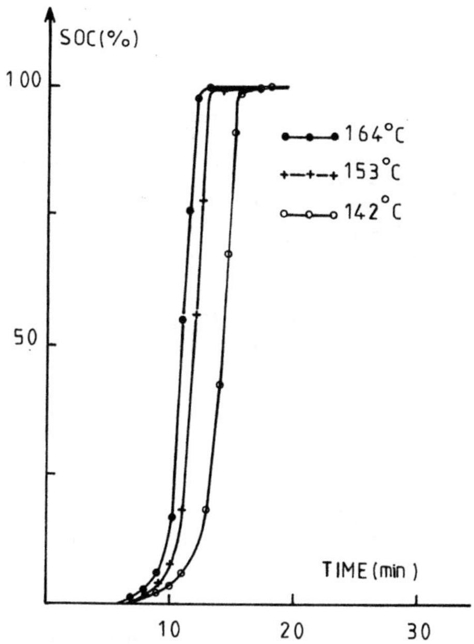

Fig. 15.23. SOC–time histories at the centre of the sample, for air temperatures of 142, 153 and 164 °C.

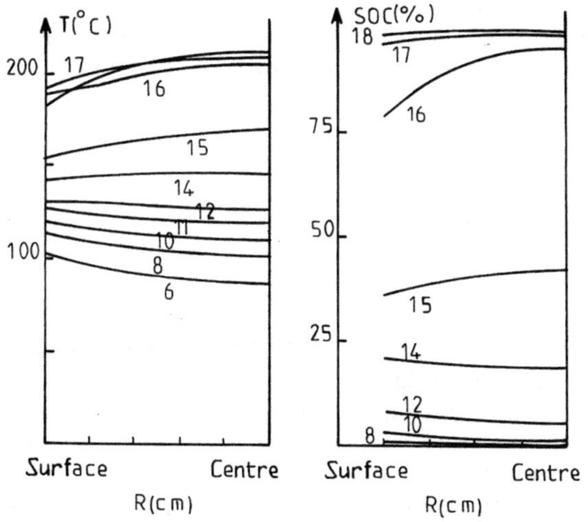

Fig. 15.24. Profiles of temperature (left) and SOC (right) developed along the radius. Temperature of heated air = 142 °C. Diameter = 1.4 cm.

higher temperature. In the same way, some gradients of temperature are developed within the resin, with the highest temperature on the surface. The following statement can be made: the higher the mould temperature, the steeper are the gradients of temperature and SOC.

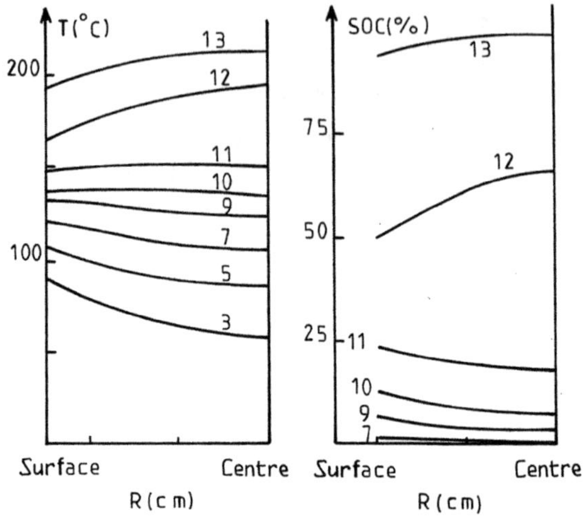

Fig. 15.25. Profiles of temperature (left) and SOC (right) developed along the radius. Temperature of heated air = 153 °C. Diameter = 1.4 cm.

2. During the curing period, other important gradients of temperature and of SOC are developed through the cross-section of the sample, with the highest value at the centre of the sample. The steepest value of these gradients appears at the maximum value of the temperature at the centre of the resin.
3. The reaction is not complete when the temperature at the centre of the resin reaches the maximum value as was sometimes supposed [16, 17], and a longer time is necessitated, especially during the cooling period.

15.5 Conclusions

The main characteristic of the mould heated by air is the rather low value of the coefficient of surface heat convection. This fact is responsible for a low rate of heat transfer between the resin and the surrounding atmosphere, which appears either during the heating period of the mould when the mould is put in contact with the heated air or during the period of reaction when the resin is hotter than the surrounding atmosphere.

The effect of the other parameters, e.g., temperature and dimension of the sample, on the process is of great importance. Because of the internal heat generated by the cure reaction, an increase in the temperature of the mould or an increase in the dimensions of the sample produces the following effects: a higher increase in temperature in the resin, and a steeper gradient of temperature and SOC developed throughout the sample.

The effect of the values determined for the kinetic and thermal parameters on the process is also of great interest. High accuracy is necessary, especially for the activation energy, the cure enthalpy and the heat capacity.

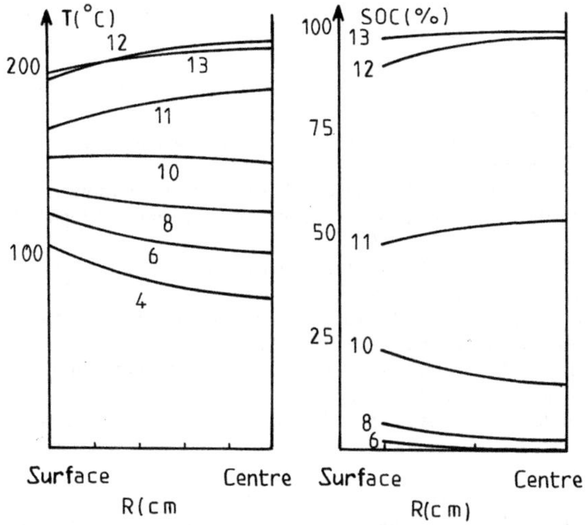

Fig. 15.26. Profiles of temperature (left) and SOC (right) developed along the radius. Temperature of heated air = 164 °C. Diameter = 1.4 cm.

Because of the presence of two simultaneous phenomena, i.e., heating and cure reaction, the process is very complex. The use of a numerical model taking into account all the known facts is thus of great importance for various reasons:

It can give a fuller insight into the process, with the profiles of temperature and SOC developed through the sample, which contribute to the heterogeneity of the sample and to the poor properties of the final material.

It is necessary for determining the best operational conditions for a given size of the mould, when it is desirable to reduce the time of cure, and to improve the productivity.

It is the most powerful technique for evaluating the new operational conditions which are to be used for a larger sample. The problem of increasing the dimensions of a sample is always difficult, but it is even more difficult and more sensitive in the case of a cure reaction with high enthalpy. By simulating the process, the model can predict the difficulty which could be encountered by the change in dimension of the sample.

References

1. Chater M, Bouzon J, Vergnaud JM. Modelling of the cure of epoxide resin in long cylindrical samples. J Polym Engng 1987; 7: 289
2. McAdams WH. Heat transmission, 3rd edn, 1954, pp 165–183 (McGraw-Hill Series in Chemical Engineering, 7)
3. Chater M, Lalart D, Michel-Dansac F, Vergnaud JM. Variation of enthalpy and kinetic parameters of the cure of epoxide resin with the composition of the binary system. Eur Polym J 1986; 22: 805
4. Chataing G. Chater M, Vergnaud JM. Determination of heat flux and profiles of temperature and SOC in isothermal calorimetry with reaction of high enthalpy. Thermochim Acta 1985; 89: 107

5. Khouider A, Rochette JB, Vergnaud JM. Study of heat flux and profiles of temperature and SOC obtained in an isothermal calorimeter with rubber cure of low enthalpy. Thermochim Acta 1985; 89: 81
6. Vergnaud JM. Modelling in calorimetry working under isothermal conditions or in scanning mode. Thermochim Acta 1987; 114: 15
7. Chater M, Vergnaud JM. Effect of kinetic and thermal parameters on the process of cure of epoxy resin in a cylindrical mould. Eur Polym J 1987; 23: 563
8. Rochette JB, Sadr A, Abdul M, Vergnaud JM. Effect of a variation in the order of the overall reaction of cure on the profiles of temperature and state of cure developed at the middle of rubber sheets. Thermochim Acta 1984; 79: 187
9. Sadr A, Hivert M, Vergnaud JM. Effect of mold temperature and cure enthalpy on the state of cure at the middle of thin rubber sheets. Thermochim Acta 1983; 71: 237
10. Khouider A, Bouzon J, Vergnaud JM. Effect of cure enthalpy on cure progress during the cooling period in motionless air. Thermochim Acta 1986; 102: 83
11. Ferradou C, Vergnaud JM. Effect of variation in the kinetic parameters on the vulcanization of rubber sheet. Thermochim Acta 1983; 68: 283
12. Ferradou C, Rochette JB, Vergnaud JM. Effect of a variation in kinetic parameters (rate constant, activation energy) on the vulcanization of rubber sheets in injection molding process. J Appl Polym Sci 1985; 30: 2663
13. Abdul M, Rochette JB, Sadr A, Vergnaud JM. Effect of variation in the values of the thermal properties of rubber on the vulcanization of thin and thick sheets. Thermochim Acta 1984; 80: 287
14. Chater M, Bouzon J, Vergnaud JM. Modelling of the cure of epoxide resin: effect of the diameter of cylindrical moulds. Plast Rubber Process Applic 1987; 7: 199
15. Chater M, Vergnaud JM. Effect of the temperature of the mold on the cure of epoxy resin. J Polym Engng 1988; 8: 1
16. Nixon JA, Hutchinson JM. Analysis of the cure of sheet moulding compound. I. Development of the model. Plast Rubber Process Applic 1985; 5: 349
17. Hutchinson JM, Nixon JA. Analysis of the cure of sheet moulding compound. II. Application of the model to various moulding geometries. Plast Rubber Process Applic 1985; 5: 359
18. Pusatcioglu SY, Hassler JC, Fricke AL, McGee HA. Effect of temperature gradients on cure and stress gradients in thick thermoset castings. J Appl Polym Sci 1980; 25: 381
19. Barone MR, Caulk DA. The effect of deformation and thermoset cure on heat conduction in a chopped-fibre reinforced polyester during compression molding. Int J Heat Mass Transfer 1979; 22: 1021
20. Progelhof RC, Throne JL. Non-isothermal curing of reactive plastics. Polym Engng Sci 1975; 15: 690
21. Mallick PK, Raghupathi N. Effect of cure cycle on mechanical properties of thick section fibre-reinforced poly-thermoset moldings. Polym Engng Sci 1979; 19: 774

Chapter 16

Cure of Epoxy Resin in a Long Cylindrical Mould Heated by Oil – Applications to Heat Dissipation and Reactive Injection Moulding

16.1 Introduction

The first objective in this chapter is to build a model able to describe the process of heating and curing a thermoset resin, and to give information on it. In fact, the process is about the same as that already studied in the preceding chapter with the mould heated by air, with the main difference that the surrounding fluid is an oil. The numerical model is thus the same as the model described in the preceding chapter. It takes into account not only the heat transfer by convection to the external surface of the mould and heat conduction through the mould and resin, but also the heat generated by the cure reaction.

A long cylinder is also considered for the mould, so that the heat transfer is radial only. As the glass mould is immersed into the heated oil in a bath, the principle is very simple, and moreover this type of mould can be applied for simulating various applications [1].

This type of mould is more suited to industrial applications, as heated oil may be used to transport heat to or from the mould in a flexible way [2].

One of the interesting applications of this mould and heating system studied is that of heat dissipation. As already shown for the mould heated by air, a large increase in temperature is observed in the resin due to the exothermic reaction and to the low heat conductivity of the resin. This can be detrimental to the quality of the final product. In order to reduce this increase in temperature, a method to extract the excess heat during the cure reaction is studied with the help of experiment and calculation using the model. It consists of varying the temperature of the oil very quickly, and in this way, the system based on a mould heated by oil is made very flexible [3].

Another application, well known for rubber, is the injection of the material in liquid state at a given temperature, which is selected at a value between the room and the mould temperature. The interest of the injection system is obvious, as it enables the time of the first step of heating in the mould to be reduced. The mould is thus coupled with a preheating system, and the curing time in the mould is shortened. However, some drawbacks may appear with the possibility for the resin to cure in the preheating system. Good knowledge of the cure process is thus necessary, and the model is of help. Of course, to simplify the difficult experimental problems, the process of resin injection is simulated in this case, by immersing the mould with the resin in it in several oil baths at various temperatures [4]. However, these results give significant results which can be successfully compared to experiments performed with an actual injection system [5].

The general plan used in this chapter is slightly different from the plan followed in the previous chapter, in order to place more emphasis on the topics of interest: the process of heat dissipation and the process of injection moulding.

16.2 Modelling of the Process

16.2.1 Mathematical Treatment [1]

Assumptions

The following general assumptions are made:

1. Only radial heat transfer by conduction through the circular cross-section of the mould and resin is considered, as the cylindrical mould is very long.
2. Heat is transferred through the liquid–mould interface either by free convection when the liquid oil is motionless, or by forced convection when the oil is stirred.
3. The cure reaction is expressed by a single Arrhenius equation, whose parameters are obtained from calorimetric measurements in scanning mode.
4. The thermal parameters may vary slightly with temperature.

Mathematical Treatment

The equation for the transient radial heat transfer through the circular cross-section is given by

$$\rho C \frac{\partial T}{\partial t} = \frac{1}{r} \frac{\partial}{\partial r}\left(r\lambda \frac{\partial T}{\partial r}\right) + \rho \frac{\partial Q}{\partial t} \tag{16.1}$$

where the contribution of the internal heat generated by the cure reaction is seen on the right-hand side.

At each position, the rate of heat evolved from the overall cure reaction is expressed as a function of the heat evolved up to time t, Q_t, and the total heat evolved after infinite time, Q_∞, by an Arrhenius equation:

$$\frac{1}{Q_\infty} \frac{dQ_t}{dt} = k_0 \left(1 - \frac{Q_t}{Q_\infty}\right)^n \exp\left(-\frac{E}{RT}\right) \tag{16.2}$$

with the pre-exponential factor k_0, the activation energy E and the order of reaction n.

16.2.2 Modelling and Numerical Analysis [1]

As in the case of the mould heated by air, the present problem for the mould heated by oil has no analytical solution, because of the presence of the internal heat evolved from the cure reaction. The problem must be solved by using a numerical model with finite differences. This model is the same as that described for the mould heated by air (Chap. 15). Sometimes, depending on the problem, a slight difference occurs for the convection heat at the external surface. The new equation is thus given in each case.

16.3 Study of the Process with Oil at Constant Temperature [1]

16.3.1 Experiment

Mould and Temperature Measurement

The mould, cylindrical in shape, was made of glass, with a length much longer than the diameter (internal diameter 1.2 cm, external diameter 1.6 cm). Because of this fact, only the radial heat is considered, especially at midlength where the thermocouple was placed. The temperature at the centre of the resin was measured and recorded throughout the process.

The mould with the resin in it and the thermocouple very well fixed was placed in heated silicone oil (Rhodorsil 47 V 100, Rhône-Poulenc).

Materials

The formulation used consisted of:

1. Epoxy resin DGEBA (Lopox 200 from CDF Chimie)
2. Hardener D 2605 (CDF Chimie) containing a mixture of an anhydride of Me THP and an accelerator, 0.75% of the aromatic amine of type A 105 (CDF Chimie).

These components were mixed at room temperature by adding the hardener with continuous stirring until a clear mixture was obtained. The mixture was thus poured into the cylindrical glass mould.

Determination of Kinetic Parameters

The kinetic parameters of the cure reaction were determined by using a differential calorimeter (DSC 111, Setaram) working in scanning mode at

$2\,^{\circ}\mathrm{C\,min^{-1}}$. A sample of about 150 mg was encapsulated in a sealed metal holder. The heat flux–temperature curves were registered and treated with the help of a computer (HP 86, Hewlett Packard) connected with the calorimeter. The validity of these kinetic parameters was tested by comparing the heat flux–temperature curves obtained from experiment and calculation [5, 6] (Chap. 13, DSC). The enthalpy of cure was determined from the same experiments by calculating the area under the exotherm and above the reference baseline obtained with the cured resin.

The heat capacity was determined by calorimetry for each component of the resin in liquid form, and for the final solid resin.

The values of the kinetic and thermal parameters were as follows:

Kinetics of cure: $k_0 = 2.3 \times 10^{11}\,\mathrm{s^{-1}}$, $n = 1.5$, $E = 27\,140\,\mathrm{cal\,m^{-1}}$
$Q_\infty = 76\,\mathrm{cal\,g^{-1}}$

Thermal properties:

Resin, $C = 0.2 + 0.003T\,\mathrm{cal\,g^{-1}\,deg^{-1}}$ (T in °C), $\lambda = 10^{-3}\,\mathrm{cal\,cm^{-1}\,s^{-1}\,deg^{-1}}$

Glass, $C = 0.3\,\mathrm{cal\,g^{-1}\,deg^{-1}}$, $\lambda = 2.66 \times 10^{-3}\,\mathrm{cal\,cm^{-1}\,s^{-1}\,deg^{-1}}$

Coefficient of surface heat transfer: $20 \times 10^{-4} \times (\Delta T)^{0.25}\,\mathrm{cal\,cm^{-2}\,s^{-1}\,deg^{-1}}$

16.3.2 Results

The following results are given for the period of cure of the resin: the temperature–time histories and the state of cure (SOC)–time histories at the centre of the resin, and the profiles of temperature and SOC developed through the cross-section of the resin.

Temperature–Time Histories at the Centre

The temperature–time histories at the centre of the resin are shown in Fig. 16.1 for three temperatures of the oil: 145.5, 152 and 162 °C. The validity of the model can be appreciated, as well as the accuracy of the kinetic and thermal parameters, as a good superimposition is obtained between the experimental and calculated curves.

As in the case of the mould heated by air, three stages can be observed by considering the temperature–time history at the centre of the resin:

1. Heating of the resin up to a temperature at which the cure reaction starts. The material in liquid state is heated by conduction but also by convection due to the presence of high gradients of temperature developed through the resin [1].
2. The cure process. An appreciable increase in the slope of the curve occurs when the rate of reaction becomes significant. The temperature at the midplane rises to a maximum value.
3. A cooling period is observed after the temperature has reached its maximum. During this stage, the heat leaving the resin by conduction exceeds the heat evolved from the cure reaction.

The effect of the temperature of the bath on the process, is of interest

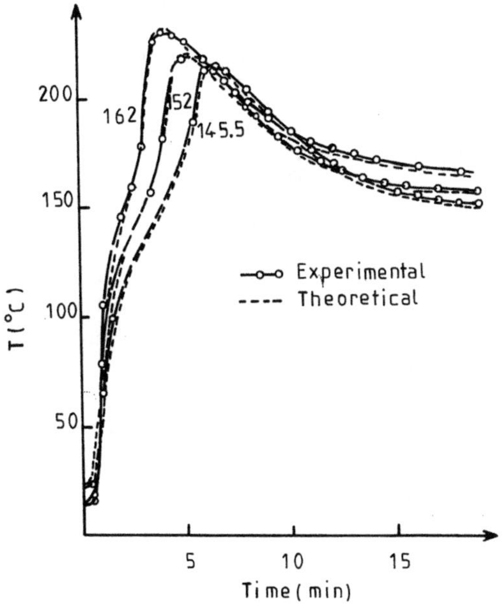

Fig. 16.1. Temperature–time histories at the centre of the resin. Sample diameter of 1.2 cm; oil temperature: 145.5, 152 and 162 °C.

especially for the time of heating and for the progress of the reaction. The following conclusions can be drawn:

1. A higher temperature of the bath is necessary for reducing the time of heating of the sample, as clearly shown in Fig. 16.1.
2. A higher temperature of the bath is also responsible for a higher increase in temperature, especially at the centre of the resin, as shown in Table 16.1.
3. A higher temperature corresponds with a shorter time for the cure of reaction, and especially with a reduction of the time at which the maximum of the temperature at the centre of the resin is attained.

SOC–Time Histories at the Centre

The SOC–time histories at the centre of the resin were calculated with the model, and plotted in Fig. 16.2 for the three temperatures of the oil. As shown by these curves, the cure at the centre of the resin takes place essentially during the second stage of cure, and it is near to completion when the maximum of temperature is attained, as shown more precisely in Table 16.1.

Profiles of Temperature and SOC

The best way of using the model is perhaps in determining the profiles of temperature and SOC expanded through the cross-section of the resin. These profiles are shown in Fig. 16.3, as calculated for the bath temperature of 145.5 °C.

Table 16.1. Characteristics of the maximum temperature

	Oil temperature (°C)		
	145.5	152	162
Maximum temperature (°C)	215	219	230
Time (s)	390	330	270
SOC (%)	97.4	98.3	99.4

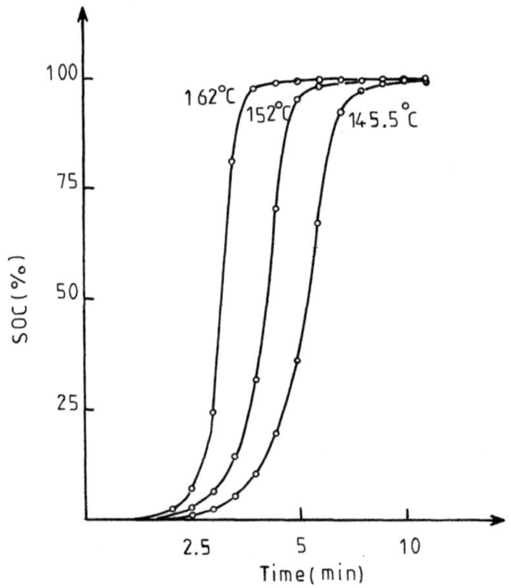

Fig. 16.2. SOC–time histories at the centre of the resin sample diameter of 1.2 cm; oil temperatures: 145.5, 152 and 162 °C.

Various conclusions can be drawn from these curves:

1. During the initial heating stage, very steep gradients of temperature within the sample are observed, with the highest temperature at the surface. The reaction starts before the temperature has reached the bath temperature. During this heating stage, a part of the reaction is thus obtained, and some small gradients of SOC are also developed within the resin with the highest value of the SOC at the surface.
2. The cure reaction increases considerably during the second stage, when the temperature exceeds 120 °C. During this stage of cure, the profiles of temperature and of SOC are completely changed. The highest temperature and SOC is at the centre of the resin, because of the internal heat generated by the

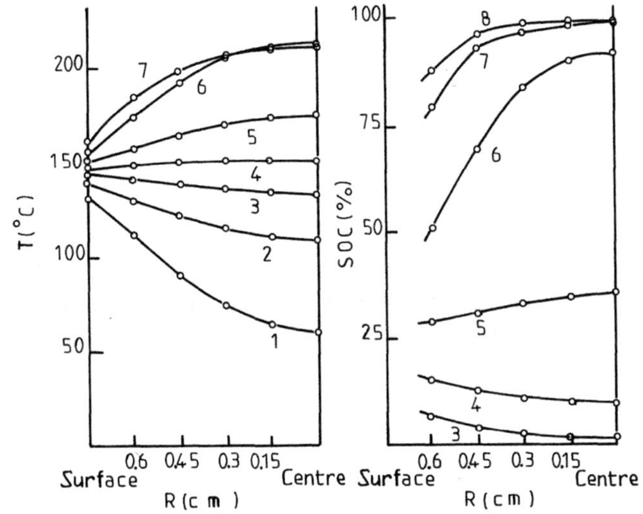

Fig. 16.3. Profiles of temperature (left) and SOC (right) developed within the cross-section of the sample at various times (min). Oil temperature 145.5 °C.

cure reaction. The steeper profile of SOC is attained when the temperature at the centre of the resin reaches the maximum value.

3. During the cooling period, the reaction goes to completion on the surface of the resin. In the centre of the resin, the reaction is complete long before the end of this stage.
4. Because of the higher coefficient of surface heat transfer for the oil than the surrounding air, higher gradients of temperature are developed with the oil, as well as steeper gradients of the SOC.

16.3.3 Conclusions

The main difference between the use of air or oil for the heated surrounding fluid is the value of the coefficient of surface heat transfer. As this coefficient is higher in the case of oil, steeper gradients of temperature and SOC are developed within the resin during the cure. Moreover, a significant increase in temperature due to the exothermic cure reaction is also observed.

Of course, a higher oil temperature allows a reduction in the heating time of the sample, but it is also responsible for a greater increase in temperature in the resin and for steeper gradients of temperature and SOC.

The problem of reducing the time of cure without greatly increasing the temperature within the resin thus arises. Two solutions are proposed for this difficult problem, one involving injection of the resin at a given temperature, and the other using a system to evacuate some heat quickly from the resin when necessary.

16.4 Study of the Process with Fibre Glass–Epoxy Resin Heated by Oil. Effect of Kinetic and Thermal Parameters

16.4.1 Experiment [2]

Material

The composite material used was a woven glass fibre impregnated with an epoxy resin (Vicotex 1454) with 49% resin by mass.

Mould and Temperature Measurement

The composite material was wrapped around itself very strongly and thus introduced into a glass test tube, cylindrical in shape, playing the role of a mould. The cylinder's length/diameter ratio is great enough for the heat to be considered as radial only. The temperature at the centre of the sample was registered with a thermocouple.

The quality of the contact between the composite and the mould is not as perfect as in the case of a pure liquid resin [1, 7]. This fact is responsible for the low coefficient of heat transfer measured on the external surface of the mould.

Kinetic and Thermal Parameters

The kinetic parameters were calculated from experiments made with a differential calorimeter scanned at a heating rate of 2 °C min^{-1} (DSC 111, Setaram). This type of apparatus is especially suitable for this kind of composite, because of the fairly large sample (around 150 mg) encapsulated in the metal holder. The heat flux–temperature history was analysed by a microcomputer (HP 86, Hewlett Packard) connected to the calorimeter. Various techniques were used for analysis, especially the Freeman–Carroll technique [6]. At the end, the experimental heat flux–temperature curve was compared with the curve calculated by using the kinetic parameters, assuming the temperature of the sample to be the same as that of the calorimeter, in order to test the accuracy of these parameters [8, 9].

The values of the thermal and kinetic parameters were as follows:

Composite resin: $k_0 = 4.5 \times 10^{19}\,\text{s}^{-1}$, $E = 43\,350\,\text{cal mol}^{-1}$,

$n = 2.3$, $Q_\infty = 29\,\text{cal g}^{-1}$ of composite,

$C = 0.23 + 10^{-4}\,T\,\text{cal g}^{-1}\,\text{deg}^{-1}$ (T in °C),

$\lambda = 10^{-3}\,\text{cal cm}^{-1}\,\text{s}^{-1}\,\text{deg}^{-1}$, $\rho = 1.48\,\text{g cm}^{-3}$

Glass: $\lambda = 2.6 \times 10^{-3}\,\text{cal cm}^{-1}\,\text{s}^{-1}\,\text{deg}^{-1}$, $\rho = 2.21\,\text{g cm}^{-3}$,

$C = 0.3\,\text{cal g}^{-1}\,\text{deg}^{-1}$

Coefficient of heat transfer: $h = 7.6 \times 10^{-4}(\Delta T)^{0.25}\,\text{cal cm}^{-2}\,\text{s}^{-1}\,\text{deg}^{-1}$

16.4.2 Cure Process and Validity of the Model [2]

The temperature–time history at the axis of the cylindrical sample is shown in Fig. 16.4 for the fresh uncured resin. The good agreement between the experimental and calculated curves is a proof of the validity of the model as well as the accuracy of the values used for the kinetics and thermal parameters.

As shown in previous papers [1, 7, 10] in the case of pure epoxy resin, three steps can be distinguished during the process:

1. The heating stage of the mould and resin when their temperature is lower than the temperature of the oil.
2. The cure reaction with the subsequent heating of the resin due to the exothermic reaction. This step starts at a temperature around 140 °C, when the increase in the slope of the temperature–time curve can be observed. An inflection point is thus obtained at 145 °C after 5.5 min and the maximum is at around 225 °C at 7.3 min. The time necessary for the temperature at the axis to reach the maximum is considerably shorter than that obtained in the case of a pure resin heated by air [7] (7.3 min instead of 12–13 min) and a little longer than that shown in the case of the pure resin heated by silicone oil [1] (7.3 min instead of 4 min).
3. A cooling period is also observed after the temperature has risen to the maximun value.

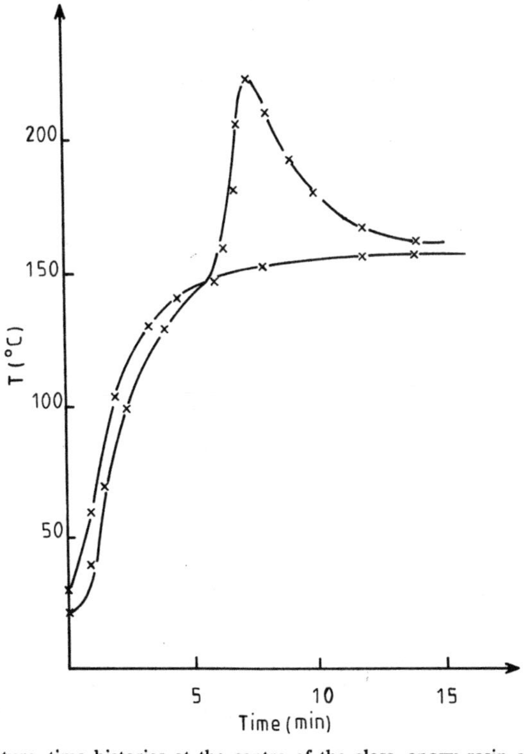

Fig. 16.4. Temperature–time histories at the centre of the glass–epoxy resin composite with the long cylindrical mould heated by silicone oil at 160 °C. × experiments.

Other results of interest are given by the model, with the profiles of temperature and SOC developed throughout the cross-section of the resin during the process. These profiles are drawn in Fig. 16.5 for the temperature (left) and for the SOC (right). The following conclusions are drawn:

1. High gradients of temperature are grown within the resin during the heating period with the highest temperature at the surface of the mould.
2. Other gradients of temperature with the highest temperature at the centre of the resin are expanded during the second step, resulting from the internal heat generated by the reaction. The steeper gradient of temperature is attained when the temperature at the axis of the cylinder reaches its maximum value (7 min 18 s).
3. During the cooling period, the gradients of temperature become flat, while the temperature of the resin is still higher than that of the oil.
4. The SOC profiles are rather flat at the end of the initial heating stage, and during the third step.
5. Very steep gradients of SOC are developed throughout the resin during the second step of cure, and especially at the maximum temperature at the centre of the resin. During a very short time, the largest part of the cure reaction is obtained.

16.4.3 Effect of the Thermal Parameters

As the model is able to simulate the whole process, it is thus possible to determine the effect of each thermal parameter on the process by varying in turn the value of each parameter.

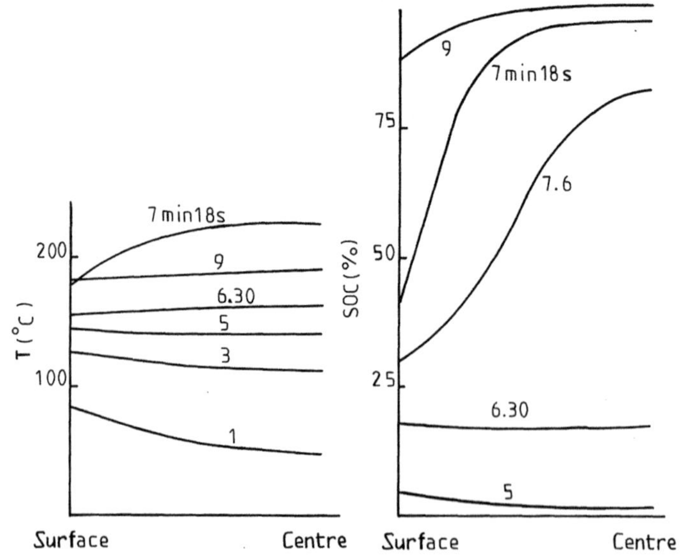

Fig. 16.5. Profiles of temperature (left) and SOC (right) in the glass–epoxy resin composite, heated by oil at 160 °C.

Three parameters of concern are considered: the thermal conductivity of the resin–glass composite, the heat capacity of the composite, and the coefficient of heat transfer to the external surface of the mould.

Thermal Conductivity of the Composite

The effect of the thermal conductivity on the process was determined by using three values: the one shown in Sect. 16.3.1, the other two are obtained by multiplying this value by 0.4 and 2.5. As shown in Fig. 16.6, where the temperature and SOC–time histories at the centre of the composite are plotted, this coefficient plays a role in the process:

1. The effect of this parameter is significant, but it must be considered that the three values of the parameter are very different.
2. The three temperature–time histories are quite different.
3. A decrease in the value of the thermal conductivity is responsible for the following facts: a shift towards longer times for the temperature–time histories during the heating period, and an increase in the maximum value of temperature.

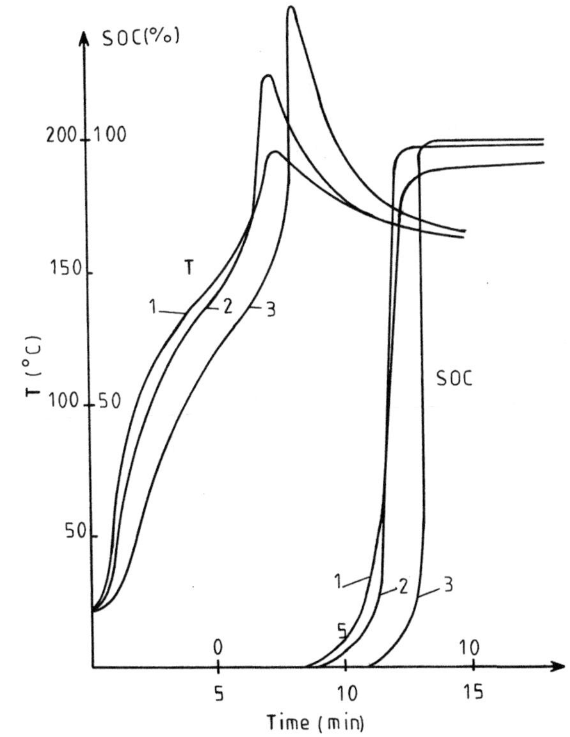

Fig. 16.6. Temperature and SOC–time histories at the centre of the glass–epoxy composite, heated by oil at 160 °C, for different values of conductivity: (1) 2.5×10^{-3}, (2) 10^{-3} and (3) 0.4×10^{-3} cal $cm^{-1} s^{-1} deg^{-1}$.

Heat Capacity of the Composite

The heat capacity does not play the same role as the thermal conductivity. The thermal conductivity acts only upon the heat transferred by conduction, while the heat capacity affects the conduction heat and the heat evolved from the cure reaction. The effect of the heat capacity of the composite on the process was determined by varying the value of the heat capacity by 10% around the mean value shown in Sect. 16.3.1:

$$C = 0.257 + 1.1 \times 10^{-4} T \qquad C = 0.21 + 0.9 \times 10^{-4} T \text{ cal g}^{-1} \text{deg}^{-1} \ (T \text{ in °C})$$

This effect of the heat capacity on the process is found to be very important, as shown in Fig. 16.7. Some conclusions are drawn:

1. The effect of this parameter is complicated, as it acts upon both the heat conduction and the heat generated by the cure reaction.
2. On the one hand, because of its action upon the conduction heat, the action of heat capacity is in inverse ratio to that of the thermal conductivity, and this fact appears clearly during the first step.
3. On the other hand, because of its action upon the heat of cure, the maximum of temperature decreases when the heat capacity is increased.

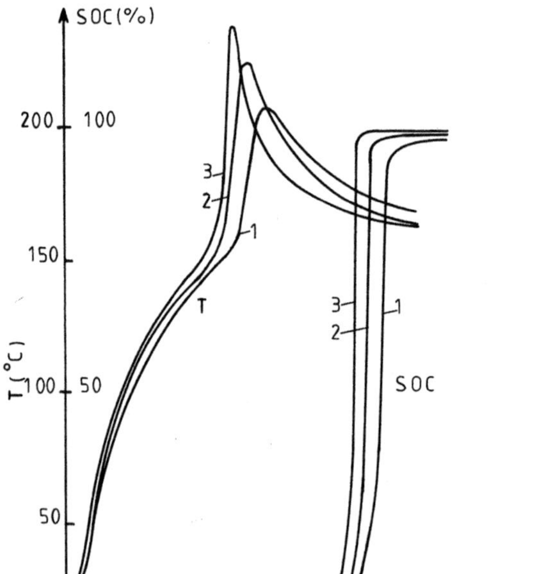

Fig. 16.7. Temperature and SOC–time histories at the centre of the glass–epoxy composite, heated by oil at 160 °C, for different values of heat capacity: (1) $0.257 + 1.1 \times 10^{-4} T$, (2) $0.23 + 10^{-4} T$ and (3) $0.21 + 0.9 \times 10^{-4}\ T \text{ cal g}^{-1} \text{deg}^{-1}$ (T in °C).

Coefficient of Surface Heat Transfer

A rather low value was found for this coefficient, compared to the value obtained with a liquid resin heated by silicone oil [1]. This result is probably due to the fact that the contact between the composite and the internal surface of the mould is not as good as in the case of the liquid resin. This result is of significant interest, and it has to be borne in mind in the case of a composite heated in a mould without high pressure.

Three values were chosen for the coefficient of heat transfer: the value determined from experiment, and two other values which are respectively half and double this first value.

As shown in Fig. 16.8, the effect of the value of the coefficient of heat transfer on the temperature–time histories is of importance. Of course, the higher this coefficient, the shorter the temperature–time history. Moreover, a higher value of h is responsible for a higher value of the temperature in the resin, and this is especially shown at the centre of the resin. The coefficient of heat transfer has also a significant effect on the SOC–time histories, an increase in its value causing an appreciable decrease in the time of cure.

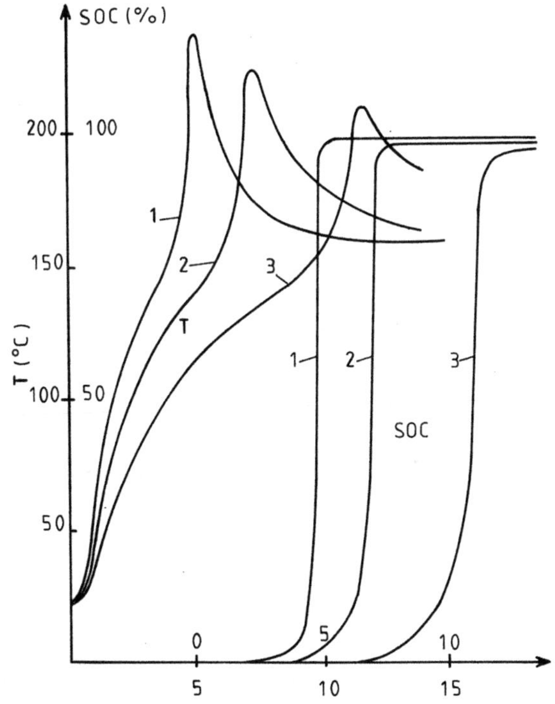

Fig. 16.8. Temperature and SOC–time histories at the centre of the glass–epoxy composite, heated by oil at 150 °C, for different coefficients of heat transfer at the surface: (1) $15.2 \times 10^{-4}\ \Delta T^{0.25}$, (2) $7.6 \times 10^{-4}\ \Delta T^{0.25}$ and (3) $3.8 \times 10^{-4}\ \Delta T^{0.25}$ cal cm^{-2}s^{-1} deg^{-1}.

16.4.4 Effect of the Kinetic Parameters

Four parameters were considered for describing the kinetics of heat generated by the cure reaction: the order of the reaction, the pre-exponential factor, the activation energy and the enthalpy of cure.

Order of the Reaction

Three values of the order of the reaction were chosen: 2.5, 2.1 and the value of 2.3 obtained from experiment. The effect of the order of the reaction on the temperature–time histories and SOC–time histories is rather small (Fig. 16.9). This fact was already shown for pure epoxy resin [11] and the cure of rubber [12]. The effect of the order on the process is negligible when the extent of cure is below 50%, and it becomes increasingly greater when the extent of cure is increased from 50% to 100%.

Pre-Exponential Factor k_0

The effect of the pre-exponential factor on the process was determined by considering the three values of this parameter: 5.4×10^{19}; 4.5×10^{19} and

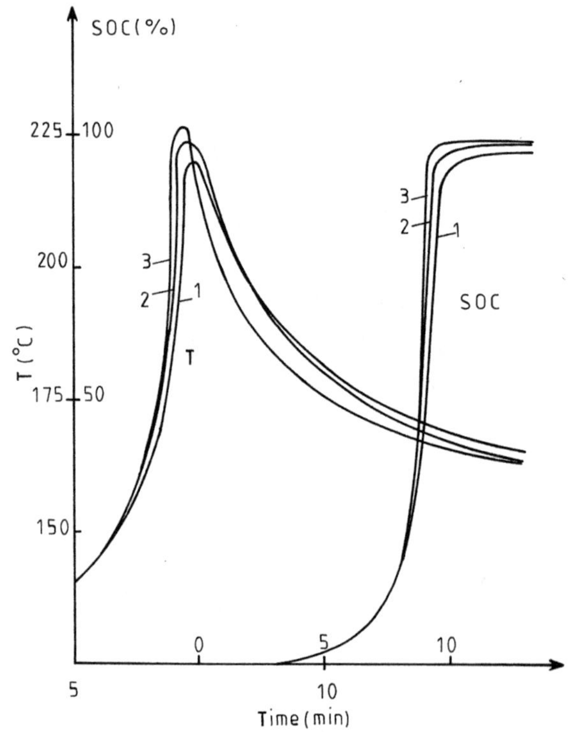

Fig. 16.9. Temperature and SOC–time histories at the centre of the glass–epoxy composite, heated by oil at 160 °C, for different orders of the overall reaction: (1) $n = 2.5$, (2) $n = 2.3$ and (3) $n = 2.1$.

$3.6 \times 10^{19} (s^{-1})$. As shown in Fig. 16.10, the effect of this parameter is of great importance, and the following conclusions are worth noting:

1. The higher the pre-exponential factor, the faster the cure reaction. A higher slope for the temperature–time history is obtained for a higher value of this parameter.
2. The maximum of the temperature is higher for the higher value of this parameter, and it is attained in a shorter time.
3. A higher value for this parameter is responsible for an increase in the SOC at a given time.

Activation Energy

The three values of the activation energy (43 500; 43 350 and 43 109 cal mol^{-1}) were calculated by keeping $k_0 \exp(-E/RT)$ constant at 138 °C, while the pre-exponential factor was varied as shown above.

The effect of a slight change in the value of the activation energy on the cure process is very important, as shown in Fig. 16.11. These curves have about the same shape as those obtained in Fig. 16.10 with the corresponding values of the pre-exponential factor. However, the fact of high importance is the very slight

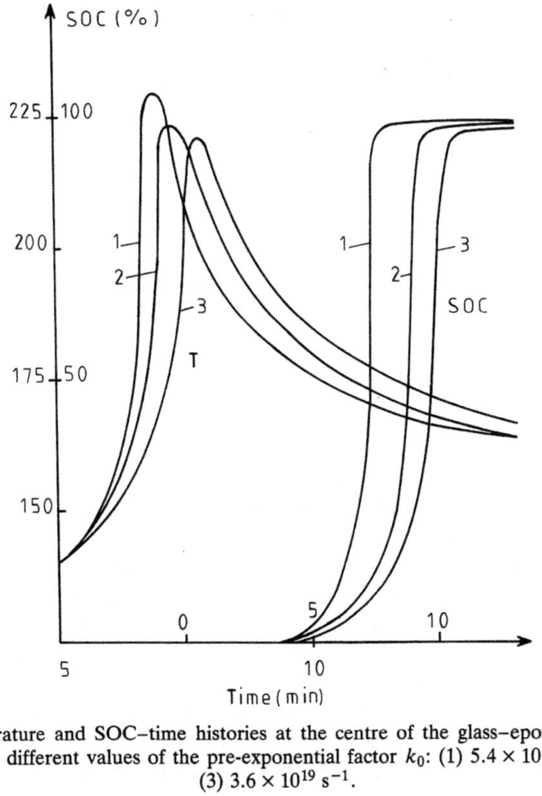

Fig. 16.10. Temperature and SOC–time histories at the centre of the glass–epoxy composite, heated by oil at 160 °C, for different values of the pre-exponential factor k_0: (1) 5.4×10^{19}, (2) 4.5×10^{19} and (3) 3.6×10^{19} s^{-1}.

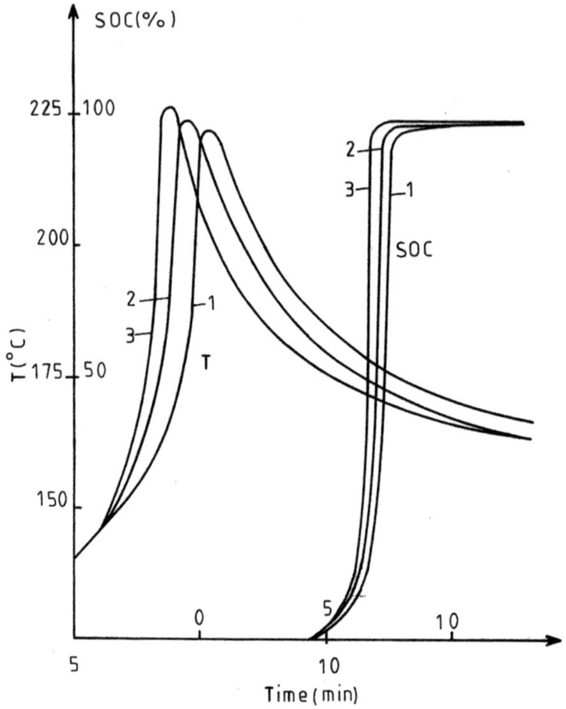

Fig. 16.11. Temperature and SOC–time histories at the centre of the glass–epoxy composite, heated by oil at 160 °C, for different values of activation energy: (1) 43 500, (2) 43 350 and (3) 43 109 kcal-mol^{-1}.

change in the activation energy (150 and 241 cal mol^{-1}) which is sufficient to produce the same effect as a change of 20% in the pre-exponential factor.

Cure Enthalpy

Three values were chosen for the cure enthalpy (32, 29 and 26 cal g^{-1}), varying by 10% around the mean value determined by calorimetry.

The effect of the cure enthalpy on the cure process is very important, as shown in Fig. 16.12. In fact, the cure enthalpy plays a role which is in inverse ratio to that of the heat capacity, at least during the cure reaction. However, the role of the enthalpy is not exactly the inverse of that of the heat capacity, because heat capacity acts not only on the heat evolved from reaction but also on heat transferred by conduction. This difference between the behaviour of the enthalpy and the heat capacity can be appreciated by comparing the curves in Fig. 16.7 and 16.12, at the beginning of the stage of reaction. In Fig. 16.12, the three temperature–time curves are separated when the reaction takes place, while in Fig. 16.7 the curves diverge from the beginning of the first step.

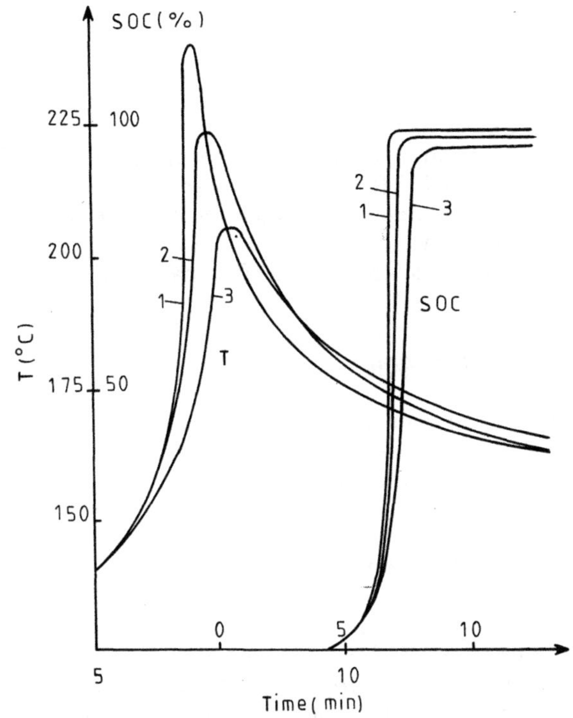

Fig. 16.12. Temperature–time and SOC–time histories at the centre of the glass–epoxy composite, heated by oil at 160 °C, for different values of cure enthalpy: (1) 32, (2) 29 and (3) 26 cal g^{-1}.

16.5 Study of the Dissipation of Heat, with the Epoxy Resin Composite Heated by Oil [3]

The low thermal conductivity of the composite and the associated high exothermicity of the cure reaction are factors of fundamental importance. Resulting from these factors, high temperatures arise within the resin during the cure, especially at the centre of the sample. These high temperatures as well as the high gradients of temperature are responsible for the discoloration or even degradation of the final material [13–16].

The main purpose of this study is to show that the high temperature gradients developed through the sample and the increase in temperature during cure can be reduced by using a mould with a controlled temperature. The mould can thus be either heated or cooled at a rather fast rate during the cure of the resin.

A complementary objective is to describe the process of heating and cure by using a model based on a numerical method with finite differences.

The mould used in this study was the long cylindrical glass mould, which was immersed, with the resin inside, in heated oil. The mould was thus immersed successively in various baths at different temperatures.

16.5.1 Experiment

Materials

The epoxy resin DGEBA (Lopox 200, CDF Chimie), the hardener (D 2605, CDF Chimie) containing a mixture of an anhydride of Me THP and an accelerator, 0.75% aromatic of type A 105 (CDF Chimie) was used.

The binary mixture was prepared at room temperature by adding the hardener with continuous stirring. The clear viscous liquid was then poured into the mould. The composition in wt% was: epoxy resin 60%, hardener 40%.

Mould and System of Heating

The glass mould (internal diameter 2 cm, external diameter 2.4 cm) with the resin in it, previously at room temperature, was introduced into silicone oil kept at a constant temperature (called the mould temperature T_m).

Kinetic and Thermal Parameters

The kinetic parameters of the cure reaction were calculated from the experiments performed with a calorimeter (DSC 111, Setaram) run in scanning mode at the heating rate of 2 °C min^{-1}. The fairly large sample (150 mg) is very convenient for mixtures such as thermosets. The heat flux–temperature history was analysed by a microcomputer (HP 86, Hewlett Packard) connected to the calorimeter, by using various methods. The experimental heat flux–temperature curve was compared to the curve obtained from calculation with the kinetic parameters by assuming that the temperature in the sample is the same as that of the calorimeter [8, 9]. Good superimposition of these two curves is significant evidence towards accepting the validity of these kinetic parameters. The thermal parameters were obtained from the literature. The parameters were as follows:

Epoxy resin: $k_0 = 3.5 \times 10^{11}\,\mathrm{s}^{-1}$, $E = 27\,750\,\mathrm{cal\,mol}^{-1}$

$n = 1.5$, $Q_\infty = 76\,\mathrm{cal\,g}^{-1}$

$C = 0.22 + 1.5 \times 10^{-3}\,T\,\mathrm{cal\,g}^{-1}\,\mathrm{deg}^{-1}$ (T in °C)

$\lambda_1 = 2.3 \times 10^{-3}\,\mathrm{cal\,cm}^{-1}\,\mathrm{s}^{-1}\,\mathrm{deg}^{-1}$ if SOC $< 94\%$

$\lambda_2 = 0.5 \times 10^{-3}$ if SOC $> 94\%$

Glass mould: $\lambda = 2.7 \times 10^{-3}\,\mathrm{cal\,cm}^{-1}\,\mathrm{s}^{-1}\,\mathrm{deg}^{-1}$, $C = 0.3\,\mathrm{cal\,g}^{-1}\,\mathrm{deg}^{-1}$,

$\rho = 2.21\,\mathrm{g\,cm}^{-3}$

Coefficient of surface heat convection: $h = 2 \times 10^{-3}\,\mathrm{cal\,cm}^{-2}\,\mathrm{s}^{-1}\,\mathrm{deg}^{-1}$

16.5.2 Process of Cure and Validity of the Model

The validity of the model was tested by comparing the temperature–time histories at the centre of the resin obtained either by experiment or calculation.

The temperature was easily determined at the axis of the cylinder. The measurements were very accurate, because the profiles of temperature through the cross-section of the resin exhibit a maximum on the axis, and so a small change in the position of the thermocouple does not alter the measurement too much.

The good agreement obtained between the calculated and experimental curves shown in Fig. 16.13 and 16.14 is proof of the validity of the numerical method as well as the accuracy of the values of the thermal and kinetic parameters. The curves in Fig. 16.13 were obtained in a conventional way [1, 2, 7, 10], by immersing the resin and the mould in the heated silicone oil kept at 155 °C. In this case, three steps are distinguished:

1. Heating of the resin, with a high slope of the temperature–time curve at the axis of the sample. This stage takes place at a temperature lower than 120 °C for time less than 4 min. Gradients of temperature are developed through the resin with the highest temperature at the surface (Fig. 16.15).
2. The cure reaction starts at around 120 °C after 4 min. Because of the high exothermicity of the reaction, an increase in temperature is observed as well as an increase in the slope of the temperature–time curve at the axis. A temperature maximum is reached at around 270 °C after 8 min. During this period, steep gradients of temperature occur through the resin (Fig. 16.15). The steepest gradient is attained between 7 and 8 min when the maximum temperature at the axis is reached.
3. After the temperature has risen to the maximum value, a cooling period occurs. Generally, because of the gelation of the resin, the thermal properties of the resin and especially the thermal conductivity are different from the values of the resin in liquid state.

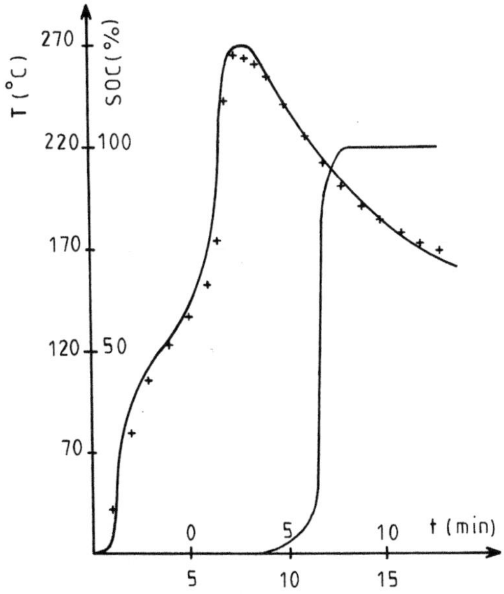

Fig. 16.13. Temperature–time and SOC–time histories at the centre of the resin. $T_m = 155$ °C, $D_i = 2$ cm, $D_e = 2.4$ cm. +, experiment; —, calculation.

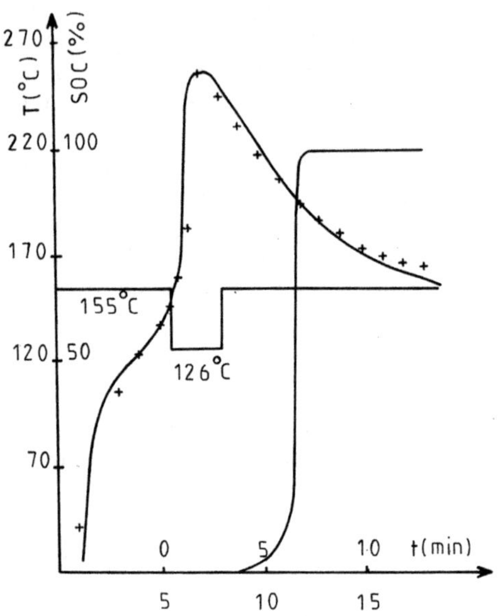

Fig. 16.14. Temperature–time and SOC–time histories at the centre of the resin. $T_m = 155$ °C for $t < 5.5$ min. $T_m = 126$ °C for $5.5 < t < 8$ min, $T_m = 155$ °C for $t > 8$ min. +, experiment; —, calculation.

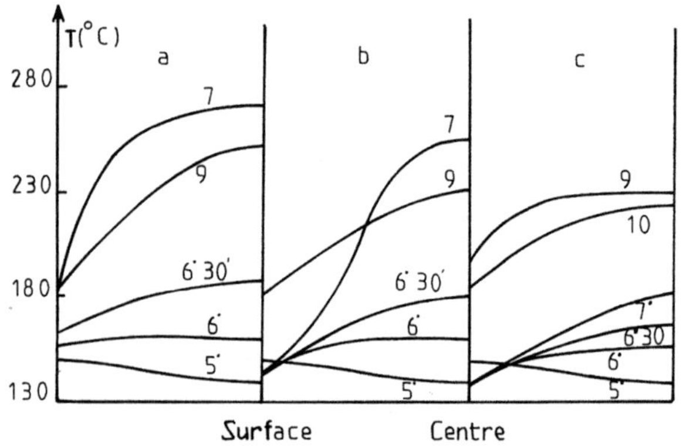

Fig. 16.15. Profiles of temperature through the resin. **a** constant heating at 155 °C; **b** $T_m = 155$ °C and 126 °C, with $t_{ch} = 5.5$ min and $t_c = 2.5$ min; **c** $T_m = 155$ °C and 126 °C, with $t_{ch} = 5$ min and $t_c = 2.5$ min.

The curves in Fig. 16.14 were obtained by immersing the filled mould in the heated silicone oil at 155 °C for 5 min 30 s, cooling the mould and resin for 2 min 30 s in silicone oil kept at 126 °C, and then reimmersing the mould in the silicone oil at 155 °C. In this more complex case, good agreement is also observed for the temperature–time curves at the axis of the resin, whether they are obtained by calculation or measurement.

Two parameters of interest are considered:

1. Time at which the cooling starts, t_{ch}
2. Length of the cooling period, t_c

16.5.3 Effect of the Length of the Cooling Period t_c

The effect of the length of the cooling stage on the temperature–time history at the axis of the resin is determined by selecting cooling times ranging from 2 to 3 min. As illustrated in Figs 16.15–16.17, this parameter is of great importance to the process.

1. Lower temperatures within the resin are obtained during the cure period (Fig. 16.15).
2. The value of the temperature maximum at the axis drops from 270 °C, obtained in the conventional case shown in Fig. 16.13, to 225–235 °C using this method.
3. The SOC–time history at the axis of the resin is not dramatically changed by the cooling period (Fig. 16.16).
4. The time necessary for the SOC to reach a value near 100% is about the same with the cooling method and the conventional technique.

16.5.4 Effect of the Time at Which the Cooling Starts, t_{ch}

The time at which the cooling period starts, t_{ch}, is also of great importance, as shown in Fig. 16.18. The temperature–time histories at the axis of the resin are determined in various cases, as well as the SOC–time histories (Fig. 16.19) at the axis of the resin.

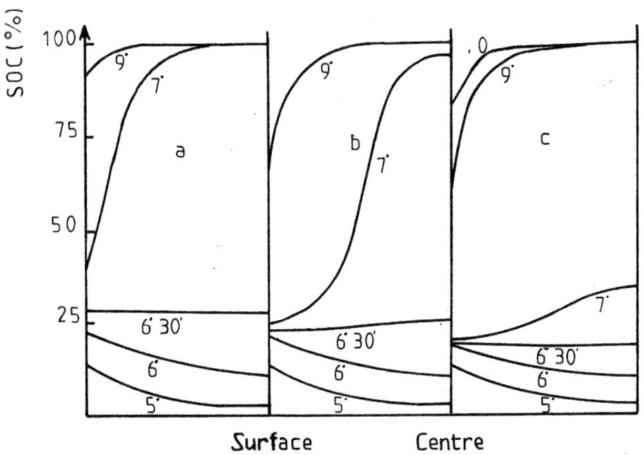

Fig. 16.16. Profiles of SOC through the resin. **a** constant heating at 155 °C; **b** $T_m = 155$ °C and 126 °C, with $t_{ch} = 5.5$ min and $t_c = 2.5$ min; **c** $T_m = 155$ °C and 126 °C, with $t_{ch} = 5$ min and $t_c = 2.5$ min.

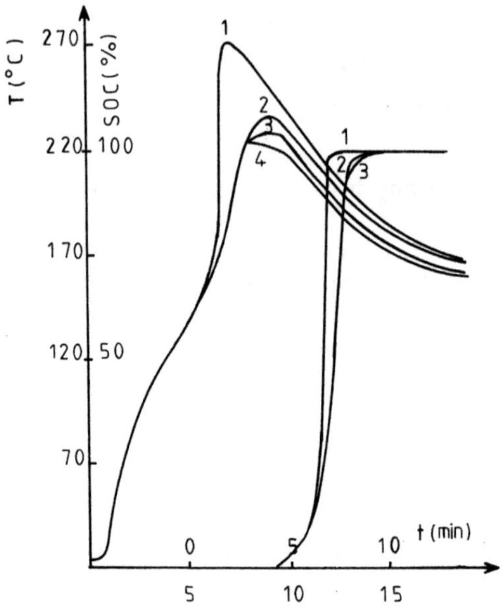

Fig. 16.17. Effect of the cooling time t_c on the temperature–time and SOC–time histories at the axis of the resin. $t_c = 0$ (1), $= 2$ min (2), $= 2.5$ min (3), $= 3$ min (4). $t_{ch} = 5$ min.

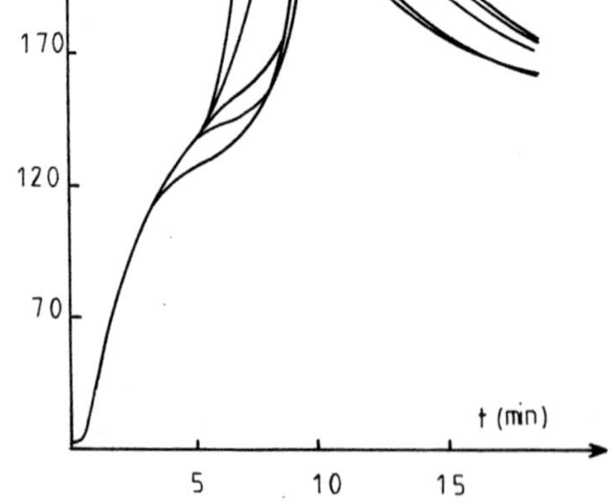

Fig. 16.18. Effect of the time at which the cooling period starts, t_{ch}, on the temperature–time history at the axis of the resin, for $T_m = 155$ °C and 126 °C and $t_c = 2.5$ min. $t_{ch} = 3$ min (6), $= 4$ min (5), $= 4.5$ min (4), $= 5$ min (3), $= 5.5$ min (2), $= 6$ min (1).

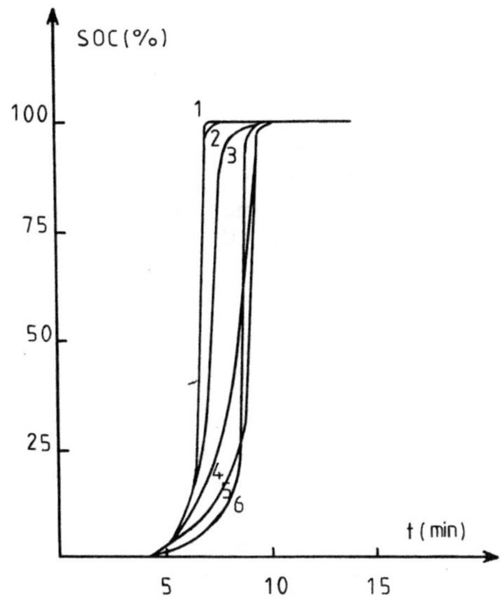

Fig. 16.19. Effect of the time at which the cooling period starts, t_{ch}, on the SOC–time histories at the axis of the resin, for $T_m = 155$ °C and 126 °C, and $t_c = 2.5$ min. $t_{ch} = 3$ min (6), = 4 min (5), = 4.5 min (4), = 5 min (3), = 5.5 min (2), = 6 min (1).

1. With no cooling period, as shown in Fig. 16.13
2. With the time t_{ch} ranging from 3 to 6 min (cooling period in each case is 2.5 min)

The effect of the value given to the time t_{ch} is clearly shown in Fig. 16.18 for the temperature–time histories and in Fig. 16.19 for the SOC–time histories at the centre of the resin. The following conclusions can be drawn:

1. When the time t_{ch} is less than 5 min, the cooling period is responsible for a decrease in the temperature maximum obtained at the axis of the sample and for a small shift toward a longer time for this maximum to remain. The longer the time t_{ch}, the lower the value of this maximum of temperature.
2. It is well known that the second-order derivation of the resin temperature with respect to time is negative during the heating period, and becomes positive when the reaction starts [1–4]. Between these two stages, there is an inflection point in the temperature–time history, at which the reaction can be assumed to start at the centre of the resin. In the classical present case, the point of inflection is shown to appear after 5 min of heating.
3. When the time t_{ch} is 5 min, a significant decrease in the maximum value of the temperature–time history at the axis is obtained (230 °C instead of 270 °C with no cooling period). The time at which this maximum occurs is around 9 min, this time being a little shorter than the corresponding times obtained with t_{ch} less than 5 min. The SOC–time history at the axis of the sample obtained when t_{ch} = 5 min, is located between the curves obtained when t_{ch} is higher and lower than 5 min, respectively. Moreover, the slope of the SOC–time history is constantly lower when t_{ch} is 5 min. The profiles of temperature

developed through the samples are also of interest (Fig. 16.15). These profiles clearly show not only that lower temperatures develop in the sample with the cooling period at $t_{ch} = 5$ min, but also that the gradients of temperature through the resin are not so steep in this case.

4. When the time t_{ch} is more than 5 min, an increase in the temperature maximum at the axis of the sample, compared to that obtained at $t_{ch} = 5$ min, is observed. The lower the time t_{ch}, the lower the temperature maximum. In fact, for $t_{ch} = 6$ min, the temperature–time history is about the same as that obtained without a cooling stage.

16.5.5 Conclusions

Due to the high exothermicity of the cure reaction of thermosets, a significant increase in temperature is observed within the resin, as well as steep temperature gradients. As a method of reducing this increase in temperature without being detrimental to the time of cure, this new way of heating the resin is efficient.

In reducing the mould temperature during cure, two parameters are of interest: the time at which the cooling stage starts and the length of this cooling stage, not including the temperature of the oil during the cooling stage. The first parameter is particularly of interest, and the best value for this time is the time at the inflection point of the temperature–time history at the axis of the sample.

The numerical model was successfully tested in a simple case, when the temperature of the oil was varied in a discontinuous way. Of course, the principle of the method remains the same with a more sophisticated apparatus able to vary the temperature continuously.

16.6 Simulation of the Process of Reactive Injection Moulding [4]

A few decades ago, the rubber industry showed great interest in the development of processes for faster vulcanisation. Two methods were found to reduce the cure time: higher mould temperatures, with some drawbacks because of the decomposition of the rubber materials; and injection moulding processes [17, 18].

Because of the low thermal conductivity of rubber compounds, heat is transferred into rubber at a low rate, and the thermal conductivity controls the speed of the process of heating and cure. The injection moulding process takes advantage of the higher temperature of the rubber at the beginning of the operation in the mould.

One aspect of fundamental importance in the case of epoxy resin or other thermosets is the exothermic nature of the cure reaction associated with the low thermal conductivity which gives rise to substantial temperature gradients and even to excessively high temperatures when they are heated in standard moulds [1, 7, 10, 19]. For instance, the enthalpy of cure is generally around 4 cal g^{-1} in the case of a rubber, but it may vary from 25 to 100 cal g^{-1}in case of thermosets, depending on the filler extent.

Then, the problem of reactive injection moulding is more difficult in the case of thermosets, especially when they have a high enthalpy of cure. However, modelling of the cure process can be of interest, because it allows calculation of the effect of the various parameters on the process and it generally gives more information than experiments [7, 10].

The first objective in this study is to show that reactive injection moulding can be described by numerical models in the case of an epoxy resin composite. In the case in hand, the sample consists of a small amount of epoxy resin and a large quantity of various kinds of fillers, with a fairly low value of cure enthalpy (18 cal g^{-1}). The model, based on an explicit numerical method with finite differences, takes into account all the facts: the heat transfer by conduction through the mould and the resin, as well as the heat transfer by convection (free or forced) between the oil and the mould surface, and the heat evolved from the overall cure reaction. This model is fully described in Chap. 15.

The mould used in this study has been described in this and the preceding chapter, and in previous papers [1–4, 7]. It consists of a long glass tube in which the resin is located. This mould can either be placed in heated air [7, 10, 11] or immersed in heated oil [1–4]. In order to simulate the process of injection with the help of this simple but convenient mould, the mould and resin are previously heated at constant temperature before being immersed in heated oil kept at a higher temperature. Of course, the temperature selected for simulating the temperature of injection must be lower than the temperature range within which the reaction starts. Experiments with a real system of injection and a mould are more difficult, essentially because of temperature measurements.

The temperature at the centre of the resin was constantly recorded; this temperature–time history of the process is very convenient and precise for testing the validity of the model.

16.6.1 Experiment

Materials

The composite used consisted of an epoxy resin with additives such as short glass fibres, fillers and other loads. The percentage of resin was quite low, as indicated by the low value of the cure enthalpy, 18 cal g^{-1}.

Mould and System of Heating

The composite was placed in a long cylindrical glass tube (internal diameter 1.2 cm, external diameter 1.6 cm). The quality of contact between the composite and mould was good, and much better than in the case of woven glass impregnated with resin [2].

The mould with the composite in it was first immersed in hot water at 75 °C until thermal equilibrium was established in the sample, and then immersed in silicone oil kept at 130 °C.

The temperature at the centre of the resin was constantly recorded during the whole process.

Kinetic and Thermal Parameters

The kinetics of the cure reaction were obtained from the heat flux–temperature curves obtained using a calorimeter run in scanning mode with a heating rate of 2 °C min^{-1} (DSC 111, Setaram), by using various methods [6].

The value of the thermal conductivity of the resin changed from 2×10^{-3} to 4×10^{-3} cal cm^{-1} s^{-1} deg^{-1} when the SOC reached the value of 68% which corresponds to the maximum temperature at the centre of the resin, and when the gelation of the resin takes place.

The kinetic and thermal parameters were as follows:

Composite: $k_0 = 9 \times 10^{21}$ s^{-1}, $E = 43\,350$ cal mol^{-1}

$n = 2.3$, $Q_\infty = 18$ cal g^{-1}, $\rho = 1.77$ g cm^{-3}

$\lambda = 2 \times 10^{-3}$ and 4×10^{-3} cal cm^{-1} s^{-1} deg^{-1}

$C = 0.27 + 10^{-4}\, T$ cal g^{-1} deg^{-1} (T in °C)

Glass mould: $\rho = 2.21$ g cm^{-3}, $C = 0.3$ cal g^{-1} deg^{-1},

$\lambda = 2.7 \times 10^{-3}$ cal cm^{-1} s^{-1} deg^{-1}

Coefficient of heat transfer: $h = 12 \times 10^{-4}\, |\Delta T|^{0.25}$ cal cm^{-2} s^{-1} deg^{-1}

16.6.2 Process and Validity of the Model

As stated previously, the temperature–time history at the centre of the resin provides the most significant information on the process of cure. The temperature is more precisely obtained here than at any other place in the resin, because of the symmetry. Moreover, the highest temperatures are attained at the centre of the resin because of the internal heat generated by the exothermic reaction.

Two temperature–time histories are plotted in Figs 16.20 and 16.21; they were obtained from experiments and calculation with the help of the model. In Fig. 16.20, the curves were obtained by immersing the filled mould successively in water at 75 °C and then in silicone oil at 130 °C. Fresh uncured resin was used in this case, and the exothermic effect resulting from the cure reaction can be determined.

The composite cured previously and cooled down to room temperature was then heated following the same temperature–time history (Fig. 16.21).

The following conclusions can be drawn:

1. Good superimposition is observed in Fig. 16.21, where the cured composite is reheated. This is a proof for the validity of the model, as well as for the accuracy of the thermal parameters.
2. Good superimposition can also be observed in Fig. 16.20, where the fresh uncured composite is heated. Thus, the model is able to describe the process either of cure, or of reheating.
3. In the process of heating and cure shown in Fig. 16.20, four steps can be distinguished.
   (a) The initial heating of the sample from room temperature to 75%, this temperature being called the injection temperature. The selection of this temperature is important, because it must be high enough so as to effect a

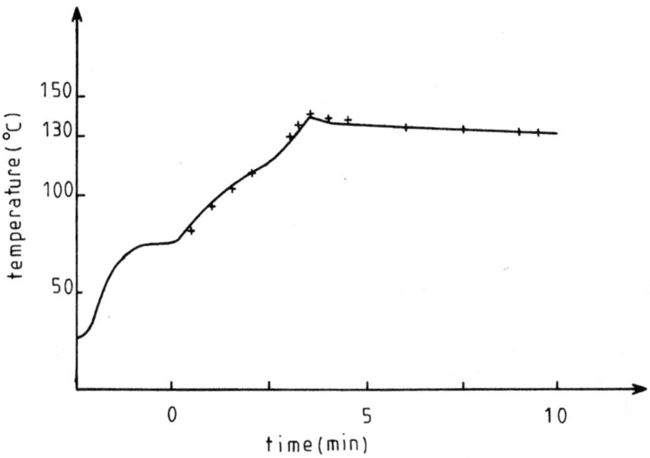

Fig. 16.20. Validity of the model in the simulation of injection moulding. Temperature–time history at the centre of the resin with reaction, for $T_m = 130$ °C and $T_{inj} = 75$ °C. +, calculation; —, experiment.

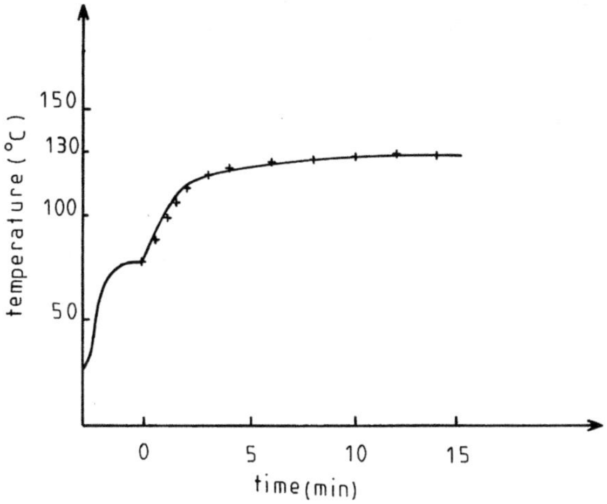

Fig. 16.21. Validity of the model. Temperature–time history in the centre of the resin without reaction, for $T_m = 130$ °C and $T_{inj} = 78$ °C. +, calculation; —, experiment.

reduction in the heating stage, and it must be low enough in order to prevent the cure reaction from taking place.

(b) The heating period of the composite which starts as soon as the mould is immersed into heated oil.

(c) At around 120 °C, an inflection point can be observed in the temperature–time history. From this time, the exothermic reaction of cure at the centre of the resin is developed. A maximum for the temperature is attained at about 140 °C.

(d) After this temperature maximum, a period of cooling down to the oil temperature is observed.

16.6.3 Effect of the Temperature of Injection

The value selected for the temperature of injection is the parameter of concern in reactive injection moulding (RIM). Three kinds of results are considered:

Temperature–time histories at the centre of the resin
SOC–time histories at the centre of the resin
Profiles of temperature and SOC through the resin

Temperature–Time Histories at the Centre

Various temperature–time histories at the centre of the resin are plotted in Fig. 16.22, by choosing four different values for the injection temperature: 22, 75, 100 and 110 °C. The following results are worth noting:

1. The lower the injection temperature, the higher the rate of heating of the resin, at the beginning of the process.
2. The higher the injection temperature, the shorter the time necessary for the temperature at the centre of the resin to rise to the maximum.
3. In all cases, the value of the highest temperature attained at the centre of the resin is not significantly changed by varying the value of the injection temperature.

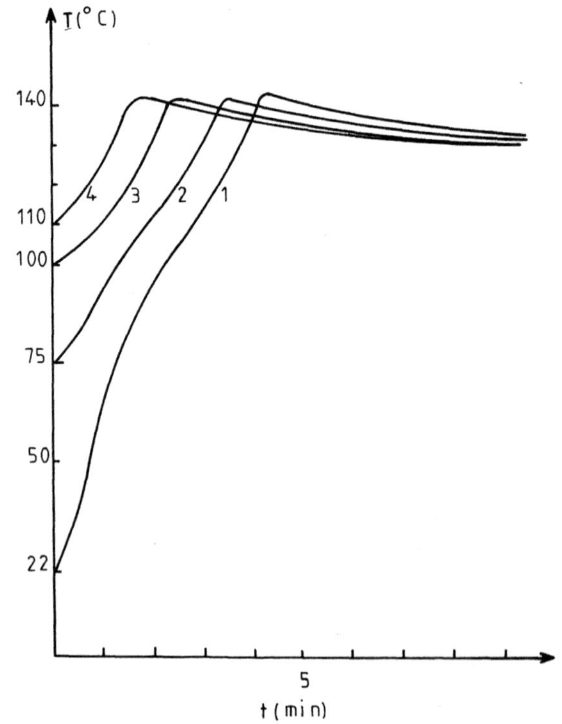

Fig. 16.22. Effect of the injection temperature on the SOC–time history at the centre of the resin. $T_m = 130$ °C. T_{inj}: (1) 22 °C, (2) 75 °C, (3) 100 °C, (4) 110 °C.

SOC–Time Histories at the Centre

The SOC–time histories at the centre of the resin are plotted in Fig. 16.23, calculated using four values of injection temperature: 22, 75, 100 and 110 °C. Some conclusions are drawn:

1. Typical S-shaped curves are obtained for the SOC–time histories at the centre of the resin.
2. The higher the injection temperature, the shorter the SOC–time history.
3. The various curves obtained for the SOC–time histories can be superimposed by a translation along the time axis. This reduction in time is of great interest, as seen in Table 16.2.

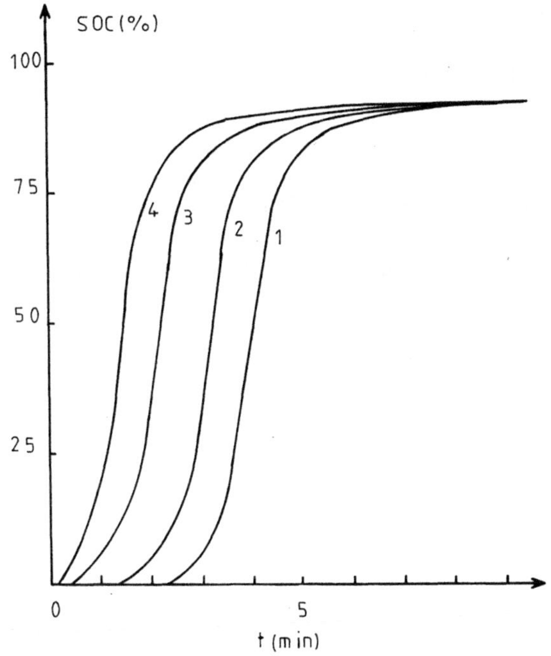

Fig. 16.23. Effect of the injection temperature on the temperature–time history at the centre of the resin. $T_m = 130$ °C. T_{inj}: (1), 22 °C, (2) 75 °C, (3) 100 °C, (4) 110 °C.

Table 16.2. Effect of the injection temperature

	Injection temperature (°C)			
	22	75	100	110
Time for T_{max}	4′18″	3′30″	2′30″	1′48″
Time for SOC = 50%	4′6″	3′12″	2′12″	1′24″
Time for SOC = 90%	6′30″	6′	5′	4′

The effect of the injection temperature on the progress of the cure reaction in the resin is also clearly shown in Fig. 16.24. The time necessary for the reaction to reach SOCs of 50%, 68% and 90% is plotted as a function of the injection temperature. A slight decrease in the time is observed when the injection temperature is lower than 75 °C. Then, a significant decrease in the time of cure is gained for higher values of injection temperature. However, a limitation for the injection temperature exists, because the reaction must not take place before the injection in the mould. The model can be of great help for determining the right value of the injection temperature.

Profiles of Temperature and SOC Through the Resin

The model is capable of calculating the profiles of temperature and SOC developed through the resin. These profiles are plotted in Fig. 16.25 for the temperature and in Fig. 16.26 for the SOC, by considering the injection temperatures of 22, 75, 100 and 110 °C.

Some results of interest are found.

1. Steep gradients of temperature are developed through the sample. During the initial heating period in oil, the temperature is highest on the surface of the resin. When the reaction takes place, other gradients are grown with the highest temperature at the centre of the resin. The steeper gradient is attained at the maximum value of the temperature at the centre of the resin.
2. The heating period is shortened by using the injection system with the obvious statement: the higher the injection temperature, the shorter the heating period.
3. Gradients of SOC are also developed during the process. Steeper gradients are obtained when the temperature at the centre of the resin rises the maximum.
4. Of course, the use of the injection system is responsible for a decrease in the heating period, and subsequently for a reduction in the cure time.

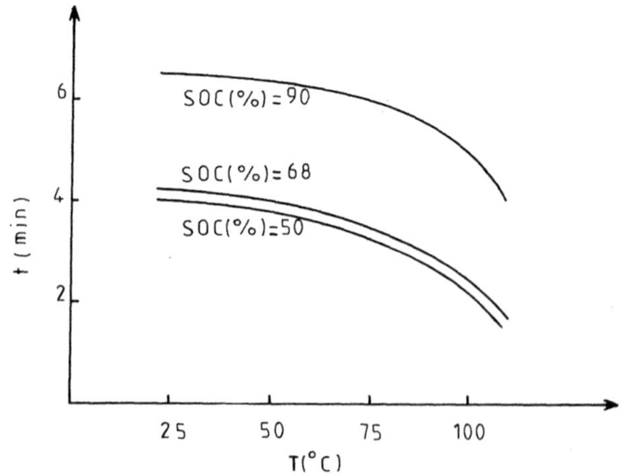

Fig. 16.24. Effect of the injection temperature on the progress of the cure reaction.

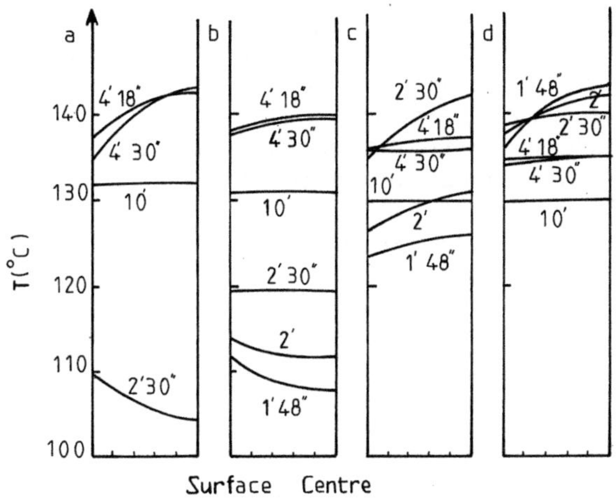

Fig. 16.25. Effect of the injection temperature on the profiles of temperature through the resin, for $T_m = 130$ °C. T_{inj}: **a** 22 °C, **b** 75 °C, **c** 100 °C, **d** 110 °C.

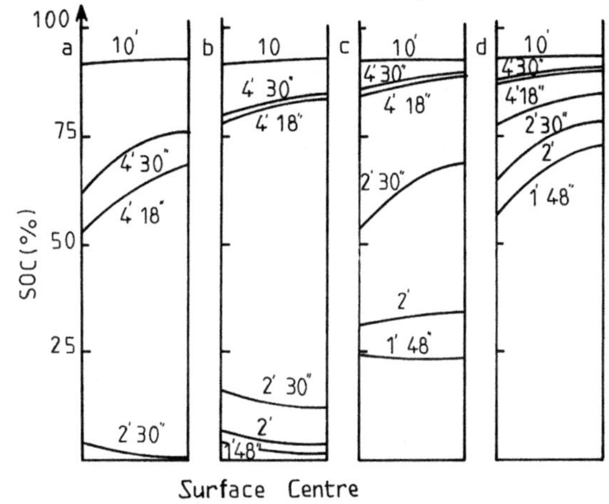

Fig. 16.26. Effect of the injection temperature on the profiles of SOC through the resin, for $T_m = 130$ °C. T_{inj}: **a** 22 °C, **b** 75 °C, **c** 100 °C, **d** 110 °C.

5. Except for the beginning of the process, the value of the SOC is highest at the centre of the resin.

16.6.4 Conclusions

The process of cure of a composite was studied by considering the temperature of injection of the resin in the heated system. A simple system with a cylindrical mould and two baths at two different temperatures was used to simulate the rather complex process. Following this principle, the mould with the resin in it was first heated up to the injection temperature, and then immersed in a bath at a higher temperature. Of course, the problems of the injection system were simplified, and shear heating neglected.

A numerical model is capable of describing the process, and the validity of this model was tested by comparing the temperature–time histories obtained at the centre of the resin.

The effect of the injection temperature on the progress of heating and cure is of great importance.

References

1. Chater M, Vergnaud, JM. Study of the process of cure of epoxide resin heated by liquid oil at constant temperature. Eur Polym J 1988; 24: 245
2. El Brouzi A, Bouzon J, Vergnaud JM. Modelling of the cure of fibre glass–epoxy resin composite heated by liquid oil at constant temperature. Effect of parameters on the process. Plast Rubber Process Applic 1989; 12: 71
3. El Brouzi A, Bouzon J, Vergnaud JM. Modelling the cure of epoxy resin heated by liquid oil. Application of the dissipation of heat generated by the reaction. Plast Rubber Process Applic 1990; 14: 95
4. El Brouzi A, Granger R, Bouzon J, Vergnaud JM. Modelling and simulation of the process of reactive injection moulding in the case of a composite with low enthalpy. Plast Rubber Process Applic 1989; 12: 7
5. Chater M, Chataing G, Vergnaud JM. Enhanced study of differential scanning calorimetry with determination of heat flux-time curves, profiles of temperature and state of cure. Thermochim Acta 1985; 90: 135
6. Kretzschmar K, Hoffman KW. Kinetic aspects arising during the differential scanning calorimetric investigation of curing epoxy resins with anhydrides. Thermochim Acta 1985; 83: 139
7. Chater M, Bouzon J, Vergnaud JM. Modelling of the cure of epoxide resin: effect of the diameter of cylindrical moulds. Plast Rubber Process Applic 1987; 7: 199
8. Liu H, Armand JY, Bouzon J, Vergnaud JM. Effect of sample size and heating rate on the DSC process for reactions of high enthalpy. Thermochim Acta 1988; 126: 81
9. Armand JY, Vergnaud JM. Effect of the value of heating rate in DSC on the kinetic parameters, when there is a high enthalpy of reaction. Thermochim Acta 1988; 131: 15
10. Chater M, Vergnaud JM. Effect of the temperature of the mould on the cure of epoxy resin. J Polym Engng 1988; 8: 1
11. Chater M, Vergnaud JM. Effect of kinetic and thermal parameters on the process of cure of epoxide resin in a cylindrical mould. Eur Polym J 1987; 23: 563
12. Rochette JB, Sadr A, Abdul M, Vergnaud JM. Effect of a variation in the order of the overall reaction of cure on the profiles of temperature and state of cure developed at the midplane of rubber sheets. Thermochim Acta 1984; 79: 187
13. Progelhof RC, Throne JL. Non-isothermal curing of reactive plastics. Polym Engng Sci 1975; 15: 690
14. Mallik PK, Raghupathi N. Effect of cure cycle on mechanical properties of thick section fiber-reinforced poly thermoset mouldings. Polym Engng Sci 1979; 19: 774
15. Barone MR, Caulk DA. The effect of deformation and thermoset cure on heat conduction in a chopped-fibre reinforced polyester during compression moulding. Int J Heat Mass Transfer 1979; 22: 1021
16. Pusatcioglu SY, Hassler JC, Fricke AL, McGee HA. Effect of temperature gradients on cure and stress gradients in thick-thermoset castings. J Appl Polym Sci 1980; 25: 381
17. Sadr A, Granger R, Vergnaud JM. Calculation of profiles of temperature and state of cure developed within the rubber mass in injection moulding processes. J. Appl Polym Sci 1984; 29: 955
18. Sadr A, Granger R, Vergnaud JM. Profiles of temperature and state of cure developed within rubber in injection molding systems. ACS Symposium Series, N. 270, Reaction Injection Moulding:

Polymer Chemistry and Engineering 1985; 18: 279
19. Nixon JA, Hutchinson JM. Analysis of the cure of sheet moulding compound. I. Development of the model. Plast Rubber Process Applic 1985; 5: 349

Chapter 17

Simultaneous Determination of Resistance to Torsion and State of Cure

17.1 Introduction

Thermosetting resin composites are transformed into three-dimensional networks by curing them in heated moulds, and the temperature as well as the time of cure are important. Good knowledge of the process is the best way to obtain final materials with the shortest possible curing time and the best properties.

The high exothermicity of the curing reaction associated with the low thermal conductivity of the resin are responsible for the development of high gradients of temperature, and for a high increase in temperature within the resin. Moreover, the viscosity of the resin generally decreases with increasing temperature until the gelation occurs. The gelation time must be known in order to apply the pressure at the proper time. Lastly, the time necessary for the cure to reach completion and for the material to attain good mechanical stability is also of concern, because it is related to the period of time over which the resin must be kept in the mould.

The main purpose of this study is to describe an experiment designed to correlate the value of the SOC with the mechanical properties of a sheet of fibre glass–resin composite during cure in an oven [1]. A Kinemat apparatus is used, modified by fitting with thermocouples to record the temperature at various places in the sheet, especially at the centre. The Kinemat apparatus is able to follow the resistance to torsion of the sample.

The second objective of this study is to determine the value of the SOC in the resin from temperature measurements at the midplane of the sample. A numerical model based on a method with finite differences is used for the purpose. It takes into account not only the heat transferred by convection to the surface of the sample and by conduction through the composite, but also the heat evolved from the cure reaction. These types of models were successfully tested with samples of various shapes: rubber sheets [2–5], or epoxy resin in cylindrical moulds [6–8].

The kinetics of heat evolved from the cure reaction is determined from measurements made using differential scanning calorimetry [9, 10].

17.2 Experiment

Material

The composite was woven glass fibre impregnated with epoxy resin with 38% resin by weight (Vicotex). Several layers were pressed together to produce a 3 mm thick sheet.

Kinetic and Thermal Parameters

The kinetic parameters were calculated from the experiments performed with a calorimeter running in scanning mode (DSC 111, Setaram) at a heating rate of 2 °C min^{-1}. A fairly large sample, around 150 mg, was encapsulated in the holder. The heat flux–temperature histories were stored and analysed using a microcomputer (HP86, Hewlett Packard) connected to the calorimeter.

The kinetic and thermal parameters were as follows:

$$n = 2.3, \qquad Q_\infty = 17.8\ \text{cal g}^{-1}, \qquad k_0 = 5 \times 10^{19}\ \text{s}^{-1},$$

$$E = 43\,350\ \text{cal mol}^{-1}, \qquad \lambda = 1.8 \times 10^{-3}\ \text{cal cm}^{-1}\,\text{s}^{-1}\,\text{deg}^{-1},$$

$$C = 0.21 + 10^{-4}\ T\ \text{cal g}^{-1}\,\text{deg}^{-1}\ (T \text{ in °C})$$

Kinemat Apparatus

The Kinemat apparatus (Prodemat, Lyon) was used for this study. The resin composite was heated by thermal convection through air in an oven kept at constant temperature. The resistance to torsion was constantly recorded during cure.

The apparatus was modified in order to measure the temperature of the air surrounding the sample and the midplane of the resin composite. Great care was taken to keep the thermocouple exactly at the midplane of the composite during cure.

17.3 Theory

Assumptions

The following assumptions are made:

1. The composite is heated by convection to the surface and by conduction through the thickness of the sample. A coefficient of heat transfer by free

convection exists through the air–composite interface. The heat transfer is unidirectional, perpendicular to the sheet.
2. The radiant heat transmission is neglected.
3. The composite is homogeneous for the heat transfer purpose.
4. The heat evolved from the cure reaction is expressed in terms of temperature and enthalpy by an Arrhenius law.
5. The process of heat transfer and cure reaction is not altered by the motion of the sample. The thermocouple is very well fixed at the centre of the composite sheet.
6. The heat conductivity λ is assumed constant.

Mathematical Treatment

The unidirectional heat transfer by conduction through the thickness of the composite is given by

$$\rho C \frac{\partial T}{\partial t} = \frac{\partial}{\partial x}\left(\lambda \frac{\partial T}{\partial x}\right) + \rho \frac{\partial Q}{\partial t} \tag{17.1}$$

where the contribution to the rate of heating of the heat transfer by conduction and the heat evolved from the reaction can be seen on the right-hand side.

At each position the rate of heat evolved from the cure reaction at time t is expressed in terms of the heat evolved up to time t, Q_t, and the total enthalpy evolved, Q_∞, by the relation

$$\frac{1}{Q_\infty}\frac{dQ_t}{dt} = k_0\left(1 - \frac{Q_t}{Q_\infty}\right)^n \exp\left(-\frac{E}{RT}\right) \tag{17.2}$$

Numerical Analysis

As no analytical solution exists for the problem, a numerical method with finite differences was used (Chap. 7). The thickness is divided into a number of equal finite slices of thickness Δx, by temperature reference planes.

Within the Solid. The heat balance during the increment of time Δt is calculated within a slice i. The new temperature after time Δt, TN_i can thus be expressed as a function of the previous temperatures obtained at the same slice i and adjacent slices $i + 1$ and $i - 1$.

$$TN_i = \frac{1}{M_i}\left[T_{i-1} + (M_i - 2)T_i + T_{i+1}\right] + \frac{1}{C_i}\Delta Q_i \tag{17.3}$$

On the Surface. The surface is heated by the warm air at instantaneous temperature T_a. The heat balance within the slice of thickness $\frac{1}{2}\Delta x$ next to the surface by considering the heat convection at the surface is evaluated during the time increment Δt. The new temperature on the surface after time Δt, TN_0, is thus expressed in terms of the previous temperature on the surface, T_0, and of the temperatures T_1 and T_a:

$$\mathrm{TN}_0 = \frac{1}{M_0}\left[2PT_a + (M_0 - 2 - 2P)T_0 + 2T_1\right] + \frac{3}{4}\frac{\Delta Q_0}{C_0} + \frac{1}{4}\frac{\Delta Q_1}{C_1} \quad (17.4)$$

by making the assumptions

$$\mathrm{TN}_{0.25} = T_{0.25} = \mathrm{TN}_0 - T_0 \quad \text{and} \quad M_{0.25} = M_0 \quad (17.5)$$

and by defining the dimensionless number P by

$$P = \frac{h\Delta x}{\lambda} \quad (17.6)$$

where h is the coefficient of heat transfer by convection at the surface. T_a and T_1 are the temperatures of air and in the sample at a distance Δx from the surface, at the previous time.

The dimensionless number M is

$$M_i = \frac{(\Delta x)^2}{\Delta t}\frac{\rho C_i}{\lambda} \quad (17.7)$$

where C_i must be expressed at each step in terms of T_i.

Cure Reaction. At each position, the heat evolved from the reaction up to time t is obtained by integrating Eq. (17.2), when the temperature is variable (Sect. 7.4.1). By introducing the function S_t,

$$S_t = \int_0^t k_0 \exp\left(-\frac{E}{RT_\theta}\right)d\theta \quad \text{with } 0 \leqslant \theta \leqslant t \quad (17.8)$$

and the recurrent relation

$$S_{t+\Delta t} = S_t + k_0\Delta t \exp\left(-\frac{E}{RT_t}\right) \quad \text{with } S_0 = 0 \quad (17.9)$$

the extent of cure Y_t at time t is thus given by

$$Y_t = \frac{Q_t}{Q_\infty} = 1 - [1 + (n-1)S_t]^{1/(1-n)} \quad (17.10)$$

17.4 Results for Cure and Resistance to Torsion

17.4.1 Variation of Mechanical Properties During Cure

The sample of fibre glass–epoxy resin composite previously at room temperature was introduced into the oven, kept at 145 °C, and its resistance to torsional motion was registered. The composite was heated by convection and conduction, as well as by the internal cure reaction.

Two curves of interest were simultaneously recorded (Fig. 17.1): one showing the temperature at the midplane of the sheet, the other the torque. The following results are noted:

1. The process of heating was rather slow, 5 min being necessary for the midplane of the sample to reach the temperature of the oven.

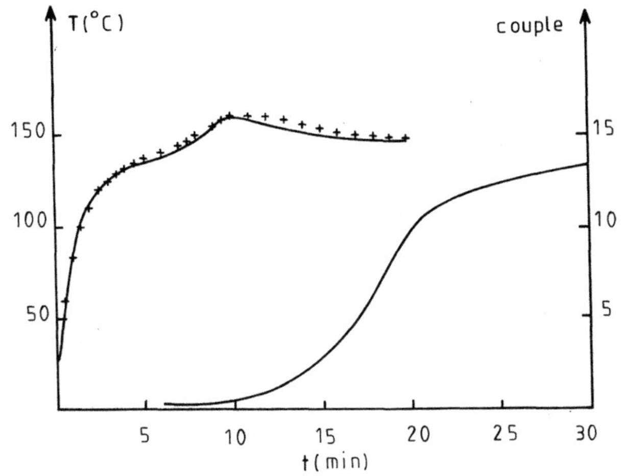

Fig. 17.1. Temperature–time history at the centre of the sheet and torque–time history of a composite during the process of cure in the Kinemat apparatus. —, experiment; +, calculation.

2. After 6 min the cure reaction starts. This is illustrated by the time at which the second-order derivative of temperature with respect to time becomes positive.
3. Due to the exothermic reaction, an increase in temperature is observed. Resulting from the heat evolved from the reaction and the heat transferred, the temperature at the midplane passes through a maximum as shown in other kinds of moulds [7, 8].
4. The resistance to torsion motion of the composite decreases a little as temperature increases up to the mould temperature.
5. When the reaction takes place, illustrated by the increase in temperature due to the exothermic reaction, the resistance to torsion of the composite starts to increase.
6. The resistance to torsion of the composite becomes higher during the cure.
7. The model describes the process of cure quite well, as the temperature–time histories at the midplane obtained by experiments and calculation are well superimposed.

At the end of the process, the motion of the apparatus was stopped. The sample was removed from the mould, cooled down to room temperature and then placed again in the oven under the same conditions as for the first heating. Good superimposition is also obtained for the experimental and calculated temperature–time histories at the midplane (Fig. 17.2).

17.4.2 Profiles of Temperature and SOC Through the Sheet

The numerical model is capable of giving more information than experiments, especially the profiles of temperature and SOC developed through the thickness of the sheet.

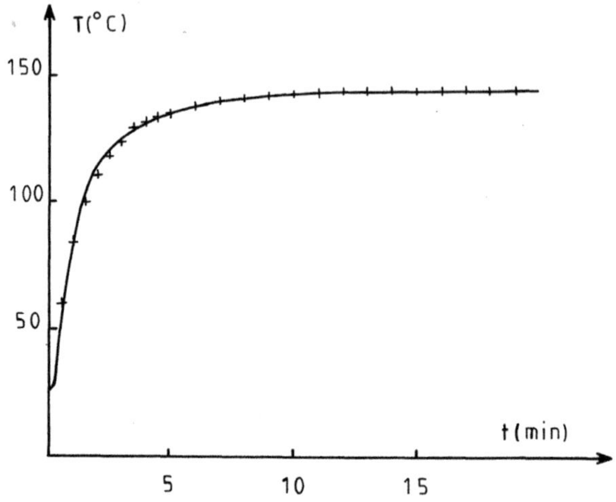

Fig. 17.2. Temperature–time history at the centre of the sheet obtained by reheating the cured sample in the Kinemat apparatus. —, experiment; +, calculation.

The profiles of temperature (left) and SOC (right) are plotted in Fig. 17.3 for the process of cure, and also for the reheating stage of the precured resin in Fig. 17.4. The following results are worth noting:

1. Three steps can be distinguished during the process of cure:
   (a) The heating stage, where some gradients of temperature are observed, especially at the beginning when the sample is strongly heated.
   (b) The stage of cure, which can be followed in Figs 17.1 and 17.3. The heat of cure is responsible for an increase in temperature at the midplane of the sheet. This temperature rises to a maximum around 10 min 30 s, and this maximum is 20 °C higher than the mould temperature.
   (c) The cooling of the sample from the maximum temperature to the mould temperature. During this stage, a significant increase in the resistance to torsion is observed.
2. In contrast with thick samples of composites heated in oil [11, 12], the profiles of temperature and SOC are almost flat for the thin sheet heated by air. This fact is perhaps of interest, showing a uniform temperature and a uniform value of the SOC within the sample.
3. From the profiles of SOC developed in Figs 17.3 and 17.4, it appears that the cure reaction is not completely achieved. This is because the time is too short for the mould temperature of 145 °C. Of course, the SOC is expressed here by the extent of heat evolved from the cure reaction.
4. From calculation, it is shown that a slight increase in SOC takes place within the sample during the reheating stage.
5. For the period of rescanning of the precured sheet, the temperature–time history at the midplane (Fig. 17.2) exhibits no significant maximum. The profiles of temperature through the thickness of the sheet, which are also flat, can be observed in Fig. 17.4.

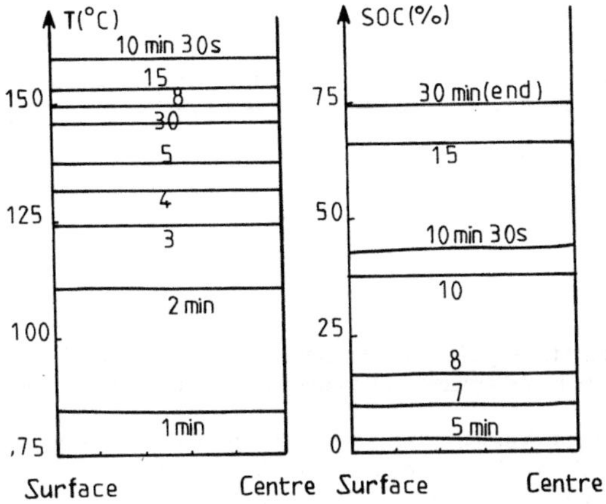

Fig. 17.3. Profiles of temperature (left) and SOC (right) within the sample during the first heating of the fresh uncured resin.

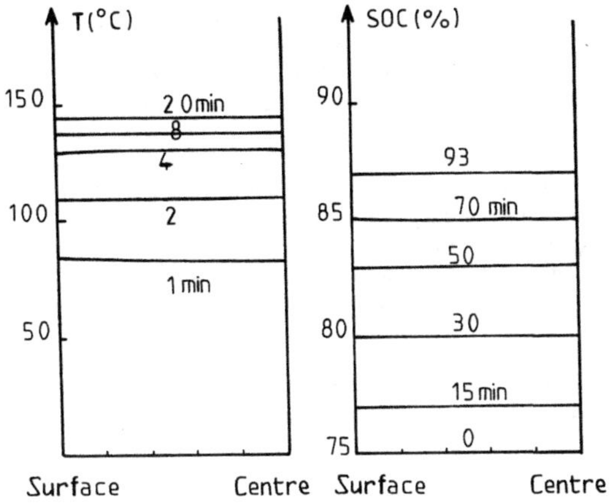

Fig. 17.4. Profiles of temperature (left) and SOC (right) within the sample during the rescanning of the precured resin.

17.5 Conclusions

The process of cure of a composite made of fibre glass and epoxy resin can be studied with the help of an apparatus measuring the mechanical properties, namely resistance to torsion of the sample during the cure.

When the sample is heated, the resistance to torsion of the sheet as well as the temperature–time history at the centre of the sheet can be simultaneously recorded.

With the help of a numerical model taking into account the convective heat to the surface and the heat conduction through the sample as well as the heat generated by the cure reaction, it is possible to determine the profiles of temperature and SOC.

As a result, simultaneous information on the change in the mechanical properties of the composite and the value of the SOC can be obtained.

References

1. El Brouzi A, Dursapt M, Vergnaud JM. Simultaneous determination of torque and state of cure of a thermoset composite by experiment and modelling of the process. Plast Rubber Compos Process Applic 1991; 5: 13
2. Abdul M, Vergnaud JM. Vulcanization progress in rubber sheets during cooling in motionless air after extraction from the mold. Thermochim Acta 1984; 76: 161
3. Khouider A, Vergnaud JM. Effect of temperature of motionless air on the cure vulcanizates after removal from the mold. J Appl Polym Sci 1986; 32: 5301
4. Khouider A, Bouzon J, Vergnaud JM. Effect of cure enthalpy on cure progress during the cooling period in motionless air. Thermochim Acta 1986; 102: 83
5. Khouider A, Turin J, Vergnaud JM. Increase in the state of cure in rubber sheets during the cooling period in water. J Polym Engng 1988; 8: 55
6. Chater M, Vergnaud JM. Effect of kinetic and thermal parameters on the process of cure of epoxy resin in a cylindrical mould. Eur Polym J 1987; 7: 563
7. Chater M, Bouzon J, Vergnaud JM. Modelling of the cure of epoxide resin: effect of the diameter of cylindrical moulds. Plast Rubber Process Applic 1987; 7: 199
8. Chater M, Vergnaud JM. Effect of the temperature of the mold on the cure of epoxy resin. J Polym Engng 1988; 8: 1
9. Liu H, Armand JY, Bouzon J, Vergnaud JM. Effect of sample size and heating rate on the DSC process for reactions of high enthalpy. Thermochim Acta 1988; 126: 81
10. Armand JY, Vergnaud JM. Effect of the value of the heating rate in DSC on the kinetic parameters, when there is a high enthalpy of reaction. Thermochim Acta 1988; 131: 15
11. El Brouzi A, Bouzon J, Vergnaud JM. Modelling of the cure of fibre glass-epoxy resin composite heated by liquid oil at constant temperature. Effect of parameters on the process. Plast Rubber Process Applic 1989; 12: 71
12. El Brouzi A, Granger R, Bouzon J, Vergnaud JM. Modelling and simulation of the process of reactive injection moulding in the case of a composite with low enthalpy. Plast Rubber Process Applic 1989; 12: 7

Chapter 18

Cure of a Loaded Polyester Resin in a Metallic Mould

18.1 Introduction

Modelling of the process of cure is necessary to afford further insight into the nature of all phenomena, and therefore to improve the quality of the final material and to reduce the time of cure in the mould. The most important aspect in the process of cure of thermosets is the high cure enthalpy associated with a low thermal conductivity, which may give rise to excessively high temperature and to substantial temperature gradients within the resin, when standard moulds are used [1–8].

Various moulds were used for testing the validity of these models, which were either placed in heated air or immersed in heated oil. The process was followed by recording the temperature at the centre of the resin, where the temperature gradient exhibits a zero value and where the highest temperature is reached. The effect of various parameters on the process of cure is thus well established either by experiment or calculation. Some relevant factors are worth noting: the diameter [5] and temperature [6] of the mould, the thermal and kinetic parameters [4, 7], and the nature of the surrounding atmosphere heating the mould [3–7]. A way of heating simulating the injection system of the resin in a mould was also studied, and the model was found to describe this method very well [8].

The first purpose of this chapter is to study the cure of an unsaturated orthophthalic poylester resin by using a metal mould able to prepare final samples which can be tested afterwards. The resin sample is a flat cylinder, and the mould is made of three parts from which the final cured resin can be easily removed. As the length of the cylindrical mould can be extended by addition of various metal annuli, cylindrical samples of various desired thicknesses can be obtained. The pressure of the resin can also be controlled. The heating system is very simple and convenient, as the mould with the resin in it is immersed in heated oil (Fig. 18.1).

The other main objective is to build and to test a numerical model in this case. The model takes into account all the known facts: heat convection (free or forced) at the oil–metal interface, heat conduction through the mould and resin, as well as heat evolved from the cure reaction. The kinetics of heat generated by

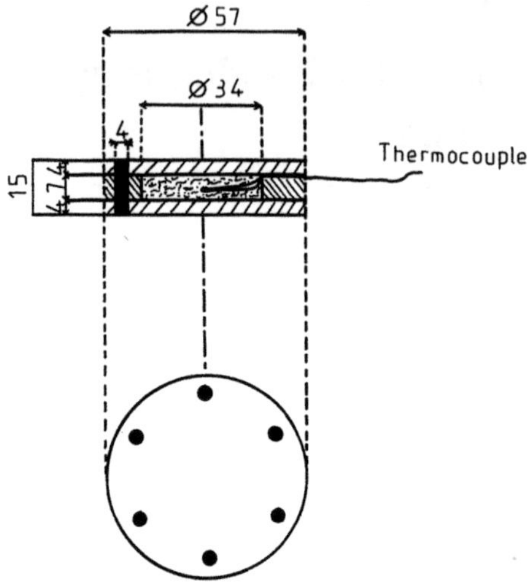

Fig. 18.1. Diagram of the mould.

the overall reaction is determined by calorimetry working in scanning mode. As this resin consists of a complex mixture, fairly large samples were used in the calorimeter in order to have a sample representative of the homogeneity of the batch from which it is taken, while the rate of heating was quite slow [9, 10].

18.2 Theory

Assumptions

The following assumptions are made:

1. One-dimensional heat transfer only is considered, as the thickness is considerably smaller than the diameter of the sample.
2. Heat is transferred by a three-step process: convection (free or forced) to the surface; conduction within the mould and the resin.
3. The rate of heat evolved from the overall cure reaction at time t is expressed in terms of temperature and heat evolved up to time t by an Arrhenius equation, whose parameters are determined by calorimetry.
4. The thermal parameters vary either with temperature or with the SOC: the heat capacity increases linearly with temperature; the thermal conductivity varies with the SOC in a discontinuous way.
5. There is perfect contact between the resin and mould.

Mathematical Treatment

The general equation for heat transfer in one-dimension with a cure reaction is expressed by

$$\rho C \frac{\partial T}{\partial t} = \frac{\partial}{\partial x}\left(\lambda \frac{\partial T}{\partial x}\right) + \rho \frac{\partial Q}{\partial t} \tag{18.1}$$

The contribution to the increase in temperature owing to heat conduction and heat generated by the cure reaction can be seen in the right-hand term.

The rate of heat evolution from the cure reaction is expressed in terms of temperature and of the partial enthalpy evolved up to time t, by the Arrhenius equation

$$\frac{1}{Q_\infty}\frac{dQ_t}{dt} = k_0\left(1 - \frac{Q_t}{Q_\infty}\right)^n \exp\left(-\frac{E}{RT}\right) \tag{18.2}$$

Numerical Analysis

No analytical solution can be obtained for the problem, because of the presence of the internal heat generated by the reaction. The problem is thus solved by using an explicit numerical method with finite differences. As shown in Fig. 18.2, the thickness of the material (mould and resin) is divided into slices of constant thicknesses either in the resin or in the mould. There are 2Nr slices in the resin and Nm − Nr slices in each metal sheet. Each plane separating two slices is associated with an integer n.

From the heat balance evaluated during the increment of time Δt within a slice, by considering transient heat conduction and heat evolved from the cure reaction in the composite, and transient heat conduction only in the mould, the new temperature after time Δt is expressed in terms of the previous temperatures obtained at the same place and adjacent places.

Composite: $0 \leq n \leq \mathrm{Nr} - 1$, *slice* $(n - 0.5)\Delta x_r \leq x \leq (n + 0.5)\Delta x_r$.

$$\mathrm{TN}_n = \frac{1}{\mathrm{Mr}}[T_{n+1} + (\mathrm{Mr} - 2)T_n + T_{n-1}] + \frac{\Delta Q_n}{C} \tag{18.3}$$

with the dimensionless number Mr for the resin

$$\mathrm{Mr} = \frac{(\Delta x_r)^2}{\Delta t}\left(\frac{\rho C}{\lambda}\right)_r \tag{18.4}$$

where ΔQ_n represents the amount of heat evolved from the cure reaction per unit mass during the increment of time Δt at the abscissa n (expressed in cal g^{-1}).

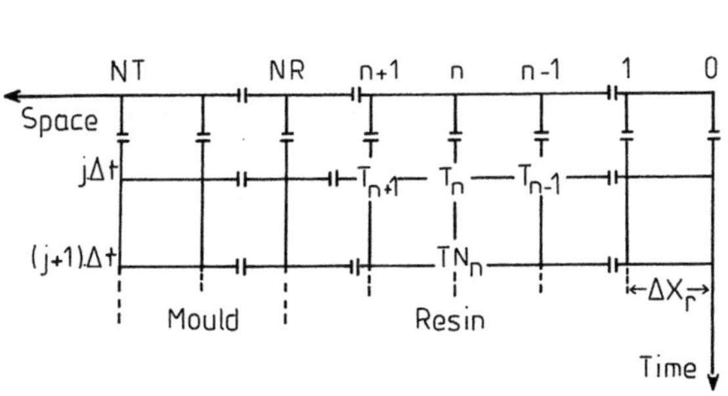

Fig. 18.2. Space–time diagram for numerical analysis.

Mould: $\mathrm{Nr} + 1 \leqslant n \leqslant \mathrm{Nm} - 1$, $x = \mathrm{Nr}\Delta x_r + (n - \mathrm{Nr})\Delta x_m$. For the slice located between the planes of abscissae $x - \frac{1}{2}\Delta x_m$ and $x + \frac{1}{2}\Delta x_m$,

$$\mathrm{TN}_n = \frac{1}{\mathrm{Mm}}\left[T_{n+1} + (\mathrm{Mm} - 2) + T_{n-1}\right] \tag{18.5}$$

with the dimensionless number Mm for the mould,

$$\mathrm{Mm} = \frac{(\Delta x_m)^2}{\Delta t}\left(\frac{\rho C}{\lambda}\right)_m \tag{18.6}$$

Composite–Mould Interface: $n = \mathrm{Nr}$. Two half slices of resin and mould on each side of the composite–mould interface are considered. The heat balance during the increment of time Δt is evaluated within these two half slices, by taking into account heat conduction in the mould and the resin and heat generated by the cure in the resin. As the resin and mould are assumed to be in perfect contact, no additional heat resistance is considered at the interface, and the temperature on each side of the interface is the same.

The new temperature after time Δt at the interface ($n = \mathrm{Nr}$) is thus expressed in terms of the previous temperatures at the interface and at the adjacent places in the resin and in the mould.

$$\mathrm{TN}_{\mathrm{Nr}} = AT_{\mathrm{Nr}+1} + (1 - A - B)T_{\mathrm{Nr}} + BT_{\mathrm{Nr}-1} + \frac{3D}{4}\Delta Q_{\mathrm{Nr}} + \frac{D}{4}\Delta Q_{\mathrm{Nr}-1} \tag{18.7}$$

with the coefficients A, B, D and F given by

$$A = \frac{\Delta t}{F}\frac{\lambda m}{\Delta x_m} \qquad B = \frac{\Delta t}{F}\frac{\lambda r}{\Delta x_r} \tag{18.8}$$

$$D = \frac{\rho_r \Delta x_r}{2F} \tag{18.8'}$$

$$F = \rho_m C_m \frac{\Delta x_m}{2} + \rho_r C_r \frac{\Delta x_r}{2} \tag{18.8''}$$

Mould–Oil Interface. The heat balance during the increment of time Δt is evaluated within half the slice of the mould next to the mould surface, by considering heat conduction through the mould and heat convection from oil to the mould surface.

The new temperature at the mould surface $\mathrm{TN}_{\mathrm{Nm}}$ is given by

$$\mathrm{TN}_{\mathrm{Nm}} = \frac{1}{\mathrm{Mm}}\left[2T_{\mathrm{Nm}-1} + (\mathrm{Mm} - 2N - 2)T_{\mathrm{Nm}} + 2NT_{\mathrm{oil}}\right] \tag{18.9}$$

where T_{oil} is the temperature of the oil far from the interface, and N is the dimensionless number

$$N = \frac{h\Delta x_m}{\lambda_m} \tag{18.10}$$

where h is the coefficient of convection heat, which is expressed by

$$h = h_0 |T_{\mathrm{oil}} - T_{\mathrm{Nm}}|^a \tag{18.11}$$

Condition of Stability. For each value of $n(0 \leqslant n \leqslant \mathrm{Nm})$, the whole coefficient of the previous temperature T_n in the equation which expresses the new temperature TN_n, must be positive. Thus the increment of time Δt must be chosen so that the following conditions are fulfilled:

$$\mathrm{Mr} > 2, \qquad \mathrm{Mm} > 2,$$
$$\mathrm{Mm} - 2N - 2 > 0, \qquad 1 - A - B > 2 \tag{18.12}$$

SOC. At each position n and each time $t = j\Delta t$, the SOC is given by

$$\mathrm{SOC} = 100Y_j \qquad \text{with} \qquad Y_j = \frac{Q_j}{Q_\infty} \tag{18.13}$$

Upon defining the function S_j,

$$S_j = \int_0^{j\Delta t} k_0 \exp\left(\frac{-E}{RT_\theta}\right) d\theta, \qquad 0 \leqslant \theta \leqslant t, \qquad t = j\Delta t \tag{18.14}$$

we have

$$S_{j+1} = S_j + k_0 \Delta t \exp\left(\frac{-E}{RT_j}\right) \tag{18.15}$$

with

$$S_0 = 0 \tag{18.1}$$

and Y_j is thus given by

$$Y_j = 1 - [1 + (n - 1)S_j]^{1/(1-n)} \tag{18.16}$$

The amount of heat generated by the reaction per unit mass, during the increment of time Δt, from t to $t + \Delta t$ is

$$\Delta Q = Q_\infty (Y_{j+1} - Y_j) \tag{18.17}$$

18.3 Experiment

Materials

The composite used was a bulk moulding compound (BMP) (Menzolit), comprising

Maleic polyester resin	16
Vinyl acetate	10.7
Hydrated alumina	66.7
Magnesia	0.1
Catalyst	0.5
Styrene	4.1
Zinc stearate	1.9
	100%
Fibre glass	20%

All the components except the fibre glass were thoroughly mixed. The glass fibres were then added and well dispersed in the viscous mixture.

Moulding and Heating System

The composite was placed in the metallic mould (Fig. 18.1), and a thin thermocouple placed at the centre of the sample. The diameter of the sample was 3.4 cm and its thickness 0.7 cm. The mould with the sample in it was then immersed in silicone oil heated at a constant temperature. The temperature at the centre of the resin was constantly recorded during the process.

Kinetic and Thermal Parameters

The kinetic parameters were previously determined from experiments performed with a DSC 111 (Setaram) run in scanning mode with a heat rate of 0.5 °C $\min^{-1}$. Large samples of 150 mg were used, because of the heterogeneity of the composite [9, 10].

The value of the coefficient of surface heat transfer h was obtained from the temperature–time histories of the precured sample when heated in oil.

The thermal and kinetic parameters were as follows:

Composite: $k_0 = 9 \times 10^{21}\,\mathrm{s}^{-1}$, $Q_\infty = 18\,\mathrm{cal\,g}^{-1}$,

$$\rho = 1.77\,\mathrm{g\,cm}^{-3},\ E = 43\,350\,\mathrm{cal\,mol}^{-1},\ n = 2.3,$$

$$C = 0.27 + 10^{-4}\,T\,\mathrm{cal\,g}^{-1}\,\mathrm{deg}^{-1}\ (T \text{ in } ^\circ\mathrm{C}),$$

$$\lambda = 1.9 \times 10^{-3}\,\mathrm{cal\,cm}^{-1}\,\mathrm{s}^{-1}\,\mathrm{deg}^{-1} \text{ when SOC} < 75,$$

$$\lambda = 3.3 \times 10^{-3}\,\mathrm{cal\,cm}^{-1}\,\mathrm{s}^{-1}\,\mathrm{deg}^{-1} \text{ when SOC} > 75$$

Mould (Duralumin, thickness 0.38 cm):

$$\rho = 2.7\,\mathrm{g\,cm}^{-3},\quad \lambda = 0.296\,\mathrm{cal\,cm}^{-1}\,\mathrm{s}^{-1}\,\mathrm{deg}^{-1},\quad C = 0.22\,\mathrm{cal\,g}^{-1}\,\mathrm{deg}^{-1}$$

Coefficient of heat transfer without motion:

$$h = 2 \times 10^{-3}|\Delta T|^{0.25}\,\mathrm{cal\,cm}^{-2}\,\mathrm{s}^{-1}\,\mathrm{deg}^{-1}$$

18.4 Results

Experimentally, it is easy to record the temperature at the centre of the resin. The model is capable of providing one with two types of results: the temperature–time history at the centre of the resin, and the profiles of temperature and SOC developed through the thickness of the resin.

18.4.1 Validity of the Model [11]

It is always of interest for a numerical model to be evaluated. The validity of this model was tested by comparing the temperature–time histories at the centre of the resin obtained either by experiment or by calculation.

Temperature–time histories at the midplane of the resin are plotted in Fig. 18.3, obtained by experiment and calculation. Two curves are shown: curve 1 was obtained with the fresh uncured resin when the resin and mould are immersed into the heated oil at 146 °C. After the resin had been cured for 300 s, the mould and resin were left to cool down to room temperature. The mould with the precured resin in it was then re-immersed in the oil bath under the same conditions, and curve 2 obtained.

The experimental and calculated temperature–time histories at the midplane are very well superimposed, with or without reaction, proving the validity of the model.

Profiles of Temperature and SOC

The profiles of temperature (Fig. 18.4) and SOC (Fig. 18.5) were obtained at 146 °C. Some conclusions are worth noting:

1. By comparing the temperature–time histories obtained with and without reaction, the portion of heat generated by the reaction appears very clearly, under exotherm 1 and above curve 2 playing the role of a reference baseline.
2. The exothermic reaction gives rise to an increase in temperature in curve 1. The temperature passes through a maximum, and finally this temperature reaches the oil temperature.

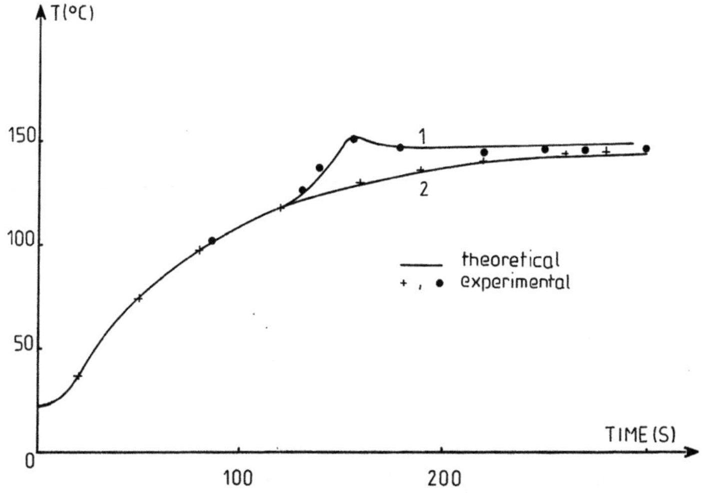

Fig. 18.3. Validity of the model. Temperature–time histories at the midplane of the resin, (1) with or (2) without reaction, with motionless oil. Mould temperature = 146 °C.

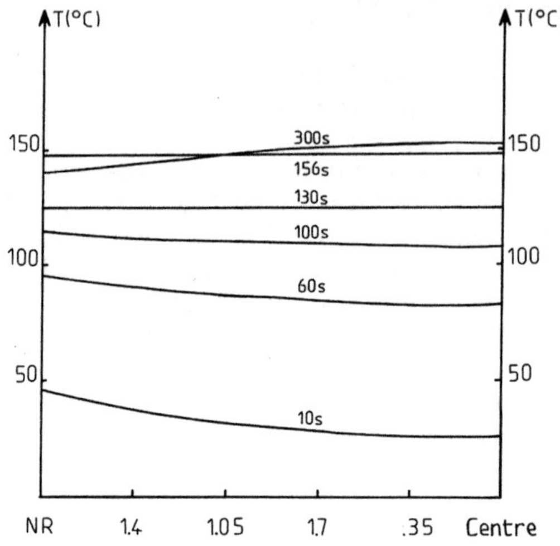

Fig. 18.4. Profiles of temperature developed within fresh uncured resin, with unstirred oil at 146 °C.

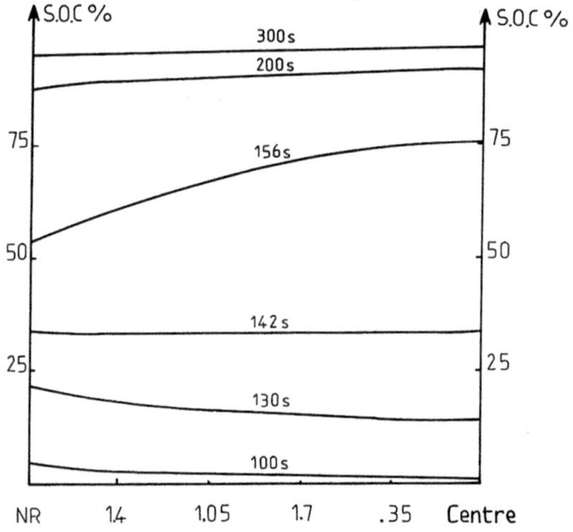

Fig. 18.5. Profiles of SOC developed within fresh uncured resin, with unstirred oil at 146 °C.

3. As for the other moulds, the temperature–time history with reaction can be divided into three stages: heating; reaction, when curves 1 and 2 separate at around 120 °C; and cooling after the maximum temperature has been attained.
4. The profiles of temperature developed within the resin give complementary information. At the beginning of the process, some gradients of temperature are developed within the resin, with the highest temperature on the surface.
5. During the cure stage, after 130 s, other gradients of temperature are developed, with the highest temperature at the centre of the resin, because of the internal heat generated by the reaction.

6. At the end of the process, when the resin and oil are at the same temperature, a flat gradient is observed.
7. At around 100 s, a small value for the SOC is observed, especially next to the surface. During the stage of reaction, other gradients of SOC are developed with a higher value at the midplane of the resin.
8. The steepest value for the gradient of SOC is observed at 156 s, exactly when the temperature at the midplane reaches it maximum value.
9. After the temperature–time history at the midplane has passed through the maximum, the profiles of SOC become less and less steep.

18.4.2 Effect of Temperature [12]

As it is always important to reduce the time of the cure cycle in order to increase the rate of production, the effect of the temperature on the process is of great concern.

Experiments and calculation are made in the same way as shown in Sect. 18.4.1, but the constant temperature of the silicone oil is selected within the range 130–166 °C.

The temperature–time histories at the centre of the resin, obtained either by calculation or experiments, are shown in Fig. 18.6 for the following four temperatures of the silicone oil: 130, 146, 157 and 166 °C. In all cases, the operational conditions are the same (thickness of the resin = 0.7 cm), except for the oil temperature.

The profiles of temperature (left) and SOC (right) are shown in Figs 18.7 (130 °C), 18.8 (157 °C) and 18.9 (166 °C), obtained by calculation by using the

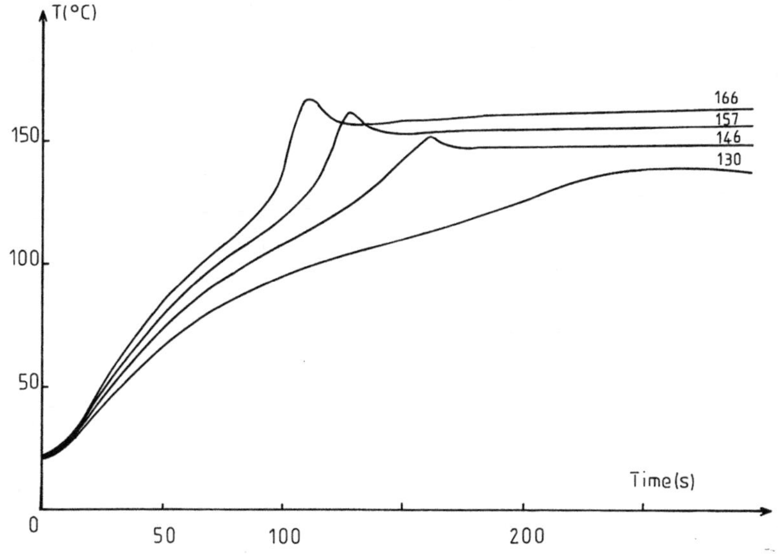

Fig. 18.6. Effect of the temperature of the oil on the temperature–time history at the centre of the resin with unstirred oil.

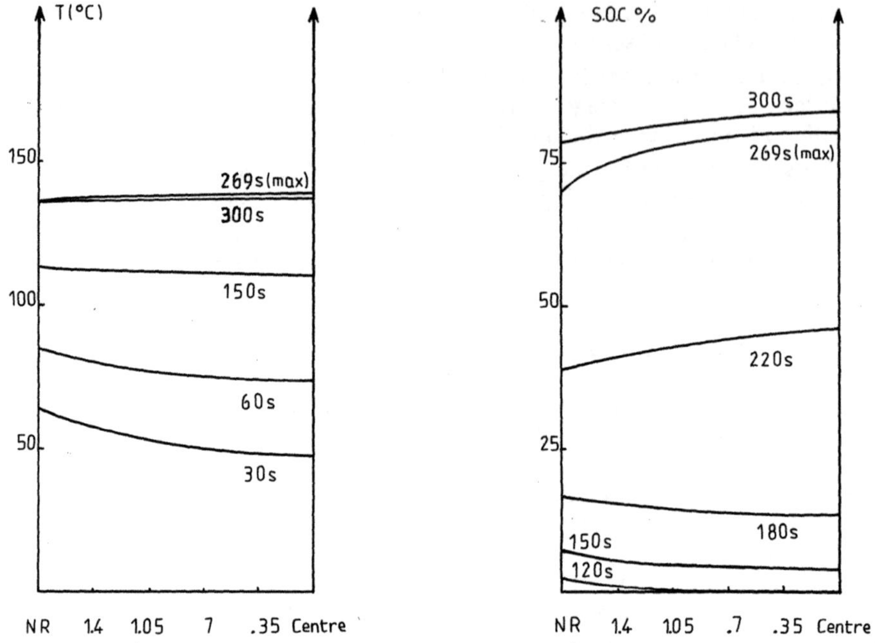

Fig. 18.7. Profiles of temperature (left) and SOC (right) developed within the resin, with unstirred oil at 130 °C.

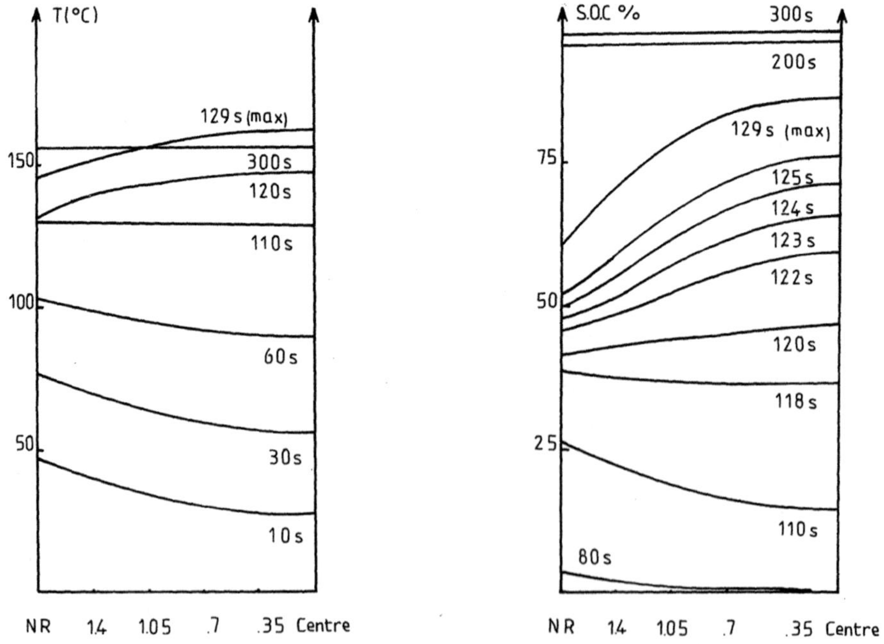

Fig. 18.8. Profiles of temperature (left) and SOC (right) developed within the resin, with unstirred oil at 157 °C.

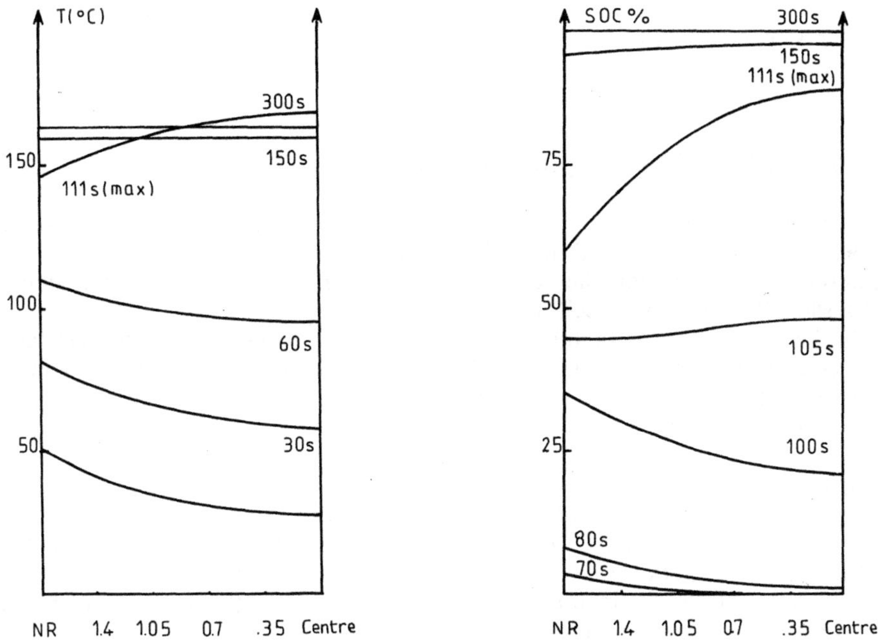

Fig. 18.9. Profiles of temperature (left) and SOC (right) developed within the resin, with unstirred oil at 166 °C.

model and the thermal and kinetic parameters given in Sect. 18.3. The following conclusions are drawn:

1. The temperature–time histories at the centre of the resin shown in Fig. 18.6 clearly illustrate the effect of the temperature of the oil on the process. At 130 °C, no maximum is observed for the temperature–time history at the centre, while this maximum appears very strongly for the higher temperatures of the oil. Of course, the higher the temperature of the oil, the higher the maximum of temperature, and the shorter the time in which it is attained.
2. Three stages can also be considered during the process: heating of the resin; the SOC; and cooling beyond the time at which the temperature at the centre of the resin has attained its maximum.
3. High gradients of temperature are observed during the first stage of heating, with the highest temperature next to the surface of the resin.
4. Other gradients of temperature are developed within the resin during the stage of cure, with the highest temperature at the centre of the resin. The steepest gradient of temperature appears at the time at which the temperature at the centre of the resin passes through its maximum.
5. Some profiles of SOC are also developed within the resin.
6. At the beginning of the process, the SOC is highest on the surface. But when the temperature at the centre of the resin attains the maximum value, the steepest gradient of SOC is observed, with the highest value at the centre of the resin.

7. The following statement can be made: the higher the temperature of the oil, the steeper the gradient of SOC at the maximum temperature at the centre of the resin.

18.4.3 Effect of Stirring the Oil [13]

The results described in the above two Sects 18.4.1 and 18.4.2 were obtained when the resin and mould were heated in motionless oil. It is thus easy to determine by calculation the coefficient of heat transfer using the dimensionless Grasshof, Prandtl and Nusselt numbers.

In the case where the liquid is stirred, it is not easy to determine by calculation the coefficient of heat transfer, as the direction of the fluid is circular and not perpendicular to the surface of the mould, and in addition, the speed of the liquid is not known with good accuracy.

However, as the stirring of the heating liquid is of great interest from a practical and industrial point of view, experiments were performed by using the same mould and the same amount of resin with the same thickness, by varying the agitation of the liquid. The coefficient of heat transfer was calculated by adjusting the temperature–time histories at the midplane of the resin obtained by calculation to those obtained by experiments, by using the precured resin and thus by considering the heating process only.

The numerical model is the same as that described with the motionless liquid, with the only difference for the coefficient of surface heat transfer. With the stirred liquid, this coefficient h is constant, and it is increased by increasing the rate of stirring.

Temperature–Time Histories at the Centre

The temperature–time history at the centre of the resin is shown in Fig. 18.10 for a given rate of stirring at 130 °C. It is difficult to express the rate of stirring by a defined value, because it depends not only on the rate of stirring of the bar magnet but also on the geometry of the system with the mould and bath. This is the reason why the stirring is expressed not by the rate but by the resulting value of the coefficient of heat transfer at the surface h. For instance, with our apparatus, a value of $h = 13 \times 10^{-3}$ cal cm^{-2} s^{-1} deg^{-1} is obtained for five revolutions of agitator per second. As shown in this figure, good superimposition is observed between the theoretical and calculated curves, proving the validity of the model.

Other temperature–time histories were calculated with various values of the coefficient of heat transfer h, ranging from 8 to 16×10^{-3} cal cm^{-2} s^{-1} deg^{-1} with the temperature of oil kept at 146 °C. In the case of the motionless oil, the value of the coefficient h is around 6.7×10^{-3} cal cm^{-2} s^{-1} deg^{-1} with $\Delta T = 126$ °C. The effect of the change in the value given to this coefficient h can thus be observed in Fig. 18.11.

The following conclusions are worth noting:

1. The comparison between the temperature–time histories at the centre of the resin at 130 °C obtained either with a motionless liquid or with a stirring

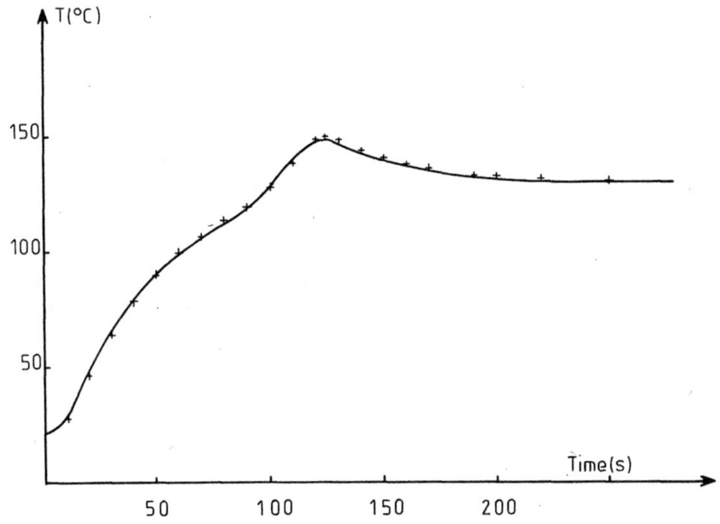

Fig. 18.10. Temperature–time history at the centre of the resin, with stirred oil at 130 °C and $h = 13 \times 10^{-3}$ cal cm^{-2} s^{-1} deg^{-1}. +, calculation; —, experiment.

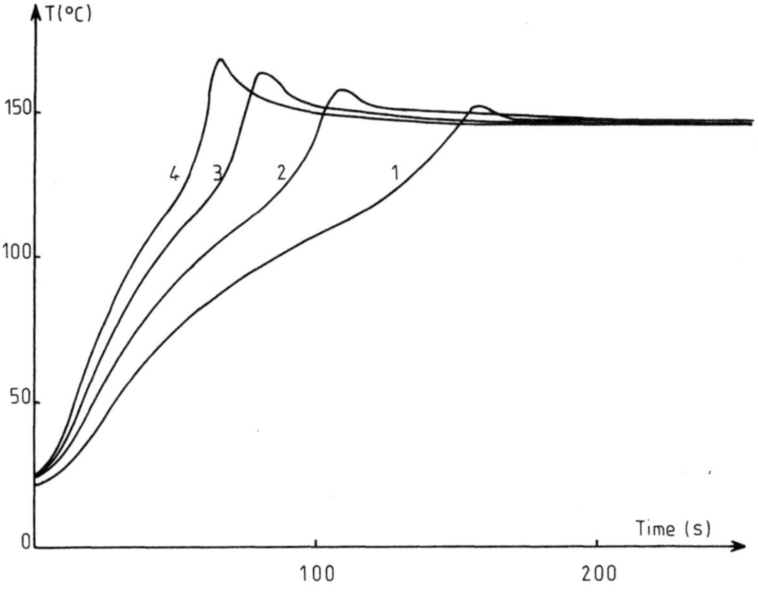

Fig. 18.11. Effect of the value of the coefficient of heat transfer h on the temperature–time history at the centre of the resin, with oil at 146 °C: (1) without stirring; (2) with stirring and $h = 8 \times 10^{-3}$; (3) with stirring and $h = 12 \times 10^{-3}$; (4) with stirring and $h = 16 \times 10^{-3}$ cal cm^{-2} s^{-1} deg^{-1}.

resulting from five revolutions of agitator per second is very informative. A very flat maximum can be observed at around 250 s with motionless liquid, while a very well established maximum is obtained at around 125 s with the stirred liquid. Moreover, it takes a rather long time for the temperature at the centre to reach the mould temperature at the end of the reaction in the case of motionless oil. However, the temperature at the centre reaches the mould temperature after not more than 200 s, when the oil is stirred.

2. The effect of the value of the coefficient of heat transfer h with stirred liquid on the temperature–time history at the centre of the resin appears very clearly in Fig. 18.12. The higher the value of the coefficient of heat transfer, the shorter the time necessary for the temperature–time history to reach the maximum.
3. The effect of the value of the coefficient of heat transfer h with a stirred liquid on the temperature–time history at the centre of the resin is also of great importance, when the value of the maximum temperature is considered. The increase in the value of the coefficient of heat transfer h is also responsible for a significant increase in the value of the maximum temperature at the centre of the resin.

Profiles of Temperature and SOC Through the Resin

The profiles of temperature and SOC developed within the resin as calculated with the model give complementary information on the process. These profiles

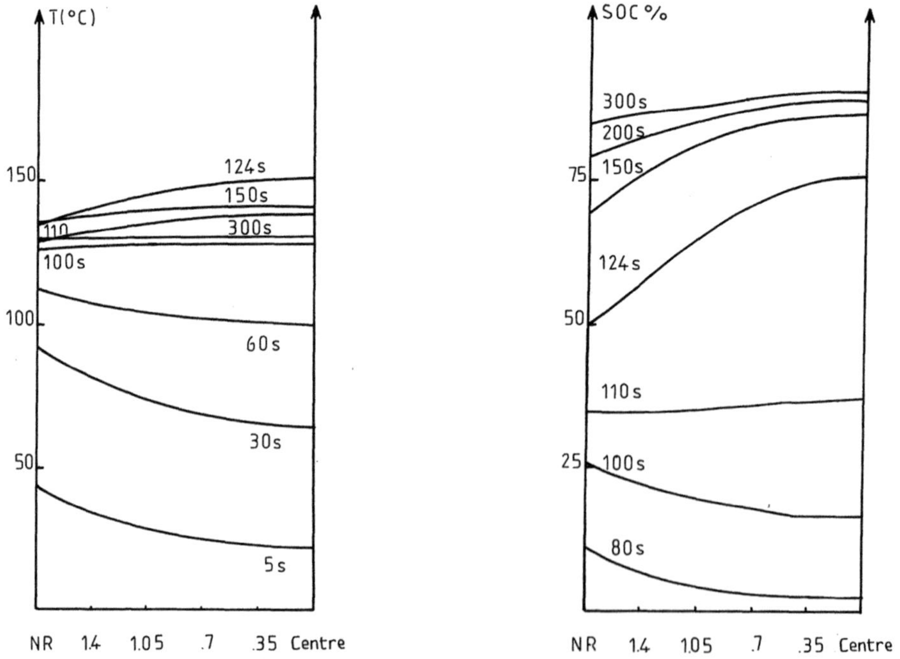

Fig. 18.12. Profiles of temperature (left) and SOC (right) developed within the resin, with stirred oil at 130 °C and $h = 13 \times 10^{-3}$ cal cm^{-2} s^{-1} deg^{-1}.

are drawn in Fig. 18.12 for the oil temperature of 130 °C and the value of h is 13×10^{-3} cal cm^{-2}s^{-1} deg^{-1}.

Other profiles are drawn in Figs 18.13–15 for the various values of the coefficient of heat transfer h, when the temperature of oil is kept at 146 °C.

Some results of note are:

1. A comparison between the profiles of temperature within the resin obtained at 130 °C with motionless oil (Fig. 18.7) and stirred oil with $h = 13 \times 10^{-3}$ cal cm^{-2} s^{-1} deg^{-1} (Fig. 18.12) is of great interest. In the case of stirred liquid, the following facts are observed: steeper gradients of temperature are developed within the resin either during the first stage of heating up to 60 s, or during the stage of cure; moreover, a significant increase in temperature at the centre of the resin of about 20 °C is attained, instead of a few degrees in motionless liquid.
2. This comparison is also of interest for the profiles of SOC. Steeper gradients of SOC are obtained in the case of the stirred liquid at 130 °C.
3. The effect of the value of the coefficient of heat transfer h on the profiles of temperature developed within the resin appears very clearly by comparing the curves drawn in Fig. 18.13–18.15 (left). Steeper gradients are developed either during the heating stage at the beginning of the process or during the curing stage when the coefficient of heat transfer is increased. Of course, the steepest gradient of temperature during the cure stage is obtained at the maximum of the temperature–time history at the centre of the resin. As already shown in Fig. 18.11 at 146 °C, the higher the coefficient of heat transfer, the higher the

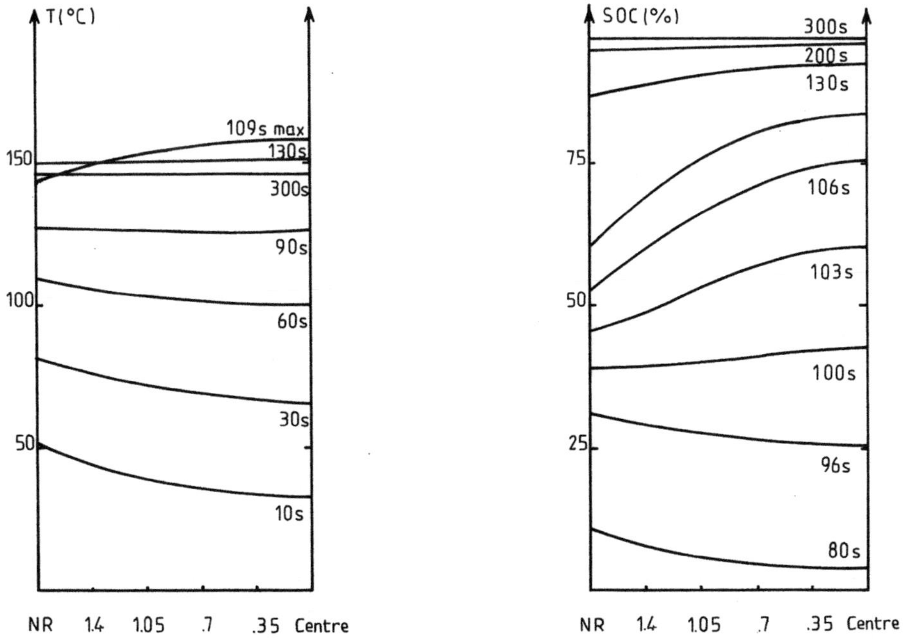

Fig. 18.13. Profiles of temperature (left) and SOC (right) developed within the resin, with stirred oil at 146 °C and $h = 8 \times 10^{-3}$ cal cm^{-2} s^{-1} deg^{-1}.

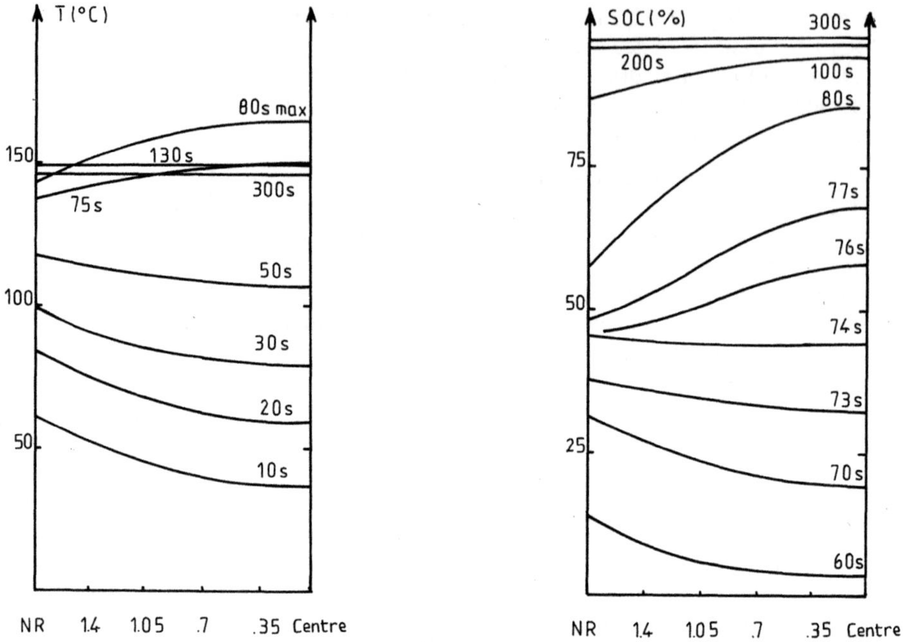

Fig. 18.14. Profiles of temperature (left) and SOC (right) developed within the resin, with stirred oil at 146 °C and $h = 12 \times 10^{-3}$ cal cm^{-2} s^{-1} deg^{-1}.

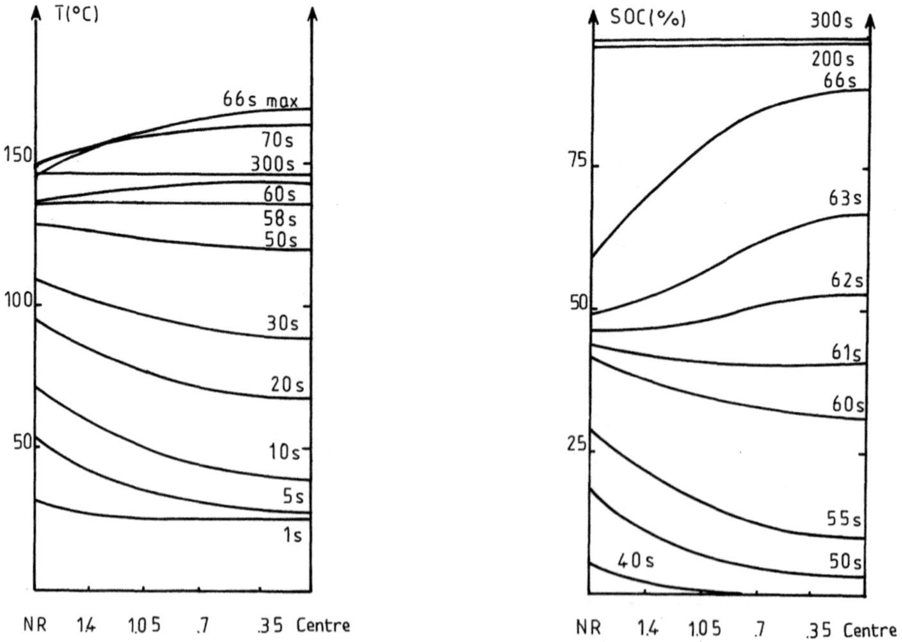

Fig. 18.15. Profiles of temperature (left) and SOC (right) developed within the resin, with stirred oil at 146 °C and $h = 16 \times 10^{-3}$ cal cm^{-2} s^{-1} deg^{-1}.

increase in temperature at the centre of the resin at the maximum value of the temperature–time history.

4. The effect of the coefficient of heat transfer *h* on the profiles of SOC developed within the resin can be appreciated by comparing the curves in Figs 18.13–18.15 (right). The reaction starts already during the heating period at the beginning of the process, and some profiles of SOC are expanded through the resin, with the highest value at the surface. During the cure period, other gradients of SOC develop with the highest value at the centre of the resin, because of the internal heat generated by the reaction. The steepest gradient of SOC is obtained at the maximum of the temperature–time history at the centre of the resin. Of course, the higher the coefficient of heat transfer, the shorter the heating period and the faster the cure period.

18.5 Conclusions

A few conclusions are drawn for the use of a metallic mould and for the process of cure:

1. This mould is of great interest because it enables the user not only to follow the process of heating and cure but also to obtain a final material which can be studied by considering either the mechanical properties or the swelling properties when the thermoset is put in contact with a liquid.
2. The operational conditions can be easily obtained. Two are of interest: the temperature of the heated oil and the stirring of the oil.
3. The numerical model is capable of describing the process by calculating the profiles of temperature and SOC developed within the resin.
4. The temperature–time history at the centre of the resin can be used to test the validity of the process.
5. The time of cure can be reduced either by increasing the temperature of the heated oil, or by stirring the oil more strongly.
6. This type of mould is very flexible for studies on a laboratory scale, as the temperature of the oil can be varied quickly by immersing the mould in various oil baths at different temperatures. Thus, the process of reactive injection moulding can be easily simulated by making inexpensive experiments. Moreover, the model is able to predict the temperature–time history of the heated oil necessary when a special cure process is desired.

References

1. Nixon JA, Hutchinson JM. Analysis of the cure of sheet moulding compound. I. Development of the model. Plast Rubber Process Applic 1985; 5: 349
2. Pusatcioglu SY, Hassler JC, Fricke AL, McGee HA. Effect of temperature gradients on cure and stress gradients in thick thermoset castings. J Appl Polym Sci 1980; 25: 381
3. Chater M, Vergnaud JM. Study of the process of cure of epoxide resin heated by liquid oil at

constant temperature. Eur Polym J 1988; 24: 245
4. El Brouzi A, Bouzon J, Vergnaud JM. Modelling of the cure of fibre glass–epoxy resin composite heated by liquid oil at constant temperature. Effect of parameters on the process. Plast Rubber Process Applic 1989; 12: 71
5. Chater M, Bouzon J, Vergnaud JM. Modelling of the cure of epoxide resin. Effect of the diameter of cylindrical moulds. Plast Rubber Process Applic 1987; 7: 199
6. Chater M, Vergnaud JM. Effect of the temperature of the mould on the cure of epoxy resin. J Polym Engng 1988; 8: 1
7. Chater M, Vergnaud JM. Effect of kinetic and thermal parameters on the process of cure of epoxide resin in a cylindrical mould. Eur Polym J 1987; 23: 563
8. El Brouzi A, Granger R, Bouzon J, Vergnaud JM. Modelling and simulation of the process of reactive injection moulding in the case of a composite with low enthalpy. Plast Rubber Process Applic 1988; 12: 7
9. Liu H, Armand JY, Bouzon J, Vergnaud JM. Effect of sample size and heating rate on the DSC process for reactions of high enthalpy. Thermochim Acta 1988; 126: 81
10. Armand JY, Vergnaud JM. Effect of the value of heating rate in DSC on the kinetic parameters, when there is a high enthalpy of reaction. Thermochim Acta 1988; 131: 15
11. Azaar K, El Brouzi A, Granger R, Vergnaud JM. Process of cure of a maleic polyester resin. Modelling and experiments with thin slab in metal mould. Eur Polym J, in press
12. Azaar K, El Brouzi A, Granger R, Vergnaud JM. Effect of temperature on the cure of a polyester resin in a thin slab. Modelling and experiments. Europ Polym J, in press
13. Azaar K, El Brouzi A, Granger R, Vergnaud JM. Modelling the cure of thermosetting polymers. Effect of the coefficient of surface heat transfer. Plast Rubber Process Applic, in press

Chapter 19

Modelling the Cure of Epoxy Resin Coating at Low Temperature (50 °C)

19.1 Introduction

Epoxy resins have been widely studied, especially with respect to their use as polymeric matrices for composites and also for high-performance coatings.

This study is concerned with the general problem of determining the best conditions of temperature and time for preparing epoxy resin coatings with good mechanical properties. Very often, when dealing with a coating deposited on a large substrate, e.g., the inside of a ship, it is not possible to heat the resin in order to reduce the cure time. The problem is how to achieve the cure of a resin coating at room temperature, or at least at a temperature around 50 °C, and to know the time required to obtain a proper coating under these conditions.

Numerical simulations of the cure process were made for epoxy resin [1–3] and for epoxy resin containing various fillers and loads [4, 5]. These models, based on numerical methods with finite differences, take into account not only the heat transferred by conduction through the resin, but also the heat evolved from the cure reaction. The kinetics of the heat generated by the cure reaction is determined using calorimetric techniques, either under isothermal conditions or in scanning mode [6–9].

The main purpose in this study is to show that coatings made of epoxy resin can be cured at low temperature, and that the state of cure (SOC) of the resin can be predicted by calculation in terms of temperature and time with the help of a kinetic equation using data obtained by calorimetry. In addition, a mechanical property, in this case hardness, was chosen to be measured following the cure, because this test is not destructive to the coating.

19.2 Experiment

Material

The resin was obtained by mixing the epoxy resin (Lopox 200, CDF Chimie) and the hardener containing an anhydride of Me THP (D 2605, CDF Chimie), with an accelerator containing 0.75% of an aromatic amine (type A 105, CDF Chimie). The binary mixture was prepared at room temperature with continuous stirring. The best composition was found to be epoxy resin 56 wt %, hardener 44 wt % [10].

Kinetic Parameters of the Resin

After mixing the two components, the mixture was applied to iron–steel sheets, with an average thickness of 1 mm.

The same mixture was encapsulated in a cylindrical steel holder and studied by calorimetry in scanning mode (DSC 111, Setaram) with a heating rate of 2 °C min^{-1}. The heat flux was recorded as a function of temperature, using a computer connected to the calorimeter. The kinetic parameters are determined using the Freeman–Carroll technique (Chap. 13).

The kinetic parameters were as follows:

$$Q_\infty = 59 \text{ cal g}^{-1}, \qquad n = 2,$$
$$E = 29 \text{ kcal mol}^{-1}, \qquad k_0 = 2 \times 10^{14} \text{ s}^{-1}$$

19.3 Theory

Assumptions

The following assumptions are made in order to simplify the problem.

1. The rate of heating in the calorimeter in scanning mode is very slow.
2. Following the first assumptions, the temperature within the sample is uniform and is the same as the temperature of the calorimeter.
3. The resin, deposited in thin layers, is heated at quite a low temperature (50 °C), so that the curing time is much longer than the heating time. As the heating time is neglected, the reaction can be considered as taking place at the constant applied temperature.
4. The rate of heat evolved from the overall cure reaction follows an Arrhenius form with constant kinetic parameters.
5. The SOC in the resin at time t is expressed by the heat evolved from the reaction up to time t as the fraction of the total heat evolved.

Mathematical Treatment

The rate of the reaction is expressed in the Arrhenius form:

$$\frac{d}{dt}\left(\frac{Q_t}{Q_\infty}\right) = k_0\left(1 - \frac{Q_t}{Q_\infty}\right)^n \exp\left(-\frac{E}{RT}\right) \tag{19.1}$$

where

n is the order of the overall reaction.

Q_t and Q_∞ are the heat evolved up to time t and after infinite time, respectively.

k_0 is the pre-exponential factor.

E is the activation energy.

19.4 Results

Two types of results are of interest: one is concerned with the determination of an equation expressing the value of the SOC in terms of temperature and time for the coating; the other with the establishment of a relationship between the hardness and the SOC [11].

19.4.1 Determination of the SOC in the Coating

Equation Expressing the SOC

The kinetic parameters were determined from the heat flux–temperature curves obtained with the calorimeter in scanning mode. The accuracy of these parameters was obtained by comparing the calculated and experimental heat flux–temperature histories. As shown in Fig. 19.1, good superimposition is observed between the curves, proving the validity of Eq. (19.1) and the accuracy of the kinetic parameters shown in Sect. 19.3.

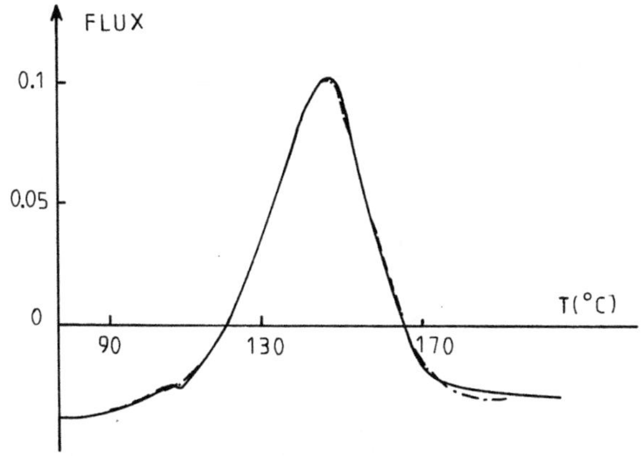

Fig. 19.1. Experimental (—) and calculated (— · — ·) heat flux–temperature curves in DSC for the cure of epoxy resin (heating rate = 2 °C min^{-1})

Measurement of SOC

After application to the metal substrate, the resin was exposed to air kept at constant temperature. As this temperature was low and the thickness of the resin layer small, thermal equilibrium was reached quickly, and the reaction can be considered as proceeding under isothermal conditions. Eq. (19.1) is thus used for determining the value of the SOC at any time for the selected constant temperature.

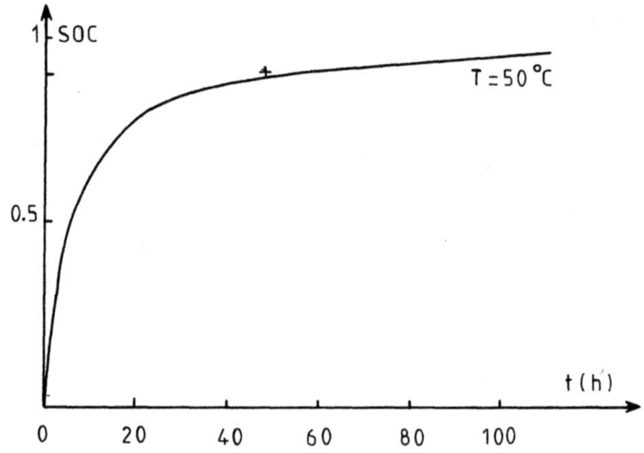

Fig. 19.2. SOC of epoxy resin coating as a function of time at 50 °C. —, calculation; +, experiment.

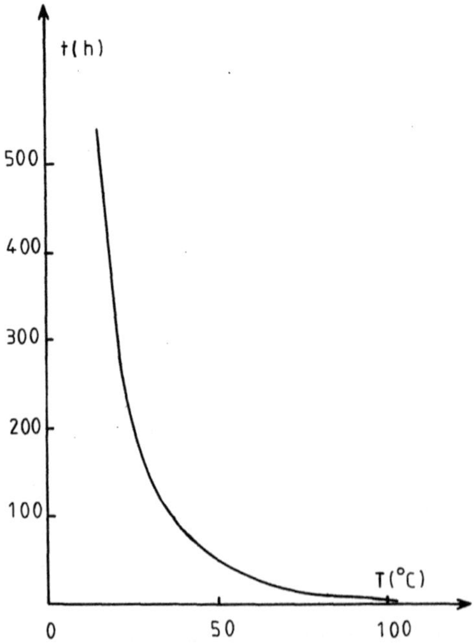

Fig. 19.3. Cure time as a function of temperature for SOC of 90%.

After curing for 44 h, a fragment of the coating was taken and placed in the calorimeter holder for determination of the value of the residual SOC. The value obtained of 6 cal g^{-1}, indicated by the + sign in Fig. 19.2 is located on the SOC–time curve calculated at 50 °C, thus supporting the validity of this method.

The time necessary for the SOC to reach 90% is expressed as a function of the constant temperature applied to the coating in Fig. 19.3. It is thus easy to appreciate the great effect of temperature on the rate of cure reaction.

19.4.2 Hardness of the Coating in Relation to the SOC

The hardness of the coating was chosen to be measured as it is easy to determine and not destructive to the sample. A small "pocket-type" hardness meter was used (Shore Durometer D, ISO R 868). In spite of the small thickness of the coating, hardness measurements were made for coatings cured at temperatures of 50, 90 and 100 °C. The hardness was measured after the coating had previously been cooled down to room temperature.

The hardness of the coating can thus be expressed as a function of the SOC cure. As shown in Fig. 19.4, there is some correlation between the hardness and SOC although it is far from perfect. Hardness values ranging from 80 to 85 were obtained when the SOC reached 90%, for various temperature conditions.

The increase in hardness as a function of time for temperatures of 50, 90 and 100 °C is illustrated in Fig. 19.5. As already shown in the case of the increase in SOC, the effect of temperature on the hardness–time curves is very significant.

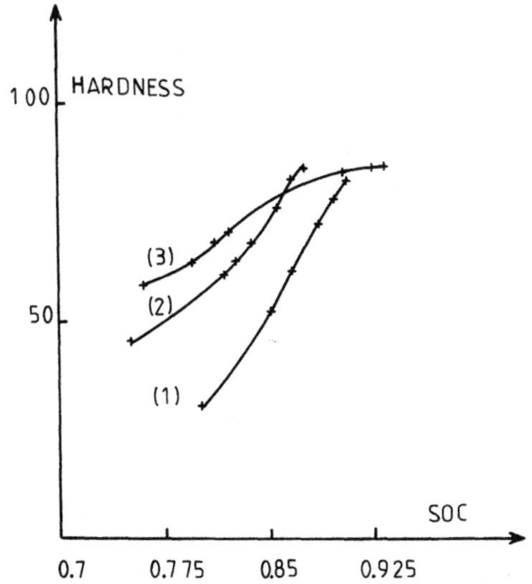

Fig. 19.4. Hardness of epoxy resin coatings as a function of the SOC, obtained at temperatures of (1) 100 °C; (2) 90 °C; (3) 50 °C.

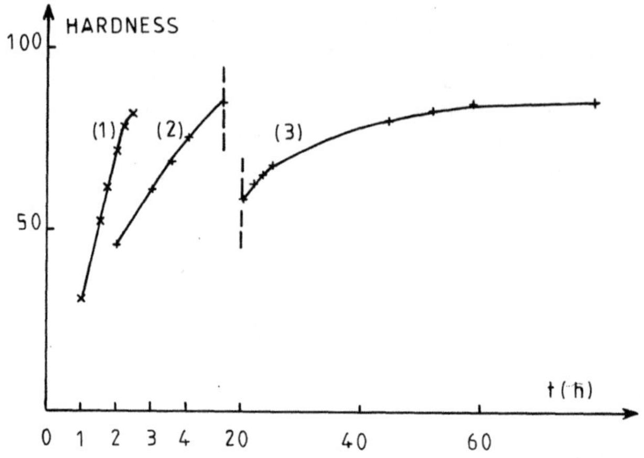

Fig. 19.5. Hardness of epoxy resin coating as a function of cure time, obtained at temperatures of (1) 100 °C, (2) 90 °C, (4) 50 °C.

19.5 Conclusions

The value of the SOC of an epoxy resin applied as a coating can be expressed in terms of temperature and time using a kinetic relationship. This expression follows an Arrhenius form, and the parameters are determined by calorimetry.

It is easy to calculate the time necessary for the SOC to reach a given value, when the coating is kept at a low and constant temperature. As the time of cure is considerably higher than the time of heating, the reaction can be considered to take place at this constant applied temperature.

The hardness of the coating can be followed during the process as it is a non-destructive test, and there is some relation between its value and the SOC.

Of course, in the case where the coating is submitted to a high temperature, the time of heating cannot be neglected, and the process must take into account the heat transferred by conduction and convection, and the heat evolved from the cure reaction.

References

1. Chater M, Vergnaud JM. Study of the process of cure of epoxide resin heated by liquid oil at constant temperature. Eur Polym J 1988; 24: 245
2. Chater M, Bouzon J, Vergnaud JM. Modelling of the cure of epoxide resin: effect of the diameter of cylindrical moulds. Plast Rubber Process Applic 1987; 7: 199
3. Chater M, Vergnaud JM. Effect of the temperature of the mould on the cure of epoxy resin. J Polym Engng 1988; 8: 1
4. Robert G, Rochette JB, Bouzon J, Vergnaud JM. Determination of the state of cure of epoxy resin used in composite with sand. Thermochim Acta 1986; 102: 93
5. El Brouzi A, Bouzon J, Vergnaud JM. Modelling of the cure of fibre glass-epoxy resin composite

heated by liquid oil at constant temperature. Effect of parameters on the process. Plast Rubber Process Applic 1989; 12: 71
6. Vergnaud JM. Modelling in calorimetry working under isothermal conditions or in scanning mode. Thermochim Acta 1987; 114: 15
7. Liu H, Armand JY, Bouzon J, Vergnaud JM. Effect of sample size and heating rate on the DSC process for reactions of high enthalpy. Thermochim Acta 1988; 126: 81
8. Armand JY, Bourgois J, Vergnaud JM. Fundamental study of heating process in DSC. Effect of the cure enthalpy and heating rate. Thermochim Acta 1987; 116: 301
9. Kretzschmar K, Hoffmann KW. Kinetic aspects arising during the differential scanning calorimetric investigation of curing epoxy resins with anhydrides. Thermochim Acta 1985; 83: 139
10. Chater M, Lalart D, Michel-Dansac F, Vergnaud JM. Variation of enthalpy and kinetic parameters of the cure of epoxide resin with the composition of the binary system. Eur Polym J 1986; 22: 805
11. Laoubi S, Vergnaud JM. Modelling the cure of epoxy resin coating at low temperature (50 °C). Thermochim Acta 1990; 162: 347

Author Index

Subject Index

GPSR Compliance

The European Union's (EU) General Product Safety Regulation (GPSR) is a set of rules that requires consumer products to be safe and our obligations to ensure this.

If you have any concerns about our products, you can contact us on ProductSafety@springernature.com

In case Publisher is established outside the EU, the EU authorized representative is:

Springer Nature Customer Service Center GmbH
Europaplatz 3
69115 Heidelberg, Germany

Batch number: 10838062

Printed by Printforce, the Netherlands